Mathematics and Its Applications

Mathematical Physics of Quantum Wires and Devices

Mathematics and Its Applications

Managing Editor:

M. HAZEWINKEL

Centre for Mathematics and Computer Science, Amsterdam, The Netherlands

Volume 506

Mathematical Physics of Quantum Wires and Devices

From Spectral Resonances to Anderson Localization

by

Norman E. Hurt

Zeta Associates, Inc.,
Fairfax, VA, U.S.A.

KLUWER ACADEMIC PUBLISHERS

DORDRECHT / BOSTON / LONDON

A C.I.P. Catalogue record for this book is available from the Library of Congress.

ISBN 978-90-481-5446-3

Published by Kluwer Academic Publishers,
P.O. Box 17, 3300 AA Dordrecht, The Netherlands.

Sold and distributed in North, Central and South America
by Kluwer Academic Publishers,
101 Philip Drive, Norwell, MA 02061, U.S.A.

In all other countries, sold and distributed
by Kluwer Academic Publishers,
P.O. Box 322, 3300 AH Dordrecht, The Netherlands.

Printed on acid-free paper

Contents

Preface

This monograph on quantum wires and quantum devices is a companion volume to the author's *Quantum Chaos and Mesoscopic Systems* (Kluwer, Dordrecht, 1997). The goal of this work is to present to the reader the mathematical physics which has arisen in the study of these systems. The course which I have taken in this volume is to juxtapose the current work on the mathematical physics of quantum devices and the details behind the work so that the reader can gain an understanding of the physics, and where possible the open problems which remain in the development of a complete mathematical description of the devices. I have attempted to include sufficient background and references so that the reader can understand the limitations of the current methods and have direction to the original material for the research on the physics of these devices.

As in the earlier volume, the monograph is a panoramic survey of the mathematical physics of quantum wires and devices. Detailed proofs are kept to a minimum, with outlines of the principal steps and references to the primary sources as required. The survey is very broad to give a general development to a variety of problems in quantum devices, not a specialty volume. The goal is to allow the reader to see the beauty of the theoretical developments in this area of study, to perceive the interrelationships of the mathematical methods in the selected topics, and to understand the pivotal results and the varied directions which the research has taken. The topics were those of interest to the author and any omission in credit or research is inadvertent.

The reader may benefit from background material including Simon's lectures on Schrödinger operators in Cycon et al. (1987), the volume on spectral theory of random Schrödinger operators by Carmona and Lacrois (1990) and the more recent volume of Stolimann (1999). In the area of quantum devices, the review of Kouwenhoven, Marcus, McEuen et al. (1997) is recommended. In the discussion of Ramanujan graphs, the primary reference is Lubotzky (1994). For additional reading on quantum chaos, the reader should note Sarnak (1993) and Zelditch (1999) and for material on random matrix theory the review of Beenakker (1997) should be consulted. Finally, for supplemental reading on automorphic forms, the reader is directed to the volume of Iwaniec (1995).

During the course of the research of this volume, the author benefitted from several communications from various researchers in this field. The author wishes to thank Professor Isaac Efrat, Professor Alex Lubotzky, Professor Moshe Morgenstern and Professor Shin-ya Koyama, who all directed me to the work of Hirofumi Nagoshi and Ortwin Scheja, when I was trying to find out about the Selberg trace formula on finite volume graphs, Professor Jonathan Bird for his many comments on the real and exciting world of mesoscopic systems, Professor Pavel Exner, for his explanations of several of his earlier results, and Professor Berggren for his supplemental comments of his research. The author wishes to especially thank Drs. Nagoshi and Scheja for sharing their doctoral research with me as it became available and answering my many questions on this area of study.

In particular I would like to express my appreciation to Professor Jonathan Bird, Professor Pavel Exner and Dr. Michael Hurt for their critical readings of preliminary drafts of chapters from this monograph and their recommendations for improvements.

Chapter 1

Quantum Wires and Devices

1.1 Quantum Waveguides, Nanotubes and Other Devices

The mathematical physics of quantum wires has developed over the last two decades. In this volume the focus will be on precise results which have been discovered regarding resonance and bound states that appear in quantum wires, Anderson localization in quantum devices, the quantum Hall effect, and graphical models of quantum wire systems. Nanotechnology today permits the fabrication of semiconductors of a variety of shapes. Using a high mobility material like $Al_x Ga_{1-x} As/GaAs$, there is a quasi two dimensional electron gas (2DEG) residing at the interface of the heterojunction. By means of a split gate deposited on the heterojunction surface, the electron gas may be confined into a quasi-one dimensional channel. The gate can be lithographically designed to shape the electron gas into other mesoscopic structures such as intersecting wires, curved wires, rings, dots, stadia and so on. In these devices, quantum mechanical effects become apparent especially at low temperatures where the electron mean free path can be quite large.

In narrow straight channels, the motion of the electron is quantized in the transverse direction and each transverse state defines a normal mode or subband. The electrons propagate in these one-dimensional subbands, of which only a few may be occupied. In short channels or electron waveguides, one encounters ballistic transport and quantization of conductance, where each conducting spin-degenerate one-dimensional subband contributes $2e^2/h$ to the differential conductance, which was observed by Wharam et al. (1988) and van Wees et al. (1988). That is, the conductance is given by $G = 2e^2 N/h$ where N is the number of occupied one-dimensional subbands.

In this survey we are interested in non straight quantum wires and related devices, e.g. bent quantum wires, wires with bumps or bulges, quantum wires with impurities, quantum wires with rough edges, intersecting quantum wires, T-shaped devices, multi-stub devices and quantum superlattices or quantum crys-

1

tals. Very interesting phenomena occur in these cases. E.g., the slightest bend in a quantum wire will produce nonpropagating states or bound states. In transport measurements one finds resonant tunneling peaks as well as anti-resonant peaks. Wu, Wybourne, Yindeepol, Weisshaar and Goodnick (1991) constructed "S-shaped" quantum wires with double right angle bends. They measured the conductance versus gate voltage (or effective width). In their tests, they observed a threshold for conductance followed by a series of plateaus. In addition, Wu et al. observed two peaks in the conductance at a gate voltage below the threshold for free electron conduction and smaller resonance peaks on the plateaus.

L-shaped devices were examined in work by Exner and coworkers, Wu and co-workers and others. Two idealized, infinite perpendicular quantum wires were modeled by Schult, Ravenhall and Wyld (1989, 1990). In their numerical work, Schult et al. found two bound states which reside at the intersection of the quantum wires; the two bound states have either even or odd parity. This work has been clarified by further research due to Berggren and coworkers and Carini and coworkers, which will be discussed below.

Electron waveguides with T-shaped structures have been studied by Berggren and Ji (1990) and Debray and coworkers. This stub structure has been examined since a transistor action is in principle achieved (called a quantum modulated transistor) for these devices. For other work on quantum interference transistors, see Datta and Bandyopadhyay (1987), Exner and Seba (1988) and Sols, Macucci, Ravaioli and Hess (1989).

Resonant and trapped states are not unique to the mathematical physics of quantum wires. In particular, the study of such resonant states appears in work on acoustics and fluid flow or water waves. In the acoustics area Lippert (1955) studied resistance and transmittance in right angled waveguides. In fluid flow Ursell (1951) showed that there are trapped modes above a long submerged horizontal cylinder of sufficiently small radius in deep water. Some forty years later, Evans, Levitin and Vassiliev (1991) proved the existence of localised, i.e. L^2, eigenfunctions whose eigenvalues are embedded in the continuous spectrum of an acoustic waveguide with certain obstacles which are symmetric with respect to the center line of the 2D waveguide. Here the Laplacian is subject to the Neumann boundary condition. For the antisymmetric problem (v.i.), they showed that there is a trapped mode below the continuous spectrum. More recently Maniar and Newman (1997) have considered trapped modes where Dirichlet boundary conditions are assumed in the water-wave problem. The analogy to quantum wires is very close in this case. In the water-wave problem linear arrays of cylinders have been studied for which the analogy to quantum superlattice structures is very similar.

One of the earliest theoretical results on bound states or trapped modes in quantum wires is due to Exner, Seba and Stovicek (1989). They showed that there exists a bound state in an L-shaped quantum waveguide below the continuous spectrum. Curvature induced bound states were shown to exist by Exner and Seba (1989, 1990). Using the Birman-Schwinger technique and the minimax principle, they showed that the Dirichlet Laplacian on a smooth curved planar strip, i.e. a quantum waveguide, which is thin enough and whose curvature decays sufficiently rapidly, has at least one isolated eigenvalue below the bottom of the continuous spectrum. As will be developed below, this result has been considerably improved.

During this same period, experimentally many exciting results were appearing on quantum wires. McEuen et al. (1990) and Wu et al. (1990) were observing sharp peaks in the conductance of quantum wires at low temperatures, viz. 70 - 250 mK which peaks are reproducible as the temperature is cycled from these low temperatures to 4.2 K. McEuen et al. attributed the two sharp peaks as due to resonant tunneling through impurity states. Transmission probabilities and thus conductances were found to undergo rapid variations just below threshold energies as well as on the conductance plateaux. These resonance structures or quasi-bound states were studied by Kirczenow (1989) in connection to quenching of the quantum Hall effect and were modelled by Sols and Maccuci (1990) for quantum wires with circular bends. Kirczenow showed that for low magnetic fields resonant electron states can give rise to quenching of the quantum Hall voltage, maxima in the longitudinal resistance and anomalous Hall plateaus.

Numerical modelling work was proceeding in an effort to develop an understanding of these features of quantum wires. Right angle quantum wires were modelled by Schult et al. (1989), Exner et al. (1989) and Weisshaar et al. (1989). Quantum wires with bends of an arbitrary angle were studied by Avishai, Bessis, Giraud and Mantica (1991) and Wu, Sprung and Martorell (1992). Using thin electromagnetic waveguides Carini, Londergan, Mullen and Murdock (1993) experimentally measured the bound state energies as well as the fields in sharply bent waveguides. Quantum wires with circular bends were examined by Sols and Maccuci (1990), Sprung, Wu and Martorell (1992), Exner and Seba (1989), Vacek, Ojiki and Kasai (1992); the latter considered the quantum wire with and without an external magnetic field. The modelling of Berggren's group has considered quantum wires and structures of various shapes. Their work has shown the interesting variation of the flow of the quantum wire near resonance features, viz. a rapid transition from "laminar" to "vortical" flow. In the vortical case the global flow directions are found to reverse direction as the energy changes from one side of a resonance to the other (v. Berggren and Ji (1993) and Berggren, Besev and Ji (1992)).

Graphical models for quantum wire systems have been examined by Exner and coworkers, although such models go back to Pauling (1936) and Ruedenberg and Scherr (1953). We briefly review the results in this area. Later in this volume we return to the question of modelling quantum wire systems, in particular we develop the Selberg trace formula or Gutzwiller trace formula for finite volume quantum wire systems. In the mathematical physics community, finite quantum graphs have been studied by Kottos and Smilansky (1998, 1999) to better understand quantum systems which are classically chaotic. Here they use an exact Gutzwiller trace formula. Kottos and Smilansky have also examined the spectral statistics of quantum graphs and the relationship to random matrix theory and they have examined some relationships to scattering theory.

In the study of quantum chaos one considers quantum ballistic structures such as squares, circles and stadia, e.g. see Ji and Berggren (1995) where they examine the energy spectrum of a stadium shaped quantum dot with and without a magnetic field. As closed structures, the theoretical results appear to agree fairly accurately with the measured data, e.g. the statistical properties of the spectra. For a review, the reader should confer Sarnak (1995). Building on work on disordered systems, semiclassical theories have been developed which provided explicit predictions regarding open quantum structures. However, as we discuss below, these predictions are not agreeing with measured data at least for instances of low number of modes entering the quantum dot. Open problems in this area are presented.

Although most of the experimental results for quantum wires and devices are related to two dimensional semiconductor heterostructures, we should also mention the recent work on nanotubes. Single wall carbon nanotubes are constructed from two dimensional graphene sheets which are rolled into nanometer diameter cylinders. These can be either one dimensional metals or semiconductors, depending on construction. For more details, see the recent papers of McEuen et al. (1999) and Wakabayashi and Sigrist (1999). We note that quantized conductance of multiwall carbon nanotubes has been observed. Wakabayashi and Sigrist have examined the transport properties through junctions in nanographite ribbons, which have a single conducting channel of edge states in the low energy regime. They have found that the conductance as a function of energy or chemical potential has an interesting structure, with sharp dips of zero conductance, i.e. zero conductance resonances.

Metallic nanowires and devices have also been examined, e.g. atomic scale Au nanowires, and electron corrals on Cu surfaces studied by Crommie, Lutz and Eigler (1997). For a review of recent work on metallic nanowires, see Kassubek, Stafford and Grabert (1998).

1.2   Anderson Localization

The question of impurities in quantum devices has been mentioned. Random impurities in crystal structures introduce what is called Anderson localization, which is pure point spectrum with exponentially decaying eigenfunctions. In one dimension any amount of impurities results in Anderson localization. In three dimensions both localized and extended, i.e. absolutely continuous spectrum, are expected with the energies of the extended states and localized states separated by the mobility edge. The Anderson model has been studied on the Bethe lattice or Cayley tree, on which one can show that there are extended states for small disorder. The Bethe lattice is an infinite connected graph with no closed loops and a fixed degree, K, or number of nearest neighbors at each vertex. Let $H_\lambda = \frac{1}{2}\Delta + \lambda V$ denote the random Hamiltonian. One finds that there is a mobility region for the Bethe lattice, with localization for energies beyond $(K + 1)/2$ for weak disorder. And one can show that for $0 < E < \sqrt{K}$ there is a $\lambda(E) > 0$ such that the spectrum of H_λ in $[-E, E]$ is pure absolutely continuous with probability one. The review of the Bethe lattice problem provides a stepping stone to the later study of the Selberg trace formula on finite volume graphs.

A second topic which is addressed in this area is Anderson localization in rough quantum wires. This is a natural extension of the work on bent quantum wires and recent results of Kleespies and Stollmann (1999) are presented.

1.3 Quantum Hall Effect

Two dimensional noninteracting electron systems subject to disorder and a strong magnetic field form the basis of study of the quantum Hall effect. The features of this class of problems include the localization-delocalization transition and plateau-to-plateau transition. Disorder is required in the quantum Hall effect. The general thinking early on in Anderson localization was that all states in two dimensions are localized due to disorder. However, for the quantum Hall systems, the center of each Landau band becomes delocalized in the presence of a strong magnetic field; viz., non zero Hall conductance means that there are delocalized states below the Fermi energy. One of the interesting problems in this area is what happens to the delocalized states as the disorder increases, where the system should become an Anderson insulator. This problem will be discussed in the chapter on the quantum Hall effect.

A theoretical basis for the quantum Hall effect has been under development since von Klitzing's discovery nearly twenty years ago. We present one of the fundamental results in this area, viz. Kunz's (1987) theorem. Connes and Bellissard have developed a more extensive theory based on non-commutative geometry. However, we only briefly mention that approach as it is fully developed in Connes (1994).

Several proposals have been made to describe bulk and edge states in the mathematical physics of the quantum Hall effect. We review recent work in this area.

1.4 Theme and Variations

A theme that is present in the results discussed so far is the question of discrete spectra lying below the bottom of the continuous spectrum. A classic example of this phenomenon arises in analytical number theory in the study of the spectra of the Laplacian on finite volume manifolds $\Gamma \backslash \mathcal{H}$ where $\mathcal{H}$ is the Poincaré upper half plane. We briefly outline the problem statement in the sections below.

1.4.1 Exceptional Spectra

Let $\mathcal{H} = \{z \in \mathbf{C} | Im(z) > 0\}$ denote the Poincaré upper half plane. Consider the functions $f(z)$ on $\mathcal{H}$ which satisfy

$$f(\gamma z) = \chi(\gamma)f(z)$$

for all $\gamma \in \Gamma$ where Γ is a cofinite, but not cocompact, discrete subgroup of $SL(2, \mathbf{R})$ and $\chi : \Gamma \to \mathbf{C}$ is a unitary character, so $|\chi| = 1$. We assume that χ is trivial on at least one of the parabolic subgroups of Γ; this means that at least one of the cusps of Γ is χ–singular. f is said to be a (Γ, χ)-function. Let $\mathbf{H} = \mathbf{H}(\Gamma, \chi)$ denote the space of all (Γ, χ)–functions on $\mathcal{H}$ with the inner product

$$(f, g) = \int_{\mathcal{F}} f(z)\overline{g(z)}\frac{dxdy}{y^2}$$

where $\mathcal{F} = \mathcal{F}_\Gamma$ is a fundamental domain for Γ acting on $\mathcal{H}$. The Laplacian $\Delta = y^2(\frac{\partial^2}{\partial x^2} + \frac{\partial^2}{\partial y^2})$ has a unique self-adjoint extension to $\mathbf{H}$. Let $\sigma(\Delta)$ denote the spectrum of this operator. In the case Γ is co-finite, then the spectrum of Δ is partly absolutely continuous and partly discrete. The structure of $\mathbf{H}$ has the form

$$\mathbf{H} = C \oplus R \oplus E$$

where $C = L^2_{cusp}(\Gamma \backslash \mathcal{H})$ is the space of cusp forms, R is the finite dimensional space spanned by residues of Eisenstein series with eigenvalues in $[0, 1/4)$ and E is the continuous spectra spanned by unitary Eisenstein series with spectrum $[1/4, \infty)$.

Let $\lambda_1(\Gamma, \chi) \leq \lambda_2(\Gamma, \chi) \leq ...$ denote the discrete eigenvalues of Δ. The discrete eigenvalues $\lambda \in (0, 1/4)$ are called exceptional.

1.4.2   The Selberg-Ramanujan Conjecture

The principal congruence subgroup of $\Gamma = PSL(2, \mathbf{Z})$ of level N is the group

$$\Gamma(N) = \{ \begin{pmatrix} a & b \\ c & d \end{pmatrix} \mid \begin{pmatrix} a & b \\ c & d \end{pmatrix} \equiv \begin{pmatrix} 1 & 0 \\ 0 & 1 \end{pmatrix} \, mod \, N \}.$$

A subgroup $\Gamma' \subset \Gamma$ is called a congruence subgroup if $\Gamma' \supset \Gamma(N)$ for some N.

The Selberg-Ramanujan conjecture asserts that $\Gamma(N)$ has no exceptional spectrum. Selberg showed originally that $\lambda_1(\Gamma(N), 1) \geq 3/16$. Huxley (1985) has since showed that $\lambda_1(\Gamma(N), 1) \geq 1/4$ for $N \leq 17$.

For the full modular group $\Gamma(1)$, Maass and Roelke had shown that there are no exceptional eigenvalues, that is

$$\lambda_1(\Gamma(1), 1) > 1/4.$$

1.4.3   Hecke Group

As another example which has been utilized in quantum chaos, consider the Hecke group $G(\mu)$ generated by $z \to z + \mu$ and $z \to -1/z$ where $\mu = 2cos(\pi/q)$, $q \geq 3, q \in \mathbf{Z}$. Sarnak (1982) has shown that $\lambda_1(G(\mu), 1) > 1/4$. For more detail, see Hejhal (1992).

1.4.4   Small Eigenvalues

For general Fuchsian groups Γ, exceptional eigenvalues can occur. E.g., Randol (1974) showed that one can devise examples where $\lambda_1(\Gamma, 1)$ is an arbitrarily small eigenvalue. See also Selberg (1965).

1.4.5 Fermat Group

Phillips and Sarnak have examined the spectrum of Fermat groups, viz. let $\Phi(N) = ker(R_N \circ \pi)$ where $R_N : \mathbf{Z}^2 \to (\mathbf{Z}/N\mathbf{Z})^2$ is reduction modulo integer N, $N \geq 1$ and $\pi : \Gamma(2) \to H_1(\Gamma(2)) \simeq \mathbf{Z}^2$, the canonical projection. Then $\Phi(N)\backslash\mathcal{H}$ is the $(\mathbf{Z}/N\mathbf{Z})^2$ Galois covering of $\Gamma(2)\backslash\mathcal{H}$ and $\overline{\Phi(N)\backslash\mathcal{H}} = \Gamma_N$, the Fermat curve $x^N + y^N = 1$. It is known that $\Phi(N)$ is a congruence group if and only if $N = 1, 2, 4, 8$. However, Phillips and Sarnak have shown that there are no exceptional eigenvalues for $N \leq 8$.

1.4.6 Weyl's Law

Weyl's law states that

$$N_{cusp}(R, \Gamma, \chi) + M(R, \Gamma, \chi) \sim (vol(\Gamma\backslash\mathcal{H})/4\pi)R^2$$

where

$$N_{cusp}(R, \Gamma, \chi) = |\{\text{cusp forms with } |r_j| \leq R\}|$$

and

$$M(R, \Gamma, \chi) = -\frac{1}{2\pi} \int_{-R}^{R} \frac{\phi'}{\phi}(1/2 + ir, \Gamma, \chi)dr.$$

Here $\lambda = 1/4 + r^2$ and ϕ is the determinant of the scattering matrix Φ. Selberg showed using his trace formula for the case $\chi = 1$ and Γ a congruence subgroup of $SL(2, \mathbf{R})$, then $M(R, \Gamma, 1) = \mathcal{O}(RlogR)$, i.e. N_{cusp} is the primary term in Weyl's law.

1.4.7 Essentially Cuspidal

For a cofinite subgroup $\Gamma \subset SL(2, \mathbf{R})$, the pair (Γ, χ) is said to be essentially cuspidal if

$$N_{cusp}(R, \Gamma, \chi) \sim (vol(\Gamma\backslash\mathcal{H})/4\pi)R^2.$$

Selberg conjectured that every (Γ, χ) is essentially cuspidal. Based on their dissolving theory for cusp forms, which can be expressed in terms of the Fermi golden rule, Phillips and Sarnak have conjectured that (Γ, χ) is essentially cuspidal if and only if Γ is arithmetic. Furthermore, in this case they conjecture that there should only be a finite number of cusp forms. These conjectures are discussed in greater depth in chapter eight.

1.4.8 Example for the Phillips-Sarnak Conjecture

Consider the principal congruence group $\Gamma(2)$. This group is freely generated by

$$A = \begin{pmatrix} 1 & 2 \\ 0 & 1 \end{pmatrix} \quad \text{and} \quad B = \begin{pmatrix} 1 & 0 \\ -2 & 1 \end{pmatrix}.$$

And in this case $\Gamma(2)\backslash\mathcal{H}$ has three cusps $\{0, 1, \infty\}$. The group of characters of $\Gamma(2)$ is given by

$$\chi_{(\xi,\eta)}(A) = exp(2\pi i \xi) \quad \text{and} \quad \chi_{(\xi,\eta)}(B) = exp(2\pi i \eta).$$

The character $\chi_{(\xi,\eta)}$ has a singular cusp if and only if (ξ, η) has the form $(\xi, 0), (0, \eta)$ or $\xi + \eta = 0$. As Phillips and Sarnak have shown, one can limit the study to χ_η for $0 < \eta < 1$ and $\xi = 0$. The spectrum of Δ for $\eta = 0$ is 0, the continuous spectrum $[1/4, \infty)$ and an infinite set of cusp forms embedded in the continuous spectrum. The spectrum $\Delta(\eta)$ for $0 < \eta < 1$ consists of at most one exceptional eigenvalue in $(0, 1/4)$, continuous spectrum $[1/4, \infty)$ and a set of embedded cusp forms in the continuous spectrum.

For this example, Phillips and Sarnak (1994) have shown that under the multiplicity hypothesis (see Phillips and Sarnak (1994)), there are at most a countable number of η's for which $(\Gamma(2), \chi_\eta)$ is essentially cuspidal. The values $\eta = j/8$ for $0 \leq j \leq 7$ are the η's for which $(\Gamma(2), \chi_\eta)$ is essentially cuspidal; and it is believed that these are the only η's for which $(\Gamma(2), \chi_\eta)$ is essentially cuspidal.

1.4.9   Summary and Questions

The following questions arise for a given discrete subgroup Γ of $SL(2, \mathbf{R})$:

- are there exceptional spectra? or is $\lambda_1(\Gamma, \chi) \geq 1/4$?

- if there are exceptional spectra, are they cuspidal?

- are there finitely many or infinitely many cuspidal eigenvalues?

- are there finitely many or infinitely many noncuspidal eigenvalues?

- what are the multiplicities of the eigenvalues?

- characterize the stability of the cusp forms to deformation of the lattice or variation of the magnetic field; i.e., characterize the Phillips-Sarnak dissolving of the cusp forms;

- does the discrete subgroup obey the conjectures of Phillips and Sarnak or the Selberg-Ramanujan conjecture?

1.5   Finite Volume Graphical Models

The final topic in this volume is the development of the Selberg trace formula on finite volume graphs as a model for quantum wire systems. A topic of recent research activity is that of Ramanujan graphs. Ramanujan graphs are K-regular, finite graphs whose nontrivial eigenvalues of the adjacency Laplacian have absolute values bounded by $2\sqrt{K-1}$. These graphs have good properties for communication networks, viz. high magnifications and small diameters. The first construction of a family of Ramanujan graphs was by Lubotzky, Phillips and Sarnak (1988) and

Margulis (1988), which construction is based on the arithmetic of quaternion algebras and the Ramanujan conjecture for Hecke operators acting on cusp forms of weight two.

Ramanujan graphs are briefly reviewed and their extension to finite volume graphs by Morgenstern, called Ramanujan diagrams, is discussed in Chapter nine. The Selberg trace formula for finite graphs has been studied several groups and is reviewed by Venkov and Nikitin (1994). This is a very interesting area of research involving work of Ihara, Weil, Serre, Bass, Hashimoto, Stark, Terras and others. The extension to finite volume graphs has occurred only recently in work of Scheja (1998) and Nagoshi (1998), although the relevant scattering theory had been examined by Li (1978).

Spectral statistics has been an active are of research in quantum chaos and random matrix theory. Various groups have examined these problems based on the spectral statistics of finite graphs, e.g. Jakobson, Miller, Rivin and Rudnick (1997) and Lafferty and Rockmore (1993, 1997). Most recently Nagoshi has used his Selberg trace formula to develop certain results on the distribution of eigenvalues on arithmetic infinite graphs. This material will be covered in the final chapter.

The analogue of the Phillips-Sarnak conjectures have been addressed by Efrat for the finite volume graphs. In this case one asks for a given discrete group Γ:

- are there exceptional eigenvalues?

- arc there finitely many or infinitely many cuspidal eigenvalues?

- are there finitely many or infinitely many noncuspidal eigenvalues?

- is there an analogue of the Phillips-Sarnak dissolving phenomena?

Results of these questions are surveyed in chapter nine.

1.6 Outline

Briefly, the outline for this volume is as follows. Although this review will focus on quantum wires, in the second chapter the recent work on acoustic and fluid flow problems will be outlined. Chapter three covers recent work on precise results for various types of bent or deformed quantum wires. Chapter four treats mathematical physics modelling related to quantum wires and devices, in particular bent and deformed quantum wires, including quantum wires in the presence of a magnetic field. Not all of the numerical results have been captured in the theorems of chapter three and we present enough details to allow the reader to identify open problems in this area of study. Selected open problems will be specifically identified throughout this volume. Chapter five deals with problems and results in open quantum devices. One of the models used for describing quantum wire systems is a graphical model which has been exploited by Exner and coworkers and Kostrykin and Schrader (1998).

Chapter six deals with Anderson localization, specifically recent results for multi-dimensional periodic structures, rough quantum wires and Bethe lattices.

Chapter seven surveys exact results for two dimensional systems in a perpendicular magnetic field, in particular the quantum Hall effect, localization-delocalization, bulk and edge states, and open problems regarding limits in low magnetic fields. Chapter eight reviews some aspects of the Selberg trace formula in background for problems in the following chapter, in particular the Selberg conjectures and the Phillips-Sarnak conjectures. Finally, chapter nine treats graphical models of quantum wires and the Selberg trace formula on finite volume graphs. We briefly survey the work on spectral statistics in this area.

Chapter 2

Acoustic and Fluid Flow

2.1 Trapped Modes in Acoustics and Fluid Flow

As noted in Chapter one, trapped modes and resonances have been studied in acoustics and fluid flow prior to their rediscovery in quantum wires. In this chapter we briefly review the mathematical methods and summarize the recent results in this area. As the reader will see, there are great similarities in these models and methods with the later work on quantum wires.

Evans and coworkers have examined the problem of trapped modes in fluid flow and the acoustic area, where they are called acoustic resonances. The existence of discrete eigenvalues of the spectrum of the Laplacian operator on unbounded domains goes back to the work of Ursell (1951) and Jones (1953); see also Rellich (1943) and the work on edgewaves. In linear water-wave theory, Ursell proved the existence of trapped modes above a long submerged horizontal circular cylinder of sufficiently small radius in deep water. Jones showed that Ursell's work extends to submerged horizontal cylindrical obstacles, (actually arbitrary symmetric cross section obstacles), which are not necessarily small, and in finite water depth; v., also Ursell (1987). In 1991 Evans and Linton returned to this subject and considered the case of a vertical cylinder extending throughout the water depth where the cylinder is located midway between two parallel channel walls. Their work suggested that there are trapped modes which are antisymmetric with respect to the midline and satisfy Neumann conditions on the boundaries. Callan, Linton and Evans (1991) using methods of Ursell proved that an antisymmetric trapped mode exists for a small circular cylinder midway between parallel walls. Ursell (1991) showed a similar result for a small sphere on the center line of a circular tube of infinite extent. For a cylinder of diameter $2a$ in the center of a channel of width $2d$, numerical work of Callan et al. suggested that there is a single trapped mode for each value of $a/d \leq 1$. The trapped modes take the form of persistent local oscillations near the cylinder at a unique frequency below the first cutoff frequency of the channel. Viz., the angular frequency is $\omega^N = g(k^N \tanh k^N h)^{1/2}$ where the wave number $k^N < \pi/2d$.

More recently, a second type of trapped mode was discovered by Maniar and

Newman (1997), which is called a Dirichlet trapped mode where Dirichlet boundary conditions are assumed for the channel walls and the center-plane. These
boundary conditions are nonphysical in the water-wave problem but have been
studied in the area of acoustic resonances; (v., Parker and Stoneman (1989)). The
Dirichlet trapped modes appear for a restricted range $0 < a/d < .6789$; however,
the wavenumber $k^D < \pi/d$ is still below the continuous spectrum, $[\pi^2/d^2, \infty)$.

2.2   Linearized Water-Wave Problem

The linear water-wave problem for a single cylinder of radius a, extending through
the entire depth of a channel, located at the center of the channel of width $2d$, is
described by a velocity potential Φ of the form

$$\Phi(x, y, z, t) = Re\{\phi(x, y)f(z)e^{-i\omega t}\}$$

where $f(z) = -igAcosh\, k(z + h)/\omega cosh\, kh$. Here the fluid domain is $-h < z < 0$.
The angular frequency satisfies the dispersion equation

$$\omega^2/g = ktanh(kh)$$

where g is the gravitational acceleration.

The function ϕ satisfies the Helmholtz equation

$$(\Delta + k^2)\phi(x, y) = 0$$

exterior to the cylinder and

$$\frac{\partial \phi}{\partial r} = 0$$

on $r = a$, expressing no-flow through the cylinder. The Neumann, resp., Dirichlet,
boundary conditions on the channel walls $y = \pm d$ are:

$$\frac{\partial \phi}{\partial y} = 0,$$

resp.,

$$\phi = 0.$$

One assumes the decay condition

$$\phi(x, y) \to 0$$

as $|x| \to \infty$ for $-d < y < d$. A nontrivial solution to this water-wave problem is
called a trapped mode.

Trapped modes can be symmetric or antisymmetric with respect to the x-axis.
For the antisymmetric case one assumes a Dirichlet condition on the centerplane

$$\phi(x, 0) = 0$$

for $-\infty < x < \infty$.

2.3 Acoustic Resonance Problem

For the acoustic resonance problem the potential has the form

$$\Phi(x,y,z,t) = Re\{\phi(x,y)e^{-i\omega t}\}$$

where $k = \omega/c$ and c is the speed of sound. Let $G = \{(x,y)||y| \leq d\}$ denote the acoustic channel with boundaries $\Gamma_\pm = \{y = \pm d\}$. Let G_+ denote the positive half, $G_+ = \{(x,y)|0 < y < d\}$. Consider an obstruction O in the duct which is symmetric about the centerline $y = 0$ of the waveguide. Let $\partial O = \partial O_+ \cup \partial O_-$ denote the piecewise smooth boundary of the obstruction parameterized as

$$\partial O_\pm = \{(x,y)|x = X(s), y = \pm Y(s), 0 \leq s \leq L\},$$

where $X(s), Y(s)$ are continuous, piecewise infinitely smooth functions such that $X'(s)^2 + Y'(s)^2 = 1, 0 \leq Y(s) < d, Y(0) = Y(L) = 0, X(0) = -a, X(L) = a$. In addition, they assume that $X'(s) \geq 0$ and $(X(s), Y(s)) \neq (X(s_1), Y(s_1))$ for all $s \neq s_1$, (i.e. no looping); and $Y'(s + 0) - Y'(s - 0) < 2$ for all $s \in (0, L)$, (i.e., no cusps protruding inside the obstacle). Set

$$O_0 = \{(x,y)|y = 0, x \in (-\infty, -a] \cup [a, +\infty)\}.$$

The symmetrical obstacle is assumed to extend from $x = -a$ to $x = a$. Like the water-wave problem ϕ satisfies the Helmholtz equation exterior to the obstacle:

$$(\Delta + k^2)\phi(x,y) = 0$$

for $(x,y) \in F = G\backslash O$. The boundary conditions are the Neumann condition

$$\frac{\partial \phi}{\partial y} = 0$$

for $(x,y) \in \Gamma_\pm$ and

$$\frac{\partial \phi}{\partial n} = 0$$

for $(x,y) \in \partial O_\pm$. Here $\partial/\partial n$ denotes the derivative with respect to the normal vector to the boundary of the obstruction. Again we assume that $\phi \to 0$ for $|x| \to \infty$ and $|y| \leq d$.

This problem will be called the full acoustic problem P.

2.4 Antisymmetric Acoustic Problem

Similar to the water-wave problem, solutions of the acoustic resonance problem can be antisymmetric about the axis $y = 0$, i.e. $\phi(x,y) = -\phi(x, -y)$. Specifically, the antisymmetric acoustic problem is:

$$(\Delta + k^2)\phi^a(x,y) = 0$$

for $(x, y) \in F_+ = G_+ \backslash O_+$. The boundary conditions are the Neumann condition

$$\frac{\partial \phi^a}{\partial y} = 0$$

for $(x, y) \in \Gamma_+$ and $\frac{\partial \phi^a}{\partial n} = 0$ for $(x, y) \in \partial O_+$ with the Dirichlet boundary condition $\phi^a = 0$ for $(x, y) \in O_0$. As before we assume the decay property that $\phi^a \to 0$ for $|x| \to \infty$ and $0 \leq |y| \leq d$. This problem will be called the antisymmetric resonance problem P^a.

2.5   Minimax Principle or Variational Estimates

Let $H_\Omega = -\Delta_\Omega^D$ denote the Friedrichs extension of the operator L which acts as the Laplacian on $D(L) = C_0^\infty(\Omega)$. The form domain of H_Ω is $H_0^1(\Omega)$ and since Ω has the segment property (v. Reed-Simon (1978)) the set $\{g \in C^\infty(\Omega) | g = 0 \text{ on } \partial\Omega\}$ is a core for H_Ω. Let $\sigma(H_\Omega)$ denote the spectrum of H_Ω. A weak version of the minimax principle says that

$$inf(\sigma(H_\Omega)) = inf_{\psi \in Q(H_\Omega)} \frac{q_\Omega(\psi, \psi)}{(\psi, \psi)}$$

where $q_\Omega(\phi, \psi) = (\phi, H_\Omega \psi)$ and $Q(H_\Omega)$ is the quadratic form domain. For a more complete discussion, see Reed and Simon (1978).

2.6   Continuous Spectra

The continuous spectra for the antisymmetric acoustic or water-wave problem is given by the interval $[(\pi/2d)^2, \infty)$ for the Neumann boundary conditions. The continuous spectra for the full acoustic problem P with Neumann boundary conditions is $[0, \infty)$.

The continuous spectra for the antisymmetric problem with Dirichlet boundary conditions is $[(\pi/d)^2, \infty)$.

2.7   Acoustic Resonances

Evans (1992) considered the case of a strip of finite length positioned midway and parallel to the sides of an infinite duct of constant width (v., also Parker (1966)). Evans provided a constructive proof to show that a trapped mode exists if the strip is sufficiently long compared to the width of the duct. Evans developed a rapidly convergent method to calculate the trapped mode frequencies. The trapped modes are symmetric or antisymmetric to the line $x = 0$. Consider the case of the solution which is even about $x = 0$. These modes are solutions of the Helmholtz equation with Neumann boundary condition on the strip and the duct boundary and a Dirichlet condition on the midline outside the strip:

$$(\Delta + k^2)\phi = 0, \text{ for } 0 < y < 1 \text{ for all } x$$

$$\phi = 0, \text{ for } y = 0 \text{ and } x > a,$$

$$\phi_y = 0, \text{ for } y = 0, 0 < x < a,$$

$$\phi_y = 0, \text{ for } y = 1, x > 0,$$

$$\phi \to 0, \text{ for } x \to \infty, 0 \le y \le 1,$$

$$\phi_r = O(r^{-1/2}) \text{where } r = [(x-a)^2 + y^2]^{1/2} \to 0,$$

and

$$\phi_x = 0, \text{for } x = 0, 0 < y < 1.$$

Here the strip is defined by $\{y = 0, |x| < a\}$ between $y = \pm 1$. The duct width is 2 and the cutoff is then $k < \pi/2$. Evans addresses the question whether there are nontrivial solutions to the trapped mode problem outlined above for $k < \pi/2$:

Theorem 1 *(Evans) The condition for a symmetric trapped mode is*

$$\frac{1}{2}\pi - \beta + \delta = ka + n\pi$$

where n is an integer and the reflection coefficient is $R = e^{-2ika}$. Here

$$\beta = \sum_{n=1}^{\infty}[\tan^{-1}(k/\kappa_n) - \tan^{-1}(k/k_n)]$$

where $p_n = n\pi$, $l_n = (n-1/2)\pi$, $k_n = (p_n^2 - k^2)^{1/2}$, $k_0 = -ik$, and $\kappa_n = (l_n^2 - k^2)^{1/2}$. One finds that $\delta \to 0$ as $a \to \infty$ so the condition is satisfied for some a sufficiently large, in which case $R = -e^{2i\beta}$.

2.8 Parker Resonances

Parker (1966) studied acoustic resonances which arise in ducted flow problems. The Parker resonances are given by a model for the duct which contains a series of equally space thin parallel plates spanning the duct and having their planes parallel to the air flow. Parker demonstrated that large pressure amplitudes can be generated at certain resonance frequencies, where the frequencies depend on the ratio of the length of each plate and the distance between adjacent plates. Based on symmetry considerations, to model Parker resonances one needs only to consider one plate. Parker considered in his paper the lowest order modes which he called α, β, γ and δ.

Evans and Linton (1994) have developed specific tools to estimate the Parker resonances. The model is given by

$$(\Delta + k^2)\phi = 0$$

for $0 < y < d$ and $x > 0$. The Parker α mode has the boundary conditions given as follows: on the plate $y = 0, 0 < x < a, \partial\phi/\partial y = 0$, on the lower boundary $y = 0, x > a, \phi = 0$, on the upper boundary $y = d, \partial\phi/\partial y = 0, x > 0$ and on the mid-chord line $x = 0, \phi = 0$. Of course, at a large distance downstream from the plate, $\phi \to 0$.

Theorem 2 *(Evans and Linton) The wavenumbers for the Parker α modes are given by*

$$ka = \pi - \chi$$

where

$$\chi = \sum_{n=1}^{\infty} \{sin^{-1}(k/p_n) - sin^{-1}(k/l_n)\};$$

here $p_n = n\pi/d$, $l_n = (n - 1/2)\pi/d$ and $kd < \pi/2$.

The paper of Evans and Linton also covers the other Parker modes. The work of Evans, Linton and Ursell (1992) shows that the Parker α and β modes exist for a plate in an open duct when the plate is off the centerline of the duct, say $y = b$ and the pressure node condition on the line $y = b, x > a$ is relaxed.

2.9   Trapped Modes Below the Continuous Spectrum

Evans, Levitin and Vassiliev (1994) extended the earlier results of Evans and coworkers to the case of a symmetric obstruction in a two dimensional acoustic waveguide of width $2d$. They have shown that trapped modes exist below the cutoff for the Neumann antisymmetric problem, i.e. there is a trapped mode having a unique wavenumber k^N satisfying $k^N < \pi/2d$.

Theorem 3 *(Evans, Levitin and Vassiliev) The antisymmetric problem P^a with Neumann boundary conditions has a trapped mode below the continuous spectra.*

The proof will be outlined in Section 2.12.

2.10   Cylindrical Obstructions

The problem of vertical cylindrical obstructions has been studied by Callan, Linton and Evans (1991), Evans and Porter (1997) and Maniar and Newman (1997), among others. These studies include the cases of single cylinders in the center of the channel or waveguide, linear arrays of cylinders along the centerline and symmetric arrays of cylinders in the channel.

For a single cylinder the trapped modes are constructed based on expressing ϕ as a sum over multipoles ψ_{2n+1}:

$$\phi(r,\theta) = \sum_{n=0}^{\infty} k^{-1} a_n \psi_{2n+1}(r,\theta)/Y'_{2n+1}(ka).$$

For background, see Twersky (1952). Applying the Neumann boundary condition on the cylinder results in an infinite set of linear real equations:

$$a_m + \sum_{n=0}^{\infty} a_n B_{mn} = 0$$

where $m = 0, 1, 2, ...$ and $B_{mn} = A_{mn} J'_{2m+1}(ka)/Y'_{2n+1}(ka)$ with A_{mn} the function

$$A_{mn} = -\frac{4}{\pi}(-1)^{m+n} \int_0^\infty \frac{e^{-\gamma d}}{\cosh \gamma d} \sinh(2n+1)v \sinh(2m+1)v\, dv$$

$$-\frac{4}{\pi} \int_0^{\pi/2} \tan \beta d \cos(2n+1)u \cos(2m+1)u\, du$$

where $\gamma = k\sinh v$ and $\beta = k\cos u$; (see, e.g., Evans and Porter (1997)). Callan, Linton and Evans show that $\sum \sum |B_{mn}| < \infty$ for $0 < ka < kd < \pi/2$ and $\coth \chi < M(ka)^{-2}$. Here $kd \cosh \chi = \pi/2$. This bound is sufficient to ensure that the determinant det_N of the truncated system

$$a_m + \sum_{n=1}^N B_{mn} a_m = 0$$

for $m = 0, 1, ..., N$ converges uniformly to det_∞.

Theorem 4 *(Callan, Linton and Evans) The infinite system of equations has a nontrivial solution with $\sum |a_m| < \infty$ if and only if $det_\infty = det(\delta_{nm} + B_{nm}(ka, kd))$ vanishes for some ka, kd satisfying the above conditions.*

They show that there is a range of ka for which

$$det_\infty(ka, 1/2) > 0$$

and

$$det_\infty(ka, 3/2) < 0;$$

for this reason, there is a $\lambda = \lambda(ka)$ for which $det_\infty(ka, \lambda(kd)) = 0$. Thus, the work of Callan et al. shows that for a sufficiently small, there is a single trapped mode for the antisymmetric problem with Neumann boundary conditions. Their numerical work suggests that there are trapped modes for all values $0 < a/d < 1$ where $2a$ is the cylinder diameter and $2d$ is the width of the channel. Although there is not a formal proof, we summarize these results as:

Theorem 5 *(Callan, Linton and Evans) For a cylindrical obstruction, there is just one trapped mode with wave number k^N below the continuous spectrum for the antisymmetric problem with Neumann boundary conditions; i.e. $k^N < \pi/2d$. The trapped modes occur for all values $0 < a/d < 1$.*

For a circular cylinder in a waveguide, Callan et al. have derived an approximation for the wave number kd, viz.

$$\frac{2kd}{\pi} \sim 1 - \frac{\pi^4 a^4}{8d^4}$$

as $a/d \to 0$.

Recent numerical modeling of Maniar and Newman (1997) and Evans and Porter (1997) has examined the antisymmetric problem with Dirichlet boundary conditions. Again they find evidence for a single trapped mode below the continuous spectrum although the Dirichlet trapped modes only occur for $0 < a/d < 0.677$. Even though a formal proof is lacking, we summarize these results as:

Theorem 6 *(Maniar, Newman, Evans, and Porter) For a cylindrical obstruction, there is just one trapped mode with wave number k^D below the continuous spectrum for the antisymmetric problem with Dirichlet boundary conditions, i.e. $k^D < \pi/d$. The trapped modes occur only in the restricted region $0 < a/d < .6789$.*

2.11   Multiple Cylinders

Evans and Porter (1997) have extended the study of trapped modes to multiple cylinders mounted along the center line of the channel. They considered both Neumann and Dirichlet modes. Their numerical work showed that in general there are less than or equal N trapped modes for any configuration of N cylinders. The exact number depends on the geometry of the configuration. In the Neumann case the modes are alternatively symmetric and antisymmetric. In particular they expanded the work of Evans, Levitin and Vassiliev to consider the Dirichlet mode for a single cylinder.

For a single cylinder were the cross section is given by $y = f(x), f(\pm a) = 0$, and $0 \leq f(x) < d$ for $|x| \leq a$, they show that Dirichlet modes exist provided

$$I = \int_{-a}^{a} sin(2\pi f(x)/d)dx > 0.$$

For a rectangular block, $f(x) = b, |x| \leq a$, this implies that $b/d < 1/2$. and if $f(x) < d/2$, then $I > 0$ and a trapped modes exists for all f. For a circular cylinder, $I = \pi a J_1(2\pi a/d) > 0$ for the region $a/d < .6098$ which agrees with the earlier work of Maniar and Newman (1997).

In general, the Neumann and Dirichlet trapped modes are given by the real zeros of the determinant of a system of equations, which set of equations is truncated to a few modes (say 3 or 4 for a single cylinder). For the case of two cylinders Evans and Porter (1997) studied the model as a function of the ratio $\mu = a/d$ and λ where the cylinders are at $(\pm\lambda a, 0)$. For sufficiently large μ there exist two trapped modes, a low frequency symmetric mode and a higher frequency antisymmetric mode for each λ. For a fixed λ, as μ decreases a value is reached where the antisymmetric mode disappears.

The work of Maniar and Newman (1997) showed that both types of trapped modes occur in the case of an infinite line of identical cylinders.

For the N-cylinder circular array problem, the array is invariant under the group C_{Nv} which can be used to develop a system of equations for each irreducible representation of C_{Nv}. This follows the approach of Gaspard and Rice (1989) in their study of the scattering problem for three discs.

2.12   Near Trapping Modes

For a finite array of vertical cylinders, Evans and Porter (1996) have examined local oscillations in the vicinity of the array at well defined frequencies which slowly decay. These modes are called near trapping modes. The near trapping arises from near vanishing (complex roots) of the determinant of an infinite system.

They follow Linton and Evans to reduce the problem to a solution of an infinite system of equations for the Fourier coefficients in the expansion of the potential near any cylinder. For the case of an array of N circular cylinders of radius a, equally space around a circle of radius R, the equation defining the trapped mode is

$$B_m^0 + \sum_{n=-\infty}^{\infty} B_n^0 K_{nm} = 0$$

where

$$K_{nm} = Z_n^0 e^{i(n-m)/2} \sum_{j=1}^{N} H_{n-m}(2kR\sin(\pi j/N))e^{i(m+n-2p)\pi j/N}$$

for the cylinder on the zeroth position on the positive x-axis. Here

$$Z_n^0 = J_n'(ka)/H_n'(ka)$$

where $H_n = J_n + iY_n$ is the Hankel function of the first kind and p is an integer. For each value of p they performed a search of the complex ka space close to the real line for a value of $a/d = 0.8$ using Newton's method and picked out complex values of ka corresponding to a zero of the determinant in this region.

Evans and Porter noted that peak forces can be associated with the occurrence of zeros of the determinant in the complex wavenumber ka space close to the real axis. As discussed above, for circular arrays one can use the symmetries of the problem to develop an infinite system of equations for each representation and look for zeros of the corresponding determinant for near trapping states.

2.13   Embedded Trapped Modes: Part I

Let M denote the n-dimensional cylinder $(-\infty, \infty) \times K$ where K is a compact manifold of dimension $n-1$ with smooth boundary ∂K. Let Δ denote the Laplacian on $F = M \setminus O$ where $O \subset$ interior (M) is a smooth bounded obstacle with Neumann boundary conditions on $\partial F = \partial M \cup \partial O$. Let $H^1(F)$ denote the Sobolev space $\{\psi \in L^2(F) | \nabla \psi \in L^2(F)\}$. The continuous spectrum of Δ is $[0, \infty)$ and since $\Delta \geq 0$, the eigenvalues of Δ are embedded in the continuous spectrum.

For a general obstacle such eigenvalues are not expected to exist. Thus, the results of Ursell, Jones and more recently Evans, Levitin and Vassiliev are of some interest. As discussed in Davies and Parnovski (1997), the result of Evans, Levitin and Vassiliev follows from the principle of symmetry which provides a decomposition

$$L^2(F) = H_1 \oplus H_2$$

into two subspaces which are invariant with respect to Δ. Assume the continuous spectrum of Δ restricted to H_1 satisfies: $\sigma_{ess}(\Delta|_{H_1}) \cap [0, p] = \emptyset$ for $p > 0$. Then by the Rayleigh-Ritz method applied to $\Delta|_{H_1}$, if there is a $\psi_0 \in H_1 \cap H^1(F)$, $\psi_0 \neq 0$ such that

$$\lambda(\psi_0) = \frac{\int_F |\nabla \psi_0|^2 dx dy}{\int_F |\psi_0|^2 dx dy} < p,$$

then the discrete spectrum of $\triangle|_{H_1}$ is nonvanishing and so also is that of $\triangle$. The specific decomposition used by Evans, Levitin and Vassiliev was for $K = [-1,1]$ and $O = \{z = (x,y) \in M | |y| < f(x)\}$ for $f \in C_0^\infty(\mathbf{R})$, $f(x) < 1$ and $f \not\equiv 0$. One takes $H_1 = H_{odd}$ and $H_2 = H_{even}$ the subspaces of odd (resp. even) functions with respect to reflection about the line $l_0 = \{z|y = 0\}$. In this case $\sigma_{ess}(\triangle|_{H_{odd}}) = [\pi^2/4, \infty)$. One notes that any eigenvalue of the antisymmetric problem is an eigenvalue of the symmetric problem. Thus, we have:

Theorem 7 *(Evans, Levitin, Vassiliev) The antisymmetric problem always has a trapped mode below the first cutoff and the symmetric problem always has a trapped moded $\lambda_0 \in (0, (\pi/2d)^2)$, i.e. embedded in the continuous spectrum.*

Davies and Parnovski (1997) has extended the Evans, Levitin and Vassiliev result to show that for a more general class of obstacles, L^2 eigenfunctions may exist or may not exist depending on the shape, size and position of the obstacle in the waveguide.

2.14   Trapped Modes Embedded in the Continuous Spectrum: Part II

In the case that the obstruction is a cylinder, the work of Maniar and Newman (1997) suggested that there are trapped modes of the antisymmetric problem above the cutoff. Evans and Porter (1997) numerically examined the region $k \in (\pi/2d, 3\pi/2d)$ for the Neumann case and $k \in (\pi/d, 2\pi/d)$ for the Dirichlet case. Evans and Porter show that for these regions the solution can be reduced to:

Theorem 8 *(Evans and Porter) Trapped modes occur at the intersection of the curves $det = 0$ and $\tilde{S} = 0$ where*

$$\tilde{S} = \sum_{n=0}^{\infty} \frac{e_n^m \cos(2n+1)u_0}{Y'_{2n+1}(ka)}.$$

Here $\{\lambda_r\}$ are the eigenvalues of the matrix $\delta_{mn} + B_{mn}$, for $m, n = 0, 1, 2...$ with eigenvector $e^r = \{e_0^r, e_1^r, ...\}$, $\lambda_m = min_r\{|\lambda_r|\}$, and $k\cos u_0 = \pi/2d$ in the Neumann case, resp. $k\cos u_0 = \pi/d$ in the Dirichlet case.

The numerical work of Evans and Porter showed that for the antisymmetric problem with Neumann boundary conditions there is a trapped mode for values $kd \in (\pi/2, 3\pi/2)$, viz.

$$kd = 1.488884\pi$$

$$a/d = .3520905,$$

and for the Dirichlet case for values of $kd \in (\pi, 2\pi)$

$$kd = 1.991867\pi$$

$$a/d = .2670474.$$

Away from the cylinder the form of the trapped mode oscillation is similar to the second cross channel eigenfunction (viz., $sin(3\pi y/2d)$ in the Neumann case and $sin(2\pi y/d)$ in the Dirichlet case) and the trapped mode solutions vanish as $|x| \to \infty$.

Again we summarize this numerical work as:

Theorem 9 *(Evans and Porter) For the case of a cylindrical obstruction there are trapped modes embedded in the continuous spectrum for the antisymmetric problem for both the Neumann and the Dirichlet boundary conditions.*

2.15 Wave Amplitudes

Maniar and Newman (1997) and Evans and Porter (1997) have studied the relative wave amplitudes of trapped modes for certain configurations of cylinders. In the water-wave problem the forces and free surface amplitudes of motion which occur in arrays of cylinders near Dirichlet frequencies are of interest to researchers. Consider the case of N equally space cylinders with radii $r_k = a_k$; define the wave amplitude to be

$$Amp(a_k/d) = max_{\theta_k \in (0,\pi)}\{|\phi(a_k, \theta_k)|\}$$

for the potential $\phi(r_k, \theta_k)$ around the kth cylinder for some trapped mode frequency; and let

$$Amp(a_k/d)/Amp(a_j/d)$$

denote the ratio of maximum wave elevations at the cylinder of radius a_k to that at cylinder of radius a_j at that trapped mode frequency.

The work of Maniar and Newman and Evans and Porter examined the maximum force on a linear array of cylinders in the channel. For the water wave problem these systems are interesting because in the head seas they will produce a force several times the force on an isolated cylinder at a frequency close to the new embedded Neumann and Dirichlet trapped mode. In their analysis of the maximum excitant force on the middle cylinder in a linear array, Evans and Porter found peaks due to the Neumann and Dirichlet modes below the cutoff but also at frequencies close to the new embedded Neumann and Dirichlet trapped modes. They also observed a peak (just below $kd = 3\pi/2$) which they identified as a near trapped mode.

2.16 Nonsymmetric Obstacles - Thin Parallel Obstacle

Evans and Linton (1993) considered the Neumann problem on the space

$$M = (-\infty, \infty) \times [-1, 1]$$

with $O = [\alpha, \beta] \times \{p\}$ with $p \in [-1, 1]$, a cut along an interval parallel to the axis of the cylinder. They showed that there is a trapped mode provided that the cut is long enough. In this case, Davies and Parnovski (1997) showed the existence of a trapped mode for cuts of arbitrary short length.

2.17   Spectral Theorem: the Neumann Case

Groves (1997) has developed the spectral theorem for the acoustic waveguide problems considered above. Let $\Omega \subset \mathbf{R}^n, n \geq 2$ denote the domain created by inserting a compact obstacle O into the uniform cylinder $M = (-\infty, \infty) \times K$ where K is a bounded, open, connected subset of $\mathbf{R}^{n-1}$ with smooth boundary. Let $A : D(A) \subset L^2(\Omega) \to L^2(\Omega)$ denote the self-adjoint operator associated with the quadratic form $q(u) = \int_\Omega |\nabla u|^2 d(x, y)$ where $u \in D(q) = H^1(\Omega)$. In the case the boundary ∂O is sufficiently smooth, the operator A is just the Friedrichs extension of the n–dimensional Laplacian $-\Delta$ on Ω with Neumann conditions on $\partial\Omega$. O is assumed to be such that Ω has the cone property (v., Adams (1975)). Groves' result is an extension of work of Goldstein (1969) and Wilcox (1975).

Let A' denote the Friedrichs extension of the $(n-1)$–dimensional Laplacian in K with Neumann boundary conditions on ∂K. Let $\nu_0, \nu_1, ...$ and $f_0, f_1, ...$ denote the eigenvalues and eigenvectors of A'. Under the conditions stated, A hs at most a countably infinite set of eigenvalues, where each eigenvalue has finite multiplicity. Let $V = \{u_1, u_2, ...\}$ be a maximal orthonormal set of eigenvectors of A with eigenvalues $\{\lambda_1, \lambda_2, ...\}$. Let $w_{n,k}^0$ denote the plane waves given by

$$w_{n,k}^0(x, y) = \frac{1}{\sqrt{2\pi}} f_n(y) e^{ikx}$$

for $(x, y) \in (-\infty, \infty) \times K$. Define the set of distorted plane waves $\{w_{n,k}\}_{(n,k)\in\Lambda}$ where

$$\Lambda = \{(n, k) \in \mathbf{N}_0 \times \mathbf{R} | k^2 + \nu_n \notin \{0, \nu_1, \nu_2, ...\} \cup \{\lambda_1, \lambda_2, ...\}\}$$

given by

$$w_{n,k}(x, y) = w_{n,k}^0(x, y) + v_{n,k}(x, y).$$

Here $v_{n,k}$ is the unique solution of

$$(-\Delta - (k^2 + \nu_n))v_{n,k}(x, y) = 0$$

for $(x, y) \in \Omega$ and

$$\frac{\partial}{\partial n} v_{n,k}(x, y) = -\frac{\partial}{\partial n} w_{n,k}^0(x, y)$$

for $(x, y) \in \partial\Omega$ and the outgoing radiation condition (v. Groves (1997)).

Theorem 10 *(Groves) The spectral decomposition of A is given by*

$$(Af)(x, y) = \lim_{\lambda \to \infty} \int \sum_{(n,k)\in N_\lambda} (k^2 + \nu_n) a_n^f(k) w_{n,k}(x, y) dk +$$

$$\lim_{M \to \infty} \sum_{n=1}^{min(|V|, \mathcal{M})} \lambda_n b_n^f u_n(x, y)$$

where

$$a_n^f(k) = \lim_{R \to \infty} \int_{\Omega_R} f(x, y) \overline{w_{n,k}(x, y)} d(x, y)$$

and

$$b_n^f = \int_\Omega f(x,y)\overline{u_n(x,y)}d(x,y).$$

Here $N_\lambda = \{(n,k) \in \mathbf{N}_0 \times \mathbf{R}|k^2 + \nu_n < \lambda\}$ *and* $\Omega_R = \Omega \cap ((-R,R) \times K)$.
  The domain of A is

$$D(A) = \{f \in L^2(\Omega)| \int_{-\infty}^{\infty} \sum_{n\in\mathbf{N}_0} (k^2 + \nu_n)^2|a_n^f(k)|^2 dk + \sum_{n=1}^{|V|} \lambda_n^2|b_n^f|^2 < \infty\}.$$

The essential spectrum of A is $\sigma_{ess}(A) = [0,\infty)$ *and the discrete spectrum is*
$\sigma_p(A) = \{\lambda_1, \lambda_2, ...\}.$

2.18 Compact Smooth Obstacles

Groves (1997) considered the case of the spectrum of the Neumann Laplacian in the infinite domain described in the last section. In this case he shows that there exists at least one embedded eigenvalue when O is piecewise smooth closed $(n-2)$-dimensional surface whose unit normal is parallel to K at each point of O for which it exists. His proof relies on the generalized eigenfunction expansion discussed in the last section.

Theorem 11 *(Groves) For the model described in the last section, A has an eigenvalue* λ *with* $\lambda < \nu_N$, *where N is the smallest number for which* $\partial f_N/\partial n$ *is not identically zero on O.*

Groves result extends the work of Evans, Linton and Ursell (1993) who considered the case $\Omega_0 = (-\infty, \infty) \times (0, f) \subset \mathbf{R}^2, O = [0, a] \times \{p\}, p \in (0, f)$ for $a, f > 0$; they had showed that there is at least one eigenvalue for sufficiently large values of a.

Like the method of Evans, Levitin and Vassiliev, one decomposes $L^2(\Omega) = H_1 \oplus H_2$ where H_1 and H_2 are A invariant subspaces:

$$H_1 = \{f|a_j^f = 0, j = 0, ..., N-1\}$$

and

$$H_2 = \{f|a_j^f = 0, j = N, N+1, ..., b_j^f = 0, j = 1, ..., |V|\}.$$

The spectral theorem shows that $\sigma_{ess}(A_1) = [\nu_N, \infty), \sigma_p(A_1) = \{\lambda_1, \lambda_2, ...\}$ and $\sigma_{ess}(A_2) = [0, \infty), \sigma_p(A_2) = \{\emptyset\}$ where $A_1 = A|_{H_1}$ and $A_2 = A|_{H_2}$. To prove that A_1 and hence A has an eigenvalue λ with $\lambda < \nu_N$, it suffices to prove that

$$inf_{u\in D(Q_1)\backslash\{0\}} \frac{Q_1(u)}{\|u\|^2} < \nu_N$$

where $Q_1(u) = \int_\Omega |\nabla u|^2 d(x,y)$ for $u \in H^1(\Omega) \cap H_1$.

Groves considered the case $O = [-a, a] \times P'$, where P' is a closed connected hypersurface of K and symmetric about $\{0\} \times K$. In this case, Groves showed that there is at least one symmetric eigenvector of A and when O is sufficiently long surface, at least one antisymmetric vector.

2.19   Perturbed Semi-infinite Cylinder

Goldstein (1969) treated the case of a pertubed semi-infinite cylinder in $\mathbf{R}^N$. Let S' denote the semi-infinite cylinder given by $x_N \geq 0, \tilde{x} = (x_1, ..., x_{N-1}) \in \ell$, where ℓ is a bounded $N - 1$ dimensional domain lying in the hyperplane $x_N = 0$. Let $A_{S'}$ denote the Friedrichs extension of the operator given by $-\Delta$ acting on $C_0^\infty(S')$. Since ℓ is a bounded $N - 1$ dimensional domain, there is a complete, countable, orthonormal set of eigenfunctions for the operator A_ℓ, with eigenvalues $\{\nu_n\}$. Define the perturbed domain Ω' which satisfies the conditions: $\Omega' \subset S', \Omega' = S'$ for some $h \leq x_N$ and $\dot{\Omega}'$ is a C^∞ surface in $\mathbf{R}^N$. Goldstein notes that there are no point eigenvalues in the interval $(-\infty, \nu_1)$; i.e., the spectrum consists of the interval $[\nu_1, \infty)$, based on the work of Jones (1953). Of course there could be point eigenvalues embedded in the continuous spectrum. However, Goldstein observes that there are no point eigenvalues if Ω' also satisfies the condition $cos\phi(s) \leq 0$ for each point $x \in \dot{\Omega}'$ where $\phi(s)$ is the angle between the exterior normal to Ω' and the positive x_N-axis. This latter result goes back to Rellich (1943).

2.20   Acoustic Operators

Weder (1991) considers the problem of perturbed acoustic operators in a stratified fluid of dimension $n + 1$. He studies the question of whether there are positive eigenvalues embedded in the essential spectrum, which in this case is $\sigma = [0, \infty)$. To obtain his result he uses a version of Mourre's estimate, v. Weder (1988). Under fairly natural conditions on the speed of propagation function there are no embedded eigenvalues, although they are found to exist for the transmission problem. Let A_0 denote the unperturbed acoustic propagator $A_0\phi = -c_0^2(y)\phi$ where $c_0(y)$ is a real valued measurable function which is bounded below and above and represents the speed of propagation of the sound waves in the fluid. The spectrum of A_0 is given by $\sigma(A_0) = [0, \infty)$ and A_0 is absolutely continuous. For the perturbed acoustic propagator, Weder shows that the essential spectrum is also $[0, \infty)$; and the question is whether there are positive eigenvalues embedded in the essential spectrum.

For Mourre's estimate the associated operator D is the self adjoint generator of the unitary representation of the dilation group given by $(U(\theta)\phi(x,y)) = e^{-\frac{n+1}{2}\theta}\phi(e^{-\theta}x, e^{-\theta}y)$ for $\phi \in L^2(\mathbf{R}^{n+1})$; viz $U(\theta) = e^{-2i\theta D}$. For the remaining development, we direct the reader to Weder's monograph.

2.21   Open Problems

For open problems in this area, we leave the reader with the formal proofs of theorems 5, 6 and 9 in this chapter.

Chapter 3

Trapped States in Bent Quantum Wires

3.1 Introduction

In this chapter the basic results on trapped states in bent quantum wires are presented. L-shaped structures have been studied in acoustics by Lippert (1954, 1955), in flat electromagnetic waveguides by Carini and coworkers (1992) following the theoretical work of Exner, Seba and Stovicek (1989), and in quantum wires by Schult, Ravenhall and Wyld (1989). Curvature induced bound states were discussed by Goldstone and Jaffe (1992), which results were later improved by Renger and Bulla (1995) and Duclos and Exner (1995). We review below also the case of quantum wires with bumps, which has been treated by a number of researchers, as well as the case of waveguides with a boundary window. The next topic of interest to the experimentalists is curvature induced resonances. The mathematical physics of resonances has been addressed by Duclos, Exner and coworkers. Quantum wire systems with discrete states embedded in the continuous spectrum are presented.

Experimentalists raised the issue of point defects in the quantum devices and we review the mathematical physics of non random point defects in this chapter. In chapter seven we turn to the case of random perturbations.

In the final sections we review the case of modelling quantum wires as one dimensional graphical models. We return to the subject of graphical models in chapter nine.

3.2 L-Shaped Quantum Structures

The work of Exner et al. showed that bound states can exist in sharply broken L-shaped structures. Specifically,

Theorem 1 *(Exner, Seba and Stovicek) Let Ω be an L-shaped strip of width d. Let $H_\Omega = -\Delta_D^\Omega$ denote the Dirichlet Laplacian. Then H_Ω has a bound state with eigenvalue $\lambda_0 = .9291\lambda_1$ where $\lambda_1 = (\pi/d)^2$ is the continuous spectrum threshold. The ground state is symmetric with respect to the symmetry axis, $(x = y)$, of Ω and is nondegenerate.*

The proof of Exner, Seba and Stovicek proceeds as follows. Since Ω has the segment property, by Dirichlet-Neumann bracketing, the continuous spectrum of H_Ω starts at $\lambda_1 = (\pi/d)^2$. The goal is to find an eigenvalue $\lambda_0 = \kappa\lambda_1$ where $\kappa < 1$. In the two arms, we have

$$\psi_I(x,y) = \sum r_j(y)\phi_j(x)$$

and

$$\psi_{II}(x,y) = \sum t_j(x)\phi_j(y)$$

where $\phi_j(t) = (2/d)^{1/2}sin(\omega_j t/d)$ with $\omega_j = j\pi$; the eigenenergies are $\lambda_j = \omega_j^2/d^2$ where $j \in \mathbf{N}$. The eigenvalue equation $H_\Omega\psi = \lambda\psi$ gives the relationships

$$\psi_I(x,y) = \sum_{j=1}^{\infty}(-1)^{j+1}r_j e^{q_j(1-y/d)}\phi_j(x)$$

and

$$\psi_{II}(x,y) = \sum_{j=1}^{\infty}(-1)^{j+1}t_j e^{q_j(1-x/d)}\phi_j(y)$$

where $q_j = \pi(j^2 - \kappa)^{1/2}$. In the square region defined by the intersecting arms

$$\psi_{III}(x,y) = \sum_{j=1}^{\infty}(-1)^{j+1}[r_j\alpha_j(y)\phi_j(x) + t_j\alpha_j(x)\phi_j(y)]$$

where $\alpha_j(x) = sinh(q_j x/d)/sinh(q_j)$. The wave function is symmetric or antisymmetric if and only if $r_j = t_j$, resp. $r_j = -t_j$. Consider the symmetric case. The matching of the normal derivatives on the boundaries of I, III and II, III gives the relationship

$$r = Cr$$

where

$$C_{jk} = \frac{1}{\pi}(1 - e^{-2\pi\sqrt{j^2-\kappa}})\frac{jk}{\sqrt{j^2 - \kappa}(j^2 + k^2 - \kappa)}.$$

Take $r_1 = a_1 = 1$ and assume that $r_j = j^{-s}a_j$ where $a = \{a_j\} \in l^\infty$ for $s > 0$. This gives rise to the equation

$$a = b + Ka$$

where $a = \{a_j\}_{j=2}^{\infty}$ and $b = \{b_j\}_{j=2}^{\infty}$ and $K = (K_{jk})_{j,k=2}^{\infty}$. Here

$$K_{jk} = \frac{1}{\pi}(1 - e^{-2\pi\sqrt{j^2-\kappa}})\frac{j^{s+1}k^{s-1}}{\sqrt{j^2 - \kappa}(j^2 + k^2 - \kappa)}$$

and

$$b_j = \frac{1}{\pi}(1 - e^{-2\pi\sqrt{j^2-\kappa}})\frac{j^{s+1}}{\sqrt{j^2 - \kappa}(j^2 + 1 - \kappa)}.$$

The equation $a = b + Ka$ has a unique solution if $\|K\|_{l^\infty} < 1$ so that

$$a = (I - K)^{-1}b.$$

This can be checked for the symmetric case. To solve for κ one finds that

$$1 = \frac{1 - e^{-2\pi\sqrt{1-\kappa}}}{\pi\sqrt{1-\kappa}}[\frac{1}{2-\kappa} + \sum_{j=2}^{\infty}\frac{a_j}{j^2 + 1 - \kappa}].$$

The method of proof leads naturally to a numerical procedure to estimate κ viz. take $a^{(0)} = b$ and use $a^{(j+1)} = b + Ka^{(j)}$. One can check that $\kappa = .9291$.

In the case of finite length arms (which open into a wider region with a lower continuum threshold), the authors point out that the bound state turns into a resonance structure. We will return to this later in this volume.

3.3 Curvature Induced Bound States

Exner and Seba (1989) showed that on a smooth curved planar strip which is thin enough and whose curvature decays sufficiently rapidly, the Dirichlet Laplacian has at least one isolated eigenvalue below the bottom of the essential spectrum. This work was expanded in Exner (1990) to curved tubes in $\mathbf{R}^3$ and in Exner (1993) to tubes of slowly decaying curvature. Goldstone and Jaffe (1992) (see also Dunne and Jaffe (1993)) argued that any bending pushes the threshold of the spectrum below the lowest transverse-mode energy, which implies that there exist bound states in curved tubes of any thickness provided the curvature is zero outside a bounded region or vanishes asymptotically. The result of Goldstone and Jaffe has been improved by Renger and Bulla (1995) and Duclos and Exner (1995).

Consider a curved strip Ω in the plane $\mathbf{R}^2$. Assume that the strip is of constant width $2d$. Let Γ denote the axis of the strip, with curvature $\gamma(s)$, where s is the arc length of Γ. Let u be the coordinate orthogonal to Γ. Thus we can represent $x = a(s) - ub'(s)$ and $y = b(s) + ua'(s)$ where $a'(s)^2 + b'(s)^2 = 1$. The signed curvature $\gamma(s)$ of Γ is $\gamma(s) = b'(s)a''(s) - a'(s)b''(s)$ and $|\gamma(s)| = (a''(s)^2 + b''^2)^{1/2}$. The curvature of a fixed u curve at the point (s, u) is

$$c(s, u) = \frac{|\gamma(s)|}{|1 + u\gamma(s)|}.$$

For more details see Exner and Seba (1989). In particular, one notes that γ is determined by a, b; and the converse is true up to Euclidean transformations.

In general, one assumes that (a1) $\gamma \in L^1_{loc}(\mathbf{R})$, (a2) $d\|\gamma\|_\infty < 1$ and (a3) Ω is not self-intersecting and (a4) γ is piecewise C^2 with $\dot{\gamma}, \ddot{\gamma}$ bounded. Let Δ_Ω^D denote the Dirichlet Laplacian on Ω. One can show that $-\Delta_\Omega^D$ is unitarily equivalent to the operator

$$H_\Omega = g^{-1/2}\partial_s g^{-1/2}\partial_s - g^{-1/2}\partial_u g^{1/2}\partial_u$$

on $L^2(\mathbf{R} \times (-d, d), g^{1/2}dsdu)$ where $g^{1/2}(s, u) = 1 + u\gamma(s)$ with Dirichlet condition at $u = \pm d$. The Goldstone-Jaffe result as modified by Duclos and Exner (1995) states that the discrete spectrum of $-\triangle_\Omega^D$ is nonempty provided the curvature does not push the essential spectrum threshold down.

Theorem 2 *("Goldstone-Jaffe") Assuming (a1)-(a3), if the strip is not straight (i.e., $\gamma \neq 0$), then $\inf \sigma(-\triangle_\Omega^D) < k_1^2$, where $k_1 = \pi/2d$. If $\inf \sigma_{ess}(-\triangle_\Omega^D) = k_1^2$, then $-\triangle_\Omega^D$ has at least one isolated eigenvalue of finite multiplicity.*

The proof follows the argument of Goldstone and Jaffe, which is based on a variational estimate and the choice of a particular trial function. Viz., define the form

$$q_\Omega(\phi, \psi) = (\phi, H_\Omega\psi)_g.$$

Then the minimax principle states that

$$\inf(\sigma(H_\Omega)) = \inf_{\psi \in Q(H_\Omega)} \frac{q_\Omega(\psi, \psi)}{(\psi, \psi)_g}$$

where $Q(H_\Omega)$ is the quadratic form domain of H_Ω. One selects a trial function $\hat{\psi}$, calculates

$$E(\hat{\psi}) = \frac{q_\Omega(\hat{\psi}, \hat{\psi})}{(\hat{\psi}, \hat{\psi})_g}$$

and shows that

$$E(\hat{\psi}) < (\pi/2d)^2.$$

Exner's methods involve a unitary transformation to get a Schrödinger operator H with some effective potential $V(s, u)$, viz.

$$U : L^2(\mathbf{R} \times (-d, d), g^{1/2}dsdu) \rightarrow L^2(\mathbf{R} \times (-d, d))$$

where

$$U\psi(s, u) = g^{1/4}\psi(s, u)$$

and

$$q_1(\phi, \psi) = g(U^{-1}\phi, U^{-1}\psi);$$

viz.,

$$H\psi = -\frac{\hbar^2}{2m}\left(\frac{\partial}{\partial s}g^{-1}\frac{\partial \psi}{\partial s} - \frac{\partial^2 \psi}{\partial u^2}\right) + V(s, u)\psi$$

where

$$V(s, u) = \frac{\hbar^2}{2m}\left(\frac{1}{2}g^{-3/2}\frac{\partial^2 \sqrt{g}}{\partial s^2} - \frac{5}{4}g^{-2}\left(\frac{\partial \sqrt{g}}{\partial s}\right)^2 - \frac{1}{4}g^{-1}\left(\frac{\partial \sqrt{g}}{\partial u}\right)^2\right).$$

Extending Exner's result, Duclos and Exner have shown:

Theorem 3 *(Duclos and Exner) If in addition to the above assumptions, γ is compact or γ is piecewise C^2 with $\dot{\gamma}, \ddot{\gamma}$ bounded and $\gamma, \dot{\gamma}, \ddot{\gamma} \in L_\epsilon^\infty(\mathbf{R})$, then $-\triangle_\Omega^D$ has at least one bound state of energy below k_1^2.*

The proof of this result divides the region at the points $s = \pm s_0$ outside the curved part with Neumann condition at $\pm s_0$: $H_\Omega^N = H_- \oplus H_0 \oplus H_+$ where the spectrum of H_0 is discrete and the tail operators have purely continuous spectrum. Thus $\sigma_{ess}(H_\Omega^N) = \sigma_{ess}(H_- \oplus H_+) = \sigma(H_- \oplus H_+) = [k_1^2, \infty)$. Since $H_\Omega^N \leq H_\Omega$, the result follows from the minimax principle.

3.3.1 Number of Bound States

The number of bound states has been estimated by Duclos and Exner as follows:

Theorem 4 *(Duclos and Exner) If conditions (a1)-(a4) hold and one assumes (a5) $\gamma, \dot{\gamma} \in L^2(\mathbf{R}, |s|ds), \ddot{\gamma} \in L^1(\mathbf{R}, |s|ds)$, then the number of isolated eigenvalues, counting multiplicity, satisfies*

$$N(H) \leq 1 + \frac{1}{8}\left(\frac{1 + a\|\gamma\|_\infty}{1 - a\|\gamma\|_\infty}\right)^2 \frac{\int_{\mathbf{R}^2} \gamma(s)^2 |s - t| \gamma(t)^2 ds dt}{\int_{\mathbf{R}} \gamma(s)^2 ds}$$

where $a < \frac{c_2}{1 + c_2 \|\gamma\|_\infty}$ where

$$c_2 = min\left\{\frac{\pi}{\|\gamma\|_\infty}, \left(\frac{\pi^2}{2\|\ddot{\gamma}\|_\infty}\right)^{1/3}, \left(\frac{\pi}{\sqrt{5}\|\dot{\gamma}\|_\infty}\right)^{1/2}\right\}.$$

The proof involves noting that $N(H) \leq N(H_0)$ and reducing the upper bound of $N(H_0)$ to a case covered by Klaus (1977).

In particular it follows that

Theorem 5 *(Exner) If Ω has a single bend, then the strip has just one bound state.*

For an improved result, see Exner and Vugalter (1997). The case of tubes in $\mathbf{R}^3$ and mildly curved tubes are covered in Duclos and Exner (1995). Locally curved layers of constant width over a smooth surface have been considered recently by Duclos, Exner and Krejcirik (1999). We direct the reader to the papers for the details.

3.3.2 Weak Coupling

The weak coupling result for mildly curved waveguides of Duclos and Exner is:

Theorem 6 *(Duclos and Exner) Assume the conditions (a1) - (a5) described above. Let $\gamma_\beta(s) = \beta\gamma(s)$ for $\beta > 0$. For all β small enough (i.e., for mildly curved tubes), the Dirichlet Laplacian, $-\Delta_\Omega^D$, corresponding to the tube generated by Γ_β, has exactly one isolated eigenvalue $E(\beta)$ and*

$$\sqrt{k_1^2 - E(\beta)} = \frac{\beta^2}{8}\left\{\|\gamma\|^2 + \frac{1}{2}\sum_{N=2}^\infty (\chi_N, u\chi_1)^2 \rho_N \int_{\mathbf{R}^2} \dot{\gamma}(s)e^{-\rho_N|s-s'|}\dot{\gamma}(s')dsds'\right\}$$

$$+O(\beta^3)$$

where $\rho_N = k_1\sqrt{N^2 - 1}$ and the sum runs over even N where k_N^2 and χ_N are the transverse mode eigenvalues and eigenfunctions.

Proof of this result depends on the general weak coupling result for Schrödinger operators which is discussed in the next section.

3.3.3 Weakly Coupled Schrödinger Operators

For weakly coupled Schrödinger operators in a cylinder, Duclos and Exner show:

Theorem 7 *(Duclos and Exner) Let M be an open connected set in $\mathbf{R}^2$ with $\bar{M}$ compact. Set $H_\lambda = -\triangle^D + \lambda V$ where $-\triangle^D$ is the Dirichlet Laplacian on $\mathbf{R} \times M$ and V is measurable and bounded. The Dirichlet Laplacian $-\triangle^D_M$ on M has purely discrete spectrum consisting of eigenvalues $0 < k_0^2 < ...$ with possibly degenerate eigenvectors χ_{N_j}, with $j = 1, ..., d_N$. Let*

$$V_{mn} = \int_M \bar{\chi}_m(y) V(., y) \chi_n(y) dy.$$

If V is nonzero and $|V|_{00} \in L^1(\mathbf{R}, |x| dx)$, then H_λ for small λ has at most one simple eigenvalue $E(\lambda) < k_0^2$ and this happens if and only if $\int_{\mathbf{R}} \lambda V_{00}(x) dx \leq 0$. In this case, one has

$$\sqrt{\kappa_0^2 - E(\lambda)} = -\frac{\lambda}{2} \int_{\mathbf{R}} V_{00}(x) dx + \mathcal{O}(\lambda^2).$$

The proof of this theorem is an application of the Birman-Schwinger principle. Here $E(\lambda)$ is a bound state of H_λ if and only if λK_α has the eigenvalue -1 for $\alpha^2 = E(\lambda)$ where

$$K_\alpha = |V|^{1/2} (-\triangle^D - \alpha^2)^{-1} V^{1/2}$$

with $V^{1/2} = |V|^{1/2} sgn V$. For $\alpha < \kappa_0$, decompose $K_\alpha = Q_\alpha + P_\alpha$ with

$$Q_\alpha(x, y, x_1, y_1) = \frac{e^{k_0(\alpha)|x|}}{2k_0(\alpha)} |V(x, y)|^{1/2} \chi_0(y) e^{k_0(\alpha)|x_1|} \chi_0(y_1) V(x_1, y_1)^{1/2}$$

and

$$P_\alpha = A_\alpha + |V|^{1/2} B_\alpha V^{1/2}.$$

Here

$$A_\alpha(x, y, x_1, y_1) =$$

$$\frac{1}{k_0(\alpha)} |V(x, y)|^{1/2} \chi_0(y) [e^{k_0(\alpha)|x|>} sinh k_0(\alpha)|x|<] \chi_0(y_1) V(x_1, y_1)^{1/2}$$

and the kernel of B_α is given by R_0 above. The notation $|x|<$ is

$$|x|_< = max\{0, min(|x|, |x_1|) sgn(xx_1)\}$$

and

$$|x|_> = max\{|x|, |x_1|\}.$$

One can show that λK_α has the eigenvalue -1 if and only if the same is true for $\lambda(I + \lambda P_\alpha)^{-1} Q_\alpha$. If $\|\lambda P_\alpha\| < 1$, then one finds in the decomposition Q_α is a rank one operator, and one can derive the following implicit function for $k_0(\alpha) = z$:

$$z = G(\lambda, z)$$

where

$$G(\lambda, z) = -\frac{\lambda}{2} \int_{\mathbf{R} \times M} e^{-z|x|} V(x,y)^{1/2} \chi_0(y) [(I + \lambda P_{\alpha(z)})^{-1} e^{-z|\cdot|} |V|^{1/2} \chi_0](x,y) dx dy.$$

If V is rapidly decreasing, G is analytic around $(0,0)$ and the first part of the theorem follows by the implicit function theorem. In addition one sees how to compute the next terms in the expansion. For the details of the general case, we refer the reader to Duclos and Exner (1995), Blanckenbecler, Goldberger and Simon (1977) and Exner (1993).

3.4 Bumps in Quantum Structures

As we have seen in the last section, bound states appear if a quantum wire of constant cross section is locally deformed or bent. Bound states also arise when straight quantum wires have protrusions, bumps, or bulges; v., Popov (1986), Itoh, Sano and Yoshii (1992), who studied the resonance structure with the tight binding lattice model, Andrews and Savage (1995), Kunze (1993) and Toyama and Nogami (1994). The precise results in this area have been derived by Bulla, Gesztesy, Renger and Simon (1995) and they show that the bump phenomenology is slightly more subtle then the bent phenomenology. Bulla et al. found that the average cross-section variation is what matters. A bound state exists in this case if the added volume is positive.

Specifically, Bulla et al. (1995) studied waveguides which are obtained by adding a small bump to the tube $\Omega_0 = \mathbf{R} \times (0,1)$. They show that this system has at least one positive eigenvalue below the essential spectrum: $\sigma_{ess}(-\triangle_\Omega^D) = \sigma_{ess}(-\triangle_{\Omega_0}^D) = [\pi^2, \infty)$ for the Dirichlet Laplacian $-\triangle_\Omega^D$.

Theorem 8 *(Bulla et al.) Let Ω be and open connected set such that (a) for some $R > 0$ $\Omega \cap \{x \in \mathbf{R}^2 || x| > R\} = \Omega_0 \cap \{x \in \mathbf{R}^2 || x| > R\}$ and (b) $\Omega_0 \subset \Omega, \Omega_0 \neq \Omega$. Then $-\triangle_\Omega^D$ has at least one eigenvalue in $(0, \pi^2)$.*

The simple proof of this theorem, as noted by Bulla et al., is to take the test function

$$\hat{\psi}_{\beta,\delta}(x,y) = \begin{cases} sin(\pi y) e^{-\delta(|x|-a)}, & \text{if } |x| > a, 0 < y < 1 \\ sin(\pi y)/(1 + \beta(1 - \frac{|x|}{a})), & \text{if } |x| \le a, 0 < y < 1/\beta \\ 0 & \text{otherwise} \end{cases}$$

where $0 < \beta < b$ and $\delta > 0$. This trial function vanishes on $\partial \Omega$ and at ∞ and so is in the form domain $Q(-\triangle_\Omega^D)$. One checks that

$$E(\hat{\psi}_{\beta,\delta}) = \pi^2 (1 - 2a\delta\beta) + O(\beta^2 \delta) + O(\delta^2).$$

Thus, for β and δ small enough

$$E(\hat{\psi}_{\beta,\delta}) < \pi^2 = inf \sigma_{ess}(-\triangle_\Omega^D).$$

Since $inf \sigma(-\triangle_\Omega^D) < E(\hat{\psi}_{\beta,\delta})$ and $-\triangle_\Omega^D > 0$, this proves the result.

3.4.1   Weak Coupling

In the case $\Omega = \Omega_\lambda$ where

$$\Omega_\lambda = \{(x,y) \in \mathbf{R}^2 | 0 < y < 1 + \lambda f(x)\}$$

where $f(x) \in C^\infty(\mathbf{R})$, compact support and $f \geq 0$.

Theorem 9 *(Bulla et al.) For Ω_λ given above with $\int f(x)dx > 0$, then for all small positive λ, the operator $-\Delta_{\Omega_\lambda}^D$ has a unique eigenvalue $E(\lambda)$ in $(0, \pi^2)$; it is simple and $E(\lambda)$ is analytic at $\lambda = 0$. Furthermore,*

$$E(\lambda) = \pi^2 - \pi^4 \lambda^2 (\int_{\mathbf{R}} f(x)dx)^2 + O(\lambda^3).$$

One notes that the value of $\lambda \int_{\mathbf{R}} f(x)dx$ is just the area of $\Omega_\lambda \backslash \Omega_0$. In other words, the last result shows that if the area $|\Omega_\lambda \backslash \Omega_0|$ is small, then the ground state is unique and we have the weak coupling result $E(\lambda) = \pi^2 - \pi^4 |\Omega \backslash \Omega_0|^2$. It follows that:

Theorem 10 *(Bulla et al.) If $\int f(x)dx < 0$, then $-\Delta_{\Omega_\lambda}^D$ has no spectra in $[0, \pi^2)$ if λ is sufficiently small.*

Theorem 11 *(Bulla et al.) If on an arbitrarily small segment of the boundary $\partial \Omega_0$ of Ω_0 the boundary condition is replaced by a Neumann boundary condition, then at least one additional eigenvalue is instantly created in $(0, \pi^2)$.*

The proof involves defining, like Exner, an operator

$$H_\lambda = U_\lambda(-\Delta_{\Omega_\lambda}^D)U_\lambda^{-1} - \pi^2$$

which acts in $L^2(\Omega_0)$ where $U_\lambda : L^2(\Omega_\lambda) \to L^2(\Omega_0)$ is the unitary operator

$$(U_\lambda \psi)(x,y) = \sqrt{1 + \lambda f(x)} \psi(x, (1 + \lambda f(x))y).$$

Second, one decomposes the operator into a rank one piece and an analytic piece. For the remainder of the proof, we direct the reader to the original paper.

3.4.2   Bumps with Volume Change Zero

Exner and Vulgater (1997) considered the case of a quantum wire with a bump where the bump has volume change of zero. Consider the Dirichlet Laplacian on the region $\Omega_\lambda = \{(x,y) \in \mathbf{R}^2 | 0 < y < a(1 + \lambda f(x))\}$ for $f \in C_0^\infty(\mathbf{R})$. Viz., they have shown the following two weak coupling results:

Theorem 12 *(Exner and Vulgater) If $\operatorname{supp} f \subset [-b, b]$ and $\int_{-b}^{b} f(x)dx = 0$, then the discrete spectrum of $-\Delta_{\Omega_\lambda}^D$ is empty for all sufficiently small $|\lambda|$ provided that $a > \frac{1}{\sqrt{3}}b$.*

Theorem 13 *(Exner and Vulgater) Under the assumptions above, $-\Delta_{\Omega_\lambda}^D$ has an isolated eigenvalue $E(\lambda)$ for any nonzero, sufficiently small $|\lambda|$ if*

$$\frac{\|f'\|^2}{\|f\|^2} < (\pi/a)^2 \frac{6}{9 + \sqrt{90 + 12\pi^2}},$$

in which case

$$-c_1\lambda^4 \le E(\lambda) - (\pi/a)^2 \le -c_2\lambda^4$$

where $c_1, c_2 > 0$.

As Exner has pointed out to the author, the first result in this section is not optimal. There is a gap between the existence and nonexistence results in these two theorems, which we leave as an open problem.

3.5 Elbow Shaped Quantum Waveguides Revisited

Exner, Seba and Stovicek (1989) showed the existence of a bound state in an L-shaped waveguide . Later results assumed that the quantum waveguide had to be C^4-smooth, so its curvature is twice differentiable. Renger and Bulla (1996) have extended the results of Exner and coworkers to include the earlier work on L-shaped devices as part of the curvature induced results. Viz., Renger and Bulla have shown that the existence of derivatives of the curvature of the strip is required only in the complement of a bounded interval of the axis, no upper bound for the width of the strip is required and the general shapes of quantum wires are permitted. Let Λ be a region which contains a curved parallel strip Ω of width $2d$, where Λ coincides with Ω in the exterior of a bounded set. Specifically Renger and Bulla assume:

(b1) $1 - d|\gamma(s)| \ge 0$ for all $s \in \mathbf{R}$ where $\gamma(s)$ is the curvature of the axis Γ of the waveguide; i.e. the edges of the waveguide do not form loops;

(b2) there is a bounded subset $M_1 \subset \mathbf{R}$ such that for all $s \in \mathbf{R}\backslash M_1$ one has $1 - d|\gamma(s)| > \epsilon > 0$;

(b3) there is a bounded subset $M_2 \subset \mathbf{R}$ such that $\gamma(s)$ is twice continuously differentiable for all s outside M_2;

(b4) Ω is injective;

(b5) there is $k_0 \in \mathbf{R}$ and interval I such that for all $s \in I$ one has $|\gamma(s)| \ge |k_0| sgn(\gamma(s)) = sgn(k_0)$; i.e., Γ has nonvanishing curvature in some interval and the sign does not change over this interval.

Theorem 14 *(Renger and Bulla) If Λ is an open set with $\Omega \subset \Lambda$ and conditions (b1) - (b5) hold, then*

$$inf\,\sigma(-\Delta_\Lambda^D) < (\pi/2d)^2.$$

Assume now decay conditions for $\gamma(s)$ and its derivatives, viz.

(b6) for every $\epsilon > 0$ there is an $M_\epsilon \subset \mathbf{R}$, bounded so that $(M_1 \cup M_2) \subset M_\epsilon$, $\gamma, \dot\gamma, \ddot\gamma < \epsilon$ and there is a bounded set Λ_ϵ such that $\Lambda\backslash[(\mathbf{R}\backslash M_\epsilon) \times (-d, d)] \subset \Lambda_\epsilon$

and $[(\mathbf{R}\backslash M_\epsilon) \times (-d,d)] \cap \Lambda_\epsilon = \emptyset$. This is the segment property which is discussed in Reed and Simon (1978).

Theorem 15 *(Renger and Bulla) Let Λ and $\Omega \subset \Lambda$ be as described above. Let $\gamma(s)$ denote the curvature of the axis of Ω and $2d$ is its width. Take the standard coordinates s, u on Ω such that $\Omega = \mathbf{R} \times (-d, d)$. Assuming (b1), (b2), (b4) and (b6), then*

$$inf\, \sigma_{ess}(-\triangle_\Lambda^D) \geq (\pi/2d)^2.$$

3.6 Curvature Induced Resonances

Curvature induced resonances in a planar two-dimensional Diriclet tube of width d have been studied by Duclos, Exner and Stovicek (1995). They showed that if the strip-axis curvature satisfies certain regularity conditions then the distances of the resonance poles from the real axis are exponentially small as $d \to 0 +$. The resonances are obtained as perturbations of an operator with eigenvalues embedded in the continuous spectrum; viz., an operator H^0 where $H^0 = A - \partial_u^2$ with $A = -\partial_s^2 + V^0$ and $V^0(s) = -\frac{1}{4}\gamma(s)^2$; we assume that Ω is not straight and does not intersect itself. Then,

$$\sigma(H^0) = \{\lambda + E, \lambda \in \sigma(A), E \in \sigma(-\partial_u^2)\}$$

where $\sigma(A) = \{\lambda_n\}_{n=1}^N \cup [0, \infty)$ and $\sigma(-\partial_u^2) = \{E_j\}_{j=1}^\infty$ with $E_j = (\pi j/d)^2$ with eigenfunctions $\chi_j(u) = \frac{2}{d}sin(\frac{\pi j u}{d})$. The condition $\int V^0(s)ds < 0$ implies that the discrete spectrum of A is nonempty, the eigenvalues λ_n are negative, simple and N is finite. (The fact that N is non zero is treated in Simon (1976) and Blankenbecler, Goldberger and Simon (1977). The fact that N is finite is covered in Klaus (1977).) Let $H = -\triangle_\Omega^D$ be the Dirichlet operator for $\Omega \subset \mathbf{R}^2$ of fixed width d. The eigenvalues

$$E_{j,n}^0 = \lambda_n + E_j$$

above E_1 are embedded in the continuous spectrum of H^0. For small enough d this occur for all $j \geq 2, n = 1, ...N$; i.e. there are N embedded eigenvalues below the threshold of the jth transverse mode for $j \geq 2$. The goal is to show that these turn into resonances for the full Hamiltonian.

The following assumptions will be used below:

(c0) Ω is not straight and does not intersect itself;

(c1) γ extends to an analytic function on $\Sigma_{\alpha_0, \eta_0} = \{z \in \mathbf{C} | arg(\pm z) < \alpha_0$ or $|Imz| < \eta_0\}$ with $\alpha_0 < \pi/2$ and $0 < \eta_0$.

(c2) for all $\alpha < \alpha_0$ and all $\eta < \eta_0$ there is a $c_{\alpha,\eta} > 0$ and $\epsilon > 0$ such that

$$|\gamma(z)| < c_{\alpha,\eta}(1 + |z|)^{-1-\epsilon}$$

holds in $\Sigma_{\alpha,\eta}$.

Let $W = H - H^0$; then Duclos, Exner and Stovicek show that the perturbation W consists of operators which couple different transverse modes, which results in embedded eigenvalues turning into resonances; and they show that the first non-real term, which is given by the Fermi golden rule, is exponentially small as $d \to 0+$:

Theorem 16 *(Duclos, Exner and Stovicek) Assume (c0) - (c2). For sufficiently small d, each eigenvalue $E_{j,n}^0$ of H^0 gives rise to a resonance $E_{j,n}(d)$ of H:*

$$E_{j,n}(d) = E_{j,n}^0 + \sum_{m=1}^{\infty} e_m^{j,n}(d)$$

where $e_m^{j,n}(d) = O(d^m)$ as $d \to 0+$. The first term of the series is real valued and the second term satisfies the bound

$$0 \leq -Im\, e_2^{j,n}(d) \leq c_{\eta,j} e^{-2\pi\eta\sqrt{2j-1}/d}$$

where $\eta \in (0, \eta_0)$.

The proof is based on the complex scaling method of Aguilar and Combes (1971). Other resonances of H are noted in their proof, but these resonances do not approach the real axis as $d \to 0$ and are not considered further.

In complex scaling one sets

$$(U_\theta \psi)(s, u) = e^{\theta/2} \psi(e^\theta s, u).$$

The operator U_θ is unitary and one defines $H_\theta = U_\theta H U_\theta$, $H_\theta^0 = U_\theta H^0 U_\theta$ and $A_\theta = U_\theta A U_\theta$. Since A_θ and $-\partial_u^2$ commute

$$\sigma(H_\theta^0) = \cup_{j=1}^\infty \{E_j + \sigma(A_\theta)\}$$

and $\sigma(A_\theta) = \sigma_{ess}(A_\theta) \cup \{\lambda_1, ..., \lambda_N\} \cup \{\nu_1, ..., \nu_M\}$ where ν_j are possible resonances; note that A cannot have embedded eigenvalues in the continuous spectrum due to assumption (c2). The essential spectrum of A_θ is $\sigma_{ess}(A_\theta) = e^{-2\theta} \mathbf{R}$. Duclos, Exner and Stovicek show that the possible resonances $E_k + \nu_m$ are not too close to the eigenvalues $E_{j,n} = E_j + \lambda_n$ so that perturbation theory can be applied to $H_\theta - H_\theta^0$. One also has $\sigma_{ess}(H_\theta) = \sigma_{ess}(H_\theta - A_\theta)$. For the proof, the reader is directed to the original paper.

Thus, Duclos, Exner and Stovicek have shown that there is a finite number of resonances in the vicinity of higher thresholds in quantum wires, which number coincides with the number of isolated eigenvalues below the bottom of the continuous spectrum. These resonances describe transitions between different transverse modes (v., Duclos, Exner and Meller (1996) for a discussion). Duclos, Exner and Stovicek (1995) also comment on the relationship to the heuristic semiclassical prediction in Landau and Lifschitz (1974); see in addition Briet, Combes and Duclos (1987).

In fact, the bound holds for the total resonance width:

Theorem 17 *(Duclos, Exner and Meller) Assume (c0) - (c2). Then for $\eta \in (0, \eta_0)$, and for $n = 1, ..., N$ there is a $C_{\eta,j} > 0$ such that for $j \geq 2$*

$$0 \leq -Im\, E_{j,n}(d) \leq C_{\eta,j} e^{-2\pi\eta\sqrt{2j-1}/d}$$

for d small enough.

The reader is also directed to the papers by Nedelec (1997), Martinez (1994) and Nakamura (1989, 1995). In particular, Nedelec has extended the results of Duclos, Exner and Meller to the case of a variable width strip. Define

$$S(a_0, b_0) = \{z \in \mathbf{C} || Imz| < a_0 | Re(z)|\} \cup \{z \in \mathbf{C} || Im(z)| < b_0\}$$

and set $\tilde{H} = -\partial_s(1 + u\gamma(s))^{-2}\partial_s - \partial_u^2 + V(s, u)$. Assume now the following conditions:

(d1) $\gamma(s)$ is analytic on $S(a_0, b_0)$ and $\gamma(s) \to 0$ as $|s| \to \infty$ for $s \in S(a_0, b_0)$;

(d2) $\gamma(s), \dot{\gamma}(s)^2, \ddot{\gamma}(s)$ are $O(|s|^{-1-\epsilon})$ for $\epsilon > 0$;

(d3) Σ is not self intersecting;

(d4) $g(s)$ is analytic on $S(a_0, b_0)$ where the width is $d(s) = dg(s), d = sup_s d(s)$ and $g(s) \to 1$ as $|s| \to \infty$ for $s \in S(a_0, b_0)$;
and

(d5) $supg^2(x) < \frac{E_k}{E_{k-1}} inf_{x \in S(a_0, b_0)}g^2(x)$ and $sup_{x \in \mathbf{R}}(x(1/g^2)'(x)| < C$, with $0 < C < 2\delta_k(E_{k-1})^{-1}$. Here $E_j = (\pi j)^2$ are eigenvalues of $H = -\partial_y^2$ with Dirichlet boundary conditions on $y = 0, y = 1$, and

$$\delta_{k-1} = sup_{x \in S(a_0, b_0)}|E_k - \frac{E_k}{g^2(x)}|.$$

Theorem 18 *(Nedelec) Assuming (d1) - (d5), for every $b < b_0$ each resonance E of $\tilde{H}$ such that $|E - E_k| < Ch$ satisfies*

$$|Im(E)| \leq C_b e^{-2b\delta/h}$$

where $k \geq 2$.

For h small enough $\delta = \sqrt{E_k - E_{k-1}}$ for the constant width case and

$$\delta = (inf_{j<k, x \in S(a_0, b_0)}Re(E_k - E_j/g^2(x))^{1/2} - (sup|E_k - E_j/g^2(x)|)^{1/2}$$

for the variable width case.

Lent (1990) and Sols and Macucci (1990) have studied scattering in quantum wires. Sols and Macucci discovered the presence of resonances associated with quasi-bound states just below the propagation thresholds. They also found that at large angles more than one bound state may develop. For other numerical work on resonances and bound states see Lin and Jaffe (1996).

3.7 Waveguides with a Boundary Window

The case of a parallel straight quantum waveguide coupled laterally through a common boundary window of length L was considered by Exner et al. (1996) following work of Kunze (1993) and Hirayama et al. (1993), who studied the case in the presence of a magnetic field. In terms of quantum chaos this system is interesting since the corresponding classical system is pseudo- integrable, where its phase space is of genus three (v. Richens and Berry (1981)). Moreover, this system illustrates a quantum mechanical system which in principle can have any number of bound states, unrelated to the classically allowed volume of phase space.

Let Ω denote the space given by a pair of parallel strips of widths d_1, d_2 placed parallel to the x-axis. Assume they are separated by a Dirichlet boundary everywhere except on the interval $(-a, a)$; set $l = 2a$.

Exner, Seba, Tater and Vanek (1996) have shown that:

Theorem 19 *(Exner, Seba, Tater and Vanek) Let $H = H(d_1, d_2, l) = -\Delta_\Omega^D$ on $L^2(\Omega)$ denote the Dirichlet Laplacian. A sufficient condition for H to have at least one bound state is that*

$$l \geq \frac{d(1 + \nu)}{\sqrt{\nu(\nu + 2)}}$$

where $\nu = min\{d_1, d_2\}/max\{d_1, d_2\}$ and $d = max\{d_1, d_2\}$. Set $D = d_1 + d_2$.
The number N of bound states of H satisfies:

$$\left[\frac{l}{d} \frac{\sqrt{\nu(\nu + 2)}}{1 + \nu}\right] \leq N \leq 1 + \left[\frac{l}{d} \frac{\sqrt{\nu(\nu + 2)}}{1 + \nu}\right].$$

Theorem 20 *(Exner, Seba, Tater and Vanek) The discrete spectrum is nonempty for any $d > 0$ and consists of a finite number of simple eigenvalues $(\pi/D)^2 < E_1(a) < ... < E_N(a) < (\pi/d)^2$; and the continuous spectrum is given by*

$$\sigma_{ess}(H(d_1, d_2, 2a)) = [(\pi/d)^2, \infty).$$

In terms of weak coupling, Exner and Vulgater (1996) have shown:

Theorem 21 *(Exner and Vulgater) If a is small enough there is just one eigenvalue and one has:*

$$-c_1 a^4 \leq E(a) - (\pi/d)^2 \leq -c_2 a^4$$

for $c_1, c_2 > 0$.

In addition one can show:

Theorem 22 *(Exner, Seba, Tater and Vanek) $H(d_1, d_2, l)$ has an isolated eigenvalue in $[\mu_D, \mu_d)$ for any $l > 0$ where $\mu_d = (\pi/d)^2$ and for $\mu_D = (\pi/D)^2$.*

This result was also shown by Bulla et al. (1995).

Exner and Vugalter (1997) have extended the above result to multiple windows and have shown the following weak coupling result for windowed coupled Dirichlet strips. Let $A = \{a_k\}_{k=1}^N$ define a set of windows $W_k = [x_k - a_k, x_k + a_k]$ along the x-axis. Set $W = \cup_{k=1}^N W_k$ and $I(W) = \sum_{k=1}^N a_k |W_k|$. Let $H(d, W)$ denote the Laplacian on $L^2(\mathbf{R} \times [0, d])$ with the Neumann boundary condition at the window part W of the x-axis and Dirichlet boundary condition along the remaining boundary.

Theorem 23 *(Exner and Vulgater) The essential spectrum is given by*

$$\sigma_{ess}(H(d_1, d_2, W)) = [(\pi/d)^2, \infty).$$

The discrete part is contained in $((\pi/D)^2, (\pi/d)^2)$, and it is finite and nonempty provided that $W \neq 0$. If $I(W)$ is sufficiently small, then $\sigma_{disc}(H(d_1, d_2, W))$ consists of just one eigenvalue $E(W) \leq (\pi/d)^2$ and there are $c_1, c_2 > 0$ such that

$$-c_1 I(W)^2 \leq E(W) - (\pi/d)^2 \leq -c_2 I(W)^2.$$

In this paper the authors have also noted that their proof extends to show that for a straight Dirichlet strip with an arbitrary small protrusion, there is a bound state.

These authors have also examined the unfolded level spacing distribution. They find that this distribution is sharply localized around one value. Thus, the level statistics and phase shift statistics show that the present system is essentially nonchaotic.

3.7.1   Mode-Matching in a Boundary Window

First we set some notation. Take the case $d_1 = d_2 = d$; let $\kappa_1 = \sqrt{\mu_d}$ and $K_1 = \sqrt{\mu_D}$. Let $\chi_j(y) = \sqrt{\frac{2}{d}}sin(\kappa_j)$ and $\phi_j(y) = \sqrt{\frac{2}{d}}sin(K_{2j-1}(d-y))$ where $\kappa_j = j\kappa_1$ and $K_{2j-1} = (2j-1)K_1$. Set

$$\psi(x, y) = \sum b_j e^{q_j(a-x)} \chi_j(y)$$

for $x \geq a$ where $q_j = \kappa_1\sqrt{j^2 - \epsilon}$. In the symmetric case

$$\psi_s(x, y) = \sum a_j \frac{cosh(p_j x)}{cosh(p_j a)} \phi_j(y)$$

for $0 \leq x \leq a$ where $p_j = \kappa_1\sqrt{(j - 1/2)^2 - \epsilon}$. Mode matching the normal derivatives at $x = a$ yields in the Neumann case the system of equations:

$$q_j b_j + \sum_{k=1}^{\infty} a_k p_k tanh(p_k a)(\chi_j, \phi_k) = 0;$$

in the Dirichlet case *tanh* becomes *coth*. Continuity at the boundary implies that

$$b_j = \sum_{k=1}^{\infty} a_k(\chi_j, \phi_k).$$

This can be rewritten as the equation

$$Ca = 0$$

where $C_{jk} = (a_j + p_k tanh(p_k a))(\chi_j, \phi_k)$. One can check that

$$(\chi_j, \phi_k) = \frac{(-1)^{j-k}}{\pi} \frac{2j}{j^2 - (k - 1/2)^2}.$$

As described above, this can be rewritten as the equation

$$b + Kb = 0$$

where

$$K_{jm} = \frac{1}{q_j} \sum_{k=1}^{\infty} (\chi_j, \phi_k) p_k \tanh(p_k a)(\phi_k, \chi_m).$$

As discussed in the section 3.2 for L-shaped waveguides, this provides a useful tool for numerically solving for the wavefunction.

Exner and coworkers also develop the scattering matrix for this case and numerically model the conductance G and the probability flow distribution $j(x)$ where vortices are noted.

3.7.2 Lateral Semitransparent Barrier

Extending the work of Popov (1986) and Exner, Seba, Tater and Vanek (1996), Exner and Krejcirik (1999) have examined the case of a quantum waveguide with a lateral semitransparent barrier, i.e. a infinite straight Dirichlet strip divided by a thin semitransparent barrier on a line parallel to the walls. The Hamiltonian in this case is

$$H_\alpha = -\Delta_\Omega + \alpha(x)\delta(y)$$

where $\Omega - \mathbf{R} \times (-d_2, d_1)$ is the double-guide strip. At the outer edge, one assumes Dirichlet boundary conditions

$$\psi(x, -d_2) = \psi(x, d_1) = 0$$

and at the barrier one assumes a δ potential:

$$\psi(x, 0+) = \psi(x, 0-) = \psi(x, 0)$$

$$\psi_y(x, 0+) - \psi_y(x, 0-) = \psi(x, 0) = \alpha(x)\psi(x, 0).$$

As above, let $d = max\{d_1, d_2\}, D = d_1 + d_2$ and

$$\nu = \frac{min\{d_1, d_2\}}{max\{d_1, d_2\}}.$$

Finally, one assumes that the barrier is a local perturbation, i.e.

$$lim_{|x| \to \infty} \alpha(x) = \alpha_0.$$

Let $\{\nu_n(\alpha)\}$ denote the eigenvalues of the transverse operator. For this model, one can show that the essential spectrum is $\sigma_{ess}(H_\alpha) = [\nu_1(\alpha_0), \infty)$. Exner and Krejcirik show that there is always a bound state below the bottom of the essential spectrum provided the effective coupling function is attractive in the mean:

Theorem 24 *(Exner and Krejcirik) If (1) $\alpha - \alpha_0 \in L^1_{loc}(\mathbf{R})$, (2) $\alpha(x) - \alpha_0 = \mathcal{O}(|x|^{-1-\epsilon})$ for some $\epsilon > 0$ as $|x| \to \infty$, and (3) $\int_{\mathbf{R}} (\alpha(x) - \alpha_0)dx < 0$, then H_α has at least one isolated eigenvalue below its essential spectrum.*

Exner and Krejcirik have estimated the number of bound states; they have numerically examined the S-matrix and the conductivity given by the Landauer formula for this model. In particular they show that in the conductivity there is a deformation from the ideal steplike shape with jumps at transverse thresholds; viz., deep resonances arise. Exner and Krejcirik also have looked at the probability flow (as in Berggren's work) and they also examined the case of an infinite cylindrical surface with a magnetic field parallel to the cylinder axis. We direct the reader to the paper for the results in these cases.

3.8   Discrete States in the Continuous Spectrum

The results cited above have focussed on the existence of discrete states below the continuous spectrum for quantum wires. The problem of bound states embedded in the continuous spectrum was mentioned in Chapter two in reference to the recent work of Evans and coworkers in acoustic and water-wave theory.

In terms of quantum waveguides, Popov and coworkers have recently derived several results for embedded bound states, in particular for the problem of waveguides coupled through a small window, waveguides with attached resonator and superlattice structures of resonators attached to quantum waveguides.

Popov's work is based on his method of "restriction-extension". Consider the case of two identical resonators $\Omega_{0_\pm}$ connected to the waveguide $\Omega_w = \{(x,y)||y| < d\}$. Let $r_{0_\pm}$ denote the connection points. Consider the Laplace operator with Dirichlet boundary condition in the space

$$L^2(\Omega_w \oplus \Omega_{0+} \oplus \Omega_{0-}).$$

Popov then constructs a model operator, $-\Delta$, based on the extension of the initial space. The extended space H can be represented in the form $H = H_s \oplus H_a$ where $H_{s(a)}$ is the corresponding subspace of symmetric (resp. antisymmetric) functions with respect to y. These subspaces are invariant for $-\Delta$ and

$$-\Delta = -\Delta|_{H_a} \oplus (-\Delta|_{H_s}).$$

Functions from the domain of $-\Delta|_{H_a}$ satisfy the Dirichlet boundary condition on the centerline $y = 0$. Thus the essential spectrum is

$$\sigma_{ess}(-\Delta|_{H_a}) = [(\pi/d)^2, \infty),$$

$$\sigma_{ess}(-\Delta|_{H_s}) = [(\pi/2d)^2, \infty),$$

and

$$\sigma_{ess}(-\Delta) = [(\pi/2d)^2, \infty).$$

Let λ_{nm} denote the eigenvalues for the resonator Ω_{0+}. Define

$$p_n(\lambda) = \begin{cases} \sqrt{(\pi n/2d)^2 - k^2}, & \text{for } (\pi n/2d)^2 > k^2 \\ i\sqrt{k^2 - (\pi n/2d)^2}, & \text{for } (\pi n/2d)^2 < k^2 \end{cases}$$

where $\lambda = k^2$; then:

Theorem 25 *(Popov) If $(\pi/2d)^2 < min_{m,n}\lambda_{mn} < (\pi/d)^2$, then for sufficiently great $|\lambda_0|$, i.e. for a sufficiently small diameter of the connection aperture, $-\triangle$ has an eigenvalue embedded in the continuous spectrum.*

The proof involves the dispersion equation for the model operator $-\triangle|_{H_a}$, where the dispersion equation is

$$\sum_{n,m=1}^{\infty} \frac{1}{(\lambda_{nm} - \lambda)(\lambda_{nm} - \lambda_0)} = \sum(\frac{1}{2p_n(\lambda)} - \frac{1}{p_n(\lambda_0)}).$$

The conditions of the theorem lead to the fact that the dispersion equation has a real root less than the lower bound of the continuous spectrum (π/d). Hence, there is an eigenvalue for $-\triangle|_{H_a}$. As $\lambda_0 \to \infty$, the root tends to $(\pi/d)^2$ and thus becomes greater than $(\pi/2d)$.

Theorem 26 *(Popov and Popova) In the case of Neuman boundary conditions, if $min_{m,n}\lambda_{mn} < (\pi/2d)^2$, then for sufficiently great $|\lambda_0|$, i.e. for a sufficiently small diameter of the connection aperture, $-\triangle$ has an eigenvalue embedded in the continuous spectrum.*

3.8.1 Waveguides Coupled Through a Window

Popov also considers the problem of waveguides coupled through a window, which we discussed above in section 3.7 in terms of the work of Exner and coworkers. Consider two identical waveguides $\Omega_{w\pm}$ coupled through a point-like window at r_0. This dispersion equation in this case is:

$$\sum_{n=1}^{\infty}(\frac{1}{p_n(\lambda)} - \frac{1}{p_n(\lambda_0)}) = 0.$$

Popov shows that if $\lambda < \lambda_1$, the minimal eigenvalue for the transverse modes, then there is an eigenvalue λ_r less than λ_1, which tends to the lower bound of the continuous spectrum at $\lambda_0 \to \infty$ (i.e., as the width of the window tends to zero). Thus, Popov's results agree with Exner and Vulgater.

For the case of three waveguides coupled through point like windows, the dispersion equation has the form:

$$\sum_{n=1}^{\infty}(\frac{1}{p_n^1(\lambda)} - \frac{1}{p_n^1(\lambda_0)}) + \sum_{n=1}^{\infty}(\frac{1}{\tilde{p}_n^2(\lambda)} - \frac{1}{\tilde{p}_n^2(\lambda_0)}) = 0$$

where

$$p_n^1(\lambda) = \begin{cases} \sqrt{(\pi n/2d_1)^2 - k^2}, & \text{for } (\pi n/2d_1)^2 > k^2 \\ i\sqrt{k^2 - (\pi n/2d_1)^2}, & \text{for } (\pi n/2d_1)^2 < k^2 \end{cases}$$

and

$$\tilde{p}_n^2(\lambda) = \begin{cases} \sqrt{(\pi n/d_2)^2 - k^2}, & \text{for } (\pi n/d_2)^2 > k^2 \\ i\sqrt{k^2 - (\pi n/d_2)^2}, & \text{for } (\pi n/d_2)^2 < k^2 \end{cases}$$

where $k^2 = \lambda$. For $d_1 < d_2 \leq 2d_1$, if $\lambda < (\pi/2d_1)^2$, then there is a root of the dispersion equation, i.e. an eigenvalue of $-\triangle|_{H_a}$. For sufficiently large $|\lambda_0|$, it is an eigenvalue embedded in the continuous spectrum.

3.8.2   Superlattice Structures

Consider the periodic system of pointlike windows at $r_{j\pm}$ for $j = 0, \pm 1, \pm 2, ...$ Let p denote the quasi-momentum associated with Bloch's condition

$$\psi(x + a, k) = exp(ipa)\psi(x, k).$$

Assume in this case that $d_2 = 2d_1$. Then the dispersion equation has the form:

$$\sum_{n=0}^{\infty}(p_n(\lambda_0) - p_n(\lambda))(2p_n(\lambda)p_n(\lambda_0))^{-1}+$$

$$\sum_{n=0}^{\infty} 2exp(-p_n(\lambda)a)\frac{cos(pa) - exp(-p_n(\lambda)a)}{1 - 2exp(-p_n(\lambda)a)cos(pa) + exp(-p_n(\lambda)a)} = 0.$$

Theorem 27 *(Popov) If $\lambda < (\pi/2d_1)^2$, then there is a band embedded in the continuous spectrum for the operator $-\Delta$.*

A similar result holds for the case of identical resonators $\Omega_{j\pm}$ attached to the waveguide Ω_w. Periodic quantum waveguides with this structure have been studied by Lent and Leng (1991) and Porod, Shao and Lent (1993). The dispersion in this case is:

$$\sum_{n,m=1}^{\infty}\frac{1}{(\lambda_{mn} - \lambda)(\lambda_{mn} - \lambda_0)} = \sum_{n=0}^{\infty}(p_n(\lambda_0) - p_n(\lambda))(2p_n(\lambda)p_n(\lambda_0))^{-1}+$$

$$\sum_{n=0}^{\infty} 2exp(-p_n(\lambda)a)\frac{cos(pa) - exp(p_n(\lambda)a)}{1 - 2exp(-p_n(\lambda)a)cos(pa) + exp(-2p_n(\lambda)a)}.$$

Theorem 28 *(Popov and Popova) The relationship $(\pi/2d)^2 \leq min_{m,n}\lambda_{m,n} < (\pi/d)^2$ is a sufficient condition in the Dirichlet case for the existence of a band embedded in the continuous spectrum.*

3.8.3   Periodic Array of Quantum Dots

Geyler and Popov (1996) have studied a specific Hubbard-like model of a periodic array of quantum dots in a uniform magnetic field. Let H_d denote the Hamiltonian of a spinless charged particle in a single quantum dot:

$$H_d = H_0 + V(r)$$

where

$$H_0 = (-i\frac{\partial}{\partial x} - \pi\xi y)^2 + (-i\frac{\partial}{\partial y} + \pi\xi x)^2$$

is the Landau operator and $V(r) = \omega_0 r^2/4$. The operator H_d has pure point spectrum with eigenvalues given by the Fock-Darwin levels:

$$E_{n_1,n_2} = \omega_1(n_1 + 1/2) + \omega_2(n_2 + 1/2)$$

where $n_1 = n + \frac{|m|+m}{2}$ and $n_2 = n + \frac{|m|-m}{2}$. Here $\omega_{1,2} = (\Omega \pm \omega_c)/2, \omega_c = 4\pi|\xi|$, is the cyclotron frequency, and $\Omega = \sqrt{\omega_c^2 + 4\omega_0^2}$. We present a more detailed treatment of the Landau operator in Chapter 7.

Let H denote the Hamiltonian of the quantum dot array (v., Geyler and Popov for the details). One can show that the Fock-Darwin levels $E_{m,n}$ for $m \neq 0$ are eigenvalues of H. Geyler and Popov consider a specific model similar to the Hubbard nearest neighbor hopping model. Using Popov's operator extension theory they show:

Theorem 29 *(Geyler and Popov) The continuous spectrum of H consists of the magneto-Bloch bands B_n where B_n is the set of all values of the function $E = \tau_n(x)$ on the spectrum of the Harper operator. Here B_n lies below $E_{0,n}$ and above $E_{0,n-1}$ for $n \geq 1$. In particular if E_{nm} is an eigenvalue distinct from every $E_{0,k}$ for $k \in \mathbf{N}$, then there is a coupling constant such that $E_{m,n}$ is embedded in the continuous spectrum of H.*

See Geyler and Popov for a discussion of $\tau_n(x)$. One notes that in this model the continuous spectrum may be absolutely continuous as well as singular continuous. See also Geyler and Popov (1994, 1995).

3.9 Zero-width Slit Models

3.9.1 Single Mode Case

Popov and Popova (1993) consider a quantum waveguide, Ω_1, of width d with a resonator attached at a point, say x_0, where the resonator, Ω_2, is $D \times L$ in size. Let $-\Delta = -(\Delta_1 \oplus \Delta_2)$ where Δ_i is the Laplace operator with Neumann boundary conditions on $\Omega_i, i = 1, 2$. Assume there is a single propagating mode say

$$u(x, k) = cos(\pi y d^{-1}) exp(i\sqrt{k^2 - \pi d^{-2}} x).$$

They show that the transmission coefficient for this mode is

$$T = (1 + \alpha_1 u(x_0, k)(\pi^2 d^{-2} - k^2)^{-1/2})^2$$

where $\alpha_1 = -u(x_0, k)(A + B)^{-1}$ with

$$A = 2d^{-1} \sum_{n=0}^{\infty} (|\pi^2 d^{-2} n^2 - k^2|^{1/2} - |\pi^2 d^{-2} - k_0|^{-1/2})$$

and

$$B = \sum_{m,n=0}^{\infty} \frac{(k^2 - k_0^2)|\psi_{nm}(x_0)|^2}{LD(\lambda_{nm} - k^2)(\lambda_{nm} - k_0^2)}.$$

Here $\psi_{nm}(x)$ is the eigenfunction of the resonator Ω_2 and

$$\lambda_{nm} = \pi^2 (m^2 D^{-2} + n^2 L^{-2})$$

and $k_0 = i\delta^{-1}exp(-\gamma)$ with $\gamma = 0.5772$, Euler's constant and δ is the radius of the opening at x_0. The resonance conditions are $k^2 = \lambda_{nm}$. Popov (1992) has shown that there is a scattering resonance $\tilde{\lambda}_{nm}$ of the open resonator near to the eigenvalue of the closed resonator. The authors develop an asymptotic form for T and study the dependence of T on the width of the resonator D. They note that is k_0 increases, i.e. δ decreases, then the resonances become narrower.

One notes that the current-voltage relationship can be calculated for this system by using Sols' (1992) formula:

$$I(V) = \frac{e}{h} \int_{-\infty}^{\infty} dE\, T(E)(f(E) - f(E+V))$$

where $E = k^2$ and $f(E)$ is the Fermi-Dirac distribution

$$f(E) = (exp[(E - \mu)/k_B T_B] + 1)^{-1}$$

where k_B is Boltzmann's constant and T_B is temperature.

3.9.2   Model of Waveguide-Resonator-Waveguide

Consider a waveguide-resonator-waveguide structure where the waveguides have width D and the resonator has width d and length L. Following the model in the last section the single mode transmission coefficient has been calculated by Popov and Popova (1994) to be:

$$T = \alpha_3^2(\alpha_1^2 + k^2 - \pi^2 d^{-2})^{-1}$$

where

$$u(x,k) = cos(\pi y D^{-1})(exp(i\sqrt{k^2 - \pi^2 D^{-2}}(x-x_1)) + exp(-i\sqrt{k^2 - \pi^2 d^{-2}}(x-x_2)).$$

Here x_1 and x_2 are the interface points of the waveguides and the resonator. One finds that

$$\alpha_1 = u(x_1,k)(A+B)(C^2 - (A+B)^2)^{-1}$$

and

$$\alpha_3 = -u(x_1,k)C(C^2 - (A+B)^2)^{-1}$$

with

$$A = 2D^{-1}\sum_{n=0}^{\infty}(|\pi D^{-2}n^2 - k^2|^{-1/2} - |\pi^2 D^{-2}n^2 - k_0|^{-1/2})$$

$$B = \sum_{n=0}^{\infty} \frac{(k^2 - k_0^2)4|\psi_{nm}(x_1)|^2}{Ld(\pi^2(m^2d^{-2} + n^2L^{-2}) - k^2)(\pi^2(m^2d^{-2} + n^2L^{-2}) - k_0^2)}$$

and

$$C = 4(dL)^{-1}\sum_{n,m=0}^{\infty}(-1)^n(\pi^2(d^{-2}m^2 + L^{-2}n^2) - k^2)^{-1}.$$

The conductance is given by Landauer's formula, which in this case is

$$G = T(1 - T)e^2/\pi\hbar$$

and the Sols' formula mentioned above can be used to calculate the I-V characteristics.

The reader should also note the other papers by Popov in this area.

3.10 Impurity Sites

3.10.1 Introduction

In the experiments of Wu et al. (1991, 1992) it was argued by these authors that the observed peaks in the conductance were due to electrons tunneling through impurity sites in the constriction as was discussed by McEuen (1990) who had measured conductance versus gate voltage in a straight quantum wire. Work by Wang, Berggren and Ji (1995) and Wang (1995) pointed out that bound states in the bent wire would also lead to electron conductance peaks. Later work of Carini, Londergan and Murdock (1997) concluded that the observed peaks are consistent with tunneling through the bound states in the quantum wire. However, conduction phenomena does take place through impurities and in this section we review the mathematical results related to impurities which can be described by point interactions in quantum wires.

The case of random potentials is treated in Chapters six and seven in our discussion of Anderson localization and the quantum Hall effect. For background reading on point interactions, the reader is directed to Albeverio et al. (1988).

3.10.2 Point Interactions in a Strip

The case of a single or multiple point interactions for a quantum strip was considered by Exner et al. (1996). Let $\Omega = \mathbf{R} \times [0, d]$ be a straight planar strip, with Dirichlet Laplacian. For simplicity, take $d = \pi$. Let $\mathbf{a} = (a, b)$ be a point perturbation with $b \in (0, \pi)$. Let $H(\alpha, \mathbf{a})$ denote the Hamiltonian where α denotes a coupling constant (with $\alpha = \infty$ corresponding to a free Hamiltonian H_0 where $H_0\psi = -\Delta\psi$). The resolvent is given by Krein's formula (v., Albeverio et al. (1988)), in terms of which spectral properties of the Hamiltonian can be studied. The free resolvent is an integral operator with kernel

$$G_0(\mathbf{x}_1, \mathbf{x}_2, z) = (H_0 - z)^{-1}(\mathbf{x}_1, \mathbf{x}_2) = \frac{i}{\pi} \sum_{n=1}^{\infty} \frac{e^{ik_n(z)|x_1 - x_2|}}{k_n(z)} sin(ny_1) sin(ny_2)$$

where $\mathbf{x}_j = (x_j, y_j)$ and $k_n(z) = \sqrt{z - n^2}$. Krein's formula then states that

$$(H(\alpha, \mathbf{a}) - z)^{-1}(\mathbf{x}_1, \mathbf{x}_2, z) = G_0(\mathbf{x}_1, \mathbf{x}_2, z) + \lambda(\alpha, \mathbf{a}, z)G_0(\mathbf{x}_1, \mathbf{a}, z)G_0(\mathbf{a}, \mathbf{x}_2, z).$$

This formula allows one to evaluate λ; viz., Exner and coworkers have shown:

Theorem 30 *(Exner et al.)* $\sigma_{ess}(H(\alpha, \mathbf{a})) = [1, \infty)$ *with multiplicity* $2[\sqrt{E/n^2}]$ *at a value* E. *The point spectrum is given by solutions* $z \in \mathbf{R}$ *to the equation* $\xi(\mathbf{a}, z) = \alpha$ *where for* $z < 1$

$$\xi(\mathbf{a}, z) = \frac{1}{\pi} \sum_{n=1}^{\infty} \left(\frac{\sin^2(nb)}{\kappa_n(z)} - \frac{1}{2n} \right).$$

Here $\kappa_n(z) = -ik_n(z) = \sqrt{n^2 - z}$. *There is just one eigenvalue* $E(\alpha, \mathbf{a})$ *in* $(-\infty, 1]$ *and there are no eigenvalues embedded in the continuous spectrum, i.e. no real eigenvalues of* H *where* $z \geq 1$ *is a solution.*

3.10.3   N Point Interactions

In the case of a finite number N of point interactions at $\mathbf{a}_j = (a_j, b_j)$ with coupling α_j, let $\mathbf{a} = \{\mathbf{a}_1, ...\}$ and $\alpha = \{\alpha_1, ...\}$. Exner et al. have shown:

Theorem 31 *(Exner et al.) The essential spectrum is* $\sigma_{ess}(H(\alpha, \mathbf{a})) = [1, \infty)$. *The discrete spectrum is determined by*

$$det\Lambda(\alpha, \mathbf{a}, z) = 0$$

where Λ *is the* $N \times N$ *matrix*

$$\Lambda_{jj} = \alpha_j - \frac{i}{\pi} \sum_{n=1}^{\infty} \left(\frac{\sin^2(nb_j)}{k_n(z)} - \frac{1}{2in} \right)$$

and

$$\Lambda_{jm} = \frac{i}{\pi} \sum_{n=1}^{\infty} \left(\frac{e^{ik_n(z)|a_j - a_m|} \sin^2(nb_j) \sin(nb_m)}{k_n(z)} \right)$$

for $j \neq m$. *In this case* $H(\alpha, \mathbf{a})$ *has at least one eigenvalue and it is nondegenerate; furthermore* $H(\alpha, \mathbf{a})$ *can have embedded eigenvalues.*

3.10.4   Strong and Weak Coupling Limits

Exner et al. have shown that in the strong coupling limit there are N eigenvalues including a possible degeneracy where $E_j(\alpha, \mathbf{a}) \sim 4e^{-4\pi\alpha_j}$ as $max_{1 \leq j \leq N} \alpha_j \to -\infty$. For weak coupling they find that $H(\alpha, \mathbf{a})$ has a single bound state.

3.10.5   Resonances Due to Point Interactions

The resonances for the single point scatterer are described as follows. Exner and coworkers have shown that $\xi(\mathbf{a}, z)$ has no real solutions for $z \geq 1$. However resonance zeros in the complex plane occur. They have shown that:

Theorem 32 *(Exner et al.) For a weak coupling there is generically one resonance pole close to each threshold with the exception of the lowest; the resonance is absent if $(N+1)b/\pi$ is an integer. Finally,*

$$z_N(\alpha) \sim (N+1)^2 - 2i\frac{|sin((N+1)b)|}{\pi^3\alpha^3}\sum_{n=1}^{N}(\frac{sin^2(nb)}{\sqrt{(N+1)^2 - n^2}} + \frac{1}{2in}) + O(\alpha^{-4}).$$

In the weak coupling limit, the resonance width is given by the expression

$$\Gamma_N(\alpha) = -2Im(z_N(\alpha)) = 4\frac{|sin((N+1)b)|}{\pi^3\alpha^2}\sum_{n=1}^{N}\frac{sin^2(nb)}{\sqrt{(N+1)^2 - n^2}} + O(\alpha^{-4}).$$

3.10.6 Scattering Matrix

The scattering matrix for the point interaction model is given by

$$S_{nm} = \sqrt{\frac{k_m}{k_n}}\begin{pmatrix} t_{nm} & r_{nm} \\ \tilde{r}_{nm} & \tilde{t}_{nm} \end{pmatrix}$$

where $\tilde{r}_{nm} = r_{nm}e^{-2i(k_n+k_m)a}$. For a single channel the reflection and transmission amplitudes are:

$$r_{nm} = \frac{i}{\pi}\frac{sin(nb)sin(mb)}{k_m(z)(\alpha - \xi(a,z))}e^{i(k_n+k_m)a}$$

and

$$t_{nm}(z) = \delta_{nm} + \frac{i}{\pi}\frac{sin(nb)sin(mb)}{k_m(z)(\alpha - \xi(a,z))}e^{i(k_n-k_m)a}.$$

These expression extend naturally to the N−channel case where

$$r_{nm}(z) = \frac{i}{\pi}\sum_{j,k=1}^{N}(\Lambda(z)^{-1})_{jk}\frac{sin(nb_j)sin(nb_k)}{k_m(z)}e^{i(k_m a_j + k_n a_k)}$$

and similarly for t_{nm}.

3.10.7 Conductance for Point Scatters

Using the scattering matrix from the last section, Exner et al. have calculated the conductance for a pair of point perturbations using Landauer's formula:

$$G(z) = \frac{2e^2}{h}\sum_{n,m=1}^{[\sqrt{z}]}\frac{k_m}{k_n}|t_{nm}(z)|^2.$$

Exner et al. have examined simple examples which show sharp resonance peaks and anti-resonance dips which approach the channel thresholds in the weak coupling limit.

3.11   Scattering Example

Consider the Hamiltonian

$$H = -\Delta_{\alpha,y} = -\frac{d^2}{dx^2} + \alpha\delta(x - y)$$

which is a δ-interaction of strength α centered at $y \in \mathbf{R}$. This case is treated in detail in Albeverio et al. (1988). We briefly review the results for this example.

Theorem 33 *Let* $-\infty < \alpha \leq \infty$. *Then, the essential spectrum of H is purely absolutely continuous with*

$$\sigma_{ess}(H) = \sigma_{ac}(H) = [0, \infty)$$

and

$$\sigma_{sc}(H) = \emptyset.$$

If $-\infty < \alpha < 0$, *then H has precisely one negative eigenvalue:*

$$\sigma_p(H) = \{-\alpha^2/4\}$$

with normalized eigenfunction

$$(-\alpha/2)^{1/2}e^{\alpha|x-y|/2}.$$

And if $\alpha \geq 0$ *or* $\alpha = \infty$, *then H has no eigenvalues,* $\sigma_p(H) = \emptyset$.

The scattering matrix for H is given by

$$S_{\alpha,y}(k) = \begin{pmatrix} t^l(k) & r^r(k) \\ r^l(k) & t^r(k) \end{pmatrix}$$

where t^l, t^r are the transmission coefficients from the left and right, and r^l, r^r are the reflection coefficients from the left and right. One can show:

Theorem 34 *The scattering matrix for H is*

$$S_{\alpha,y}(k) = \frac{1}{(2k + i\alpha)}\begin{pmatrix} 2k & -i\alpha e^{-2iky} \\ -\alpha e^{2iky} & 2k \end{pmatrix}.$$

$S_{\alpha,y}(k)$ *has a meromorphic continuation to all of* $\mathbf{C}$ *such that the pole of* $S_{\alpha,y}(k)$ *coincides with the bound state of H if* $\alpha < 0$ *or to the resonance of H if* $\alpha > 0$.

3.11.1   Infinite Cylinder in an Axial Magnetic Field

Exner et al. considered the case of a cylinder placed in a homogeneous magnetic field parallel to the cylinder axis. The eigenvalues of the free Hamiltonian are

$$E_{mn} = (m + A)^2$$

where $m \in \mathbf{Z}$ and $n = 1, 2, \dots$. The transverse eigenfunctions are η_m where $\eta_m(x) = \frac{1}{\sqrt{2\pi}} e^{i(m+A)x}$ and $\chi_n(y) = \sqrt{\frac{2}{\pi}} \sin(ny)$. Here $\phi = \pi B = 2\pi A$.

The Hamiltonian $H(\alpha, \mathbf{a}, \phi)$ is formed by adding a finite number of point perturbations to H_0. The resolvent and the scattering amplitudes are given as before; viz.,

$$r_{mn}(z) = \frac{i}{4\pi} \sum_{j,k=1}^{N} (\Lambda(z)^{-1})_{jk} \frac{e^{-i(m+A)b_j} e^{i(n+A)b_k}}{k_m(z)}$$

and similarly for $t_{nm}(z)$.

3.11.2 Periodic Array of Perturbations

Consider a single periodic array of point perturbations of strength α and spacing l. The eigenvalues are given by

$$\xi(\mathbf{a}, \theta, z) = \alpha.$$

Exner et al. show that for a suitable choice of parameters, the spectrum can have an arbitrary finite number of gaps in its spectrum if l is chosen large enough. The relationship to the Bethe-Sommerfeld conjecture and the work of Skriganov (1985) is discussed in Exner, Gawlista, Seba and Tater (1996). Viz., the number of gaps of a one dimensional periodic system is generically infinite (e.g., there are one dimensional systems with a single gap) while for higher dimensions the Bethe-Sommerfeld conjecture states that this number should be finite.

3.12 Point Interactions in a Tube

The extension of the case of point interactions to a straight Dirichlet tube has been developed by Exner (1999). The model is given by $\Omega = \mathbf{R} \times M$ where $M \subset \mathbf{R}^2$ is a close, compact set which is path connected. Assume that the boundary ∂M has the segment property. The operator $-\Delta_M^D$ has purely discrete spectrum which we denote as χ_n, ν_n for the eigenfunctions and eigenvalues. The free resolvent is an integral operator with kernel

$$G_0(\mathbf{x}_1, \mathbf{x}_2, z) = (H_0 - z)^{-1}(\mathbf{x}_1, \mathbf{x}_2) = \frac{i}{2} \sum_{n=0}^{\infty} \frac{e^{ik_n(z)|x_1 - x_2|}}{k_n(z)} \chi_n(\mathbf{y}_1) \chi_n(\mathbf{y}_2)$$

where $z \in \mathbf{C} \backslash [\nu_0, \infty)$, $\mathbf{x}_j = (x_j, \mathbf{y}_j)$ and $k_n(z) = \sqrt{z - \nu_n}$.

If there is a point interaction located at $\mathbf{a} = (a, \mathbf{b}) \in M^0$, the Hamiltonian is the self-adjoint extension $H(\alpha, \mathbf{a})$ of the operator $-\Delta_D^\Omega | C_0^\infty(\Omega \backslash \{\mathbf{a}\})$ with boundary value

$$L_1(\psi, \mathbf{a}) + 4\pi\alpha L_0(\psi, \mathbf{a}) = 0.$$

Of course the case $\alpha = \infty$ or $L_0(\psi, \mathbf{a}) = 0$ corresponds to the free Hamiltonian H_0. Here $L_0(\psi, \mathbf{a}) = \lim_{\mathbf{x} \to \mathbf{a}} \psi(\mathbf{x}) |\mathbf{x} - \mathbf{a}|$ and $L_1(\psi, \mathbf{a}) = \lim_{\mathbf{x} \to \mathbf{a}} [\psi(\mathbf{x}) - L_0(\psi, \mathbf{a}) / |\mathbf{x} - \mathbf{a}|]$.

The resolvent of $H(\alpha, \mathbf{a})$ is given by Krein's formula. Exner (1999) has shown:

Theorem 35 *(Exner) The operator $H(\alpha, \mathbf{a})$ has for $\alpha \in \mathbf{R}$ a single eigenvalue $\epsilon(\alpha, \mathbf{a}) \in (-\infty, \nu_0)$. The corresponding eigenfunction is*

$$\psi(x, \alpha, \mathbf{a}) = \sum_{n=0}^{\infty} \frac{e^{-\kappa_n(\epsilon)|x-a|}}{2\kappa_n(\epsilon)} \chi_n(\mathbf{y}) \chi_n(\mathbf{y})$$

where $\kappa_n(z) = -ik_n(z) = \sqrt{\nu_n - z}$. The eigenvalue is strictly increasing and has the weak coupling behavior

$$\epsilon(\alpha, \mathbf{a}) = \nu_0 - (\frac{|\chi_0(\mathbf{b})|^2}{2\alpha})^2 + \mathcal{O}(\alpha^{-3})$$

as $\alpha \to \infty$. There are no eigenvalues embedded in the continuous spectrum

$$\sigma_c(H(\alpha, \mathbf{a})) = [\nu_0, \infty).$$

Following Exner, Gawlista, Seba and Tater (1996), the on shell S-matrix at energy $z = k^2$ is a $2N_{open} \times 2N_{open}$ unitary matrix given by:

Theorem 36 *(Exner)*

$$S_{nm} = \sqrt{\frac{k_m}{k_n}} \left(\begin{array}{cc} t_{nm} & r_{nm} \\ \tilde{r}_{nm} & \tilde{t}_{nm} \end{array} \right)$$

where $n, m = 1, ..., N_{open}$ and $N_{open} = |\{\nu_n|\nu_n < z\}|$. Here the tilde denotes switching the sign of the longitudinal component of $\mathbf{a}$, $a \to -a$.

$$r_{nm} e^{-ik_m a} = (t_{nm} - \delta_{nm}) e^{ik_m a} = \frac{i}{2k_m} \frac{e^{ik_n a}}{\alpha - \xi(\mathbf{a}, z)} \chi_n(\mathbf{b}) \chi_m(\mathbf{b}).$$

Here $\xi(\mathbf{a}, z)$ is the regularized Green's function at $\mathbf{a}$ and one can show that

$$\xi(\mathbf{a}, z) = \sum_{n=0}^{\infty} [\frac{|\chi_n(\mathbf{b})|^2}{2\kappa_n(z)} + \frac{\sqrt{n} - \sqrt{n+1}}{2\sqrt{\pi M}}].$$

For the case of a finite number of perturbations $a = \{\mathbf{a}_1, ..., \mathbf{a}_N\}$ with $\mathbf{a}_j = (a_n, \mathbf{b}_j)$ and $\alpha = \{\alpha_1, ..., \alpha_N\}$ Exner has shown:

Theorem 37 *(Exner) The spectrum of $H(\alpha, \mathbf{a})$ consists for any $\alpha \in \mathbf{R}^N$ of absolutely continuous part $[\nu_0, \infty)$ and eigenvalues $\epsilon_1 < \epsilon_2 \leq ... \leq \epsilon_n < \nu_0$ with $1 \leq m \leq N$. The eigenvalues are given by the equation*

$$det\Lambda(\alpha, \mathbf{a}, z) = 0$$

where $\Lambda_{jj} = \alpha_j - \xi(\mathbf{a}_j, z)$ and $\Lambda_{jk} = -G_0(\mathbf{a}_j, \mathbf{a}_k, z)$ for $j \neq k$. In this case $H(\alpha, \mathbf{a})$ can have embedded eigenvalues if the family $(\Omega, \alpha, \mathbf{a})$ has suitable symmetry conditions.

The embedded eigenvalues are consistent with the work of Evans, Levitin and Vassiliev (1994) discussed in chapter 2.

Exner also addresses the case of a periodic array of point interactions. In the case in which each cell has a single point interaction, Exner shows that $H(\{\alpha, \mathbf{a}\}_{per})$ is absolutely continuous and has an arbitrary finite number of gaps. The Bethe-Sommerfeld conjecture on the finiteness of the number of open gaps for this model is still an open problem.

3.12.1 Random Point Perturbations and Random Matrix Ensembles

Exner et al. have examined the case of a set of point perturbations chosen at random in a interval of length L. Based on random matrix theory the mean conductance and variance of the conductance should be

$$< G >= \begin{cases} \frac{M}{2} - \frac{M}{4M+2} & \text{COE} \\ \frac{M}{2} & \text{CUE} \end{cases}$$

and

$$Var(G) = \begin{cases} \frac{M(M+1)^2}{(2M+1)^2(2M+3)} & \text{COE} \\ \frac{M^2}{4(4M^2-1)} & \text{CUE} \end{cases}$$

where M is the number of open channels in the quantum wire.

Exner et al. compared $Var(G)$ for the case of 25 impurities distributed at random inside a waveguide segment of length 25. They found that $Var(G)$ and $< G >$ coincide with the random matrix prediction inside the universality window being suppressed by localization for small k and enhanced by direct transmission for large k.

Resonant tunneling due to point scatterers is also discussed by Mosk et al. In addition these authors discuss the issue of subband bottom transparency, i.e. when the energy is close to the bottom of the subband, the $t-$matrix is small and there is almost no scattering, as noted by Chu and Sorbello and Bagwell. For other work on modelling impurities in quantum wires, see Chu and Sorbello (1989), Bagwell (1990), Tekman and Ceraci (1991), Levinson, Lubin and Sukhorukov (1992), Kirman and Bagwell (1991), Mosk, Nieuwenhuizen and Barnes (1996) and Ji and Berggren (1995).

3.13   Quantum Systems on Graphs

Quantum mechanics for a system whose configuration space is a graph has been considered by Pauling (1936) and Ruedenberg and Scherr (1953) in terms of free-electron models of organic molecules. More recent work on graph systems includes Adamyan (1992), Avishai and Luck (1992), Bulla and Trenckler (1990), Exner and Seba (1989), Gratus et al (1994), Montroll (1970), and Sadun and Avron (1995). Graph systems as limits of quantum wires have been studied by Exner. Exner and coworkers have shown that any branched (star shaped) system of infinitely long waveguides with Dirichlet boundary conditions has at least one bound state. Does this result extend to graph systems? Exner showed that the existence of a bound state is preserved in the zero diameter limit with appropriate coupling constants. However, Exner's simple graph systems cannot accommodate multiple bound states.

For two dimensional lattice systems, Exner and coworkers have shown that certain graph systems have infinitely many gaps in their spectra, so these systems do not obey the Bethe-Sommerfeld conjecture and behave more like one dimensional systems; however there are lattices also with no gaps if the coupling is weak enough which is not typical one dimensional behavior.

For other work on graph systems, see Exner and Seresova (1994), Exner (1997), Exner and Gawlista (1996), Avron, Raveh and Zur (1988), Exner (1996), Roth (1983), Kottos and Smilansky (1997, 1999) and Kostrykin and Schrader (1998). Carlson (1997) has treated Hill's equation on a homogeneous tree. We return to the subject of graphical models in Chapter 9.

3.13.1   Graph Systems

Exner (1996) examined several types of contact interaction on graphs. Consider a spinless particle with configuration space a graph, Γ_n, consisting of n halflines where the end points are connected at a single point. The Hilbert space for this system is $\mathcal{H} = \oplus_{j=1}^n L^2(0, \infty)$ and $H\{f_j\} = \{f_j''\}$. In the case Γ_2, i.e. a line with a single point interaction, consider two boundary conditions: the δ' interaction:

$$f_+ - f_- = \beta f'$$

for $\beta \in \mathbf{R}$, and the symmetrized δ'-interaction where the boundary conditions are:

$$f'_+ + f'_- = 0$$

$$f_+ + f_- = Df'$$

where $D \in \mathbf{R}$. More generally for Γ_n, the boundary conditions are

$$\sum f'_j = 0$$

and

$$f_j - f_k + C(f'_j - f'_k) = 0$$

for $j, k = 1, ... n$ and in the δ'_s case

$$f'_1 = ... = f'_n = f'$$

and

$$\sum_{j=1}^n f_j = Df'.$$

One finds there is a bound state of energy $E(C) = -C^{-2}$ for $-C < 0$ in the first case and $E(D) = -(n/D)^2$ if $D < 0$ in the symmetrized case.

The reflection and transmission amplitudes for the δ'_s coupling was determined by Exner and Seba (1989) to be

$$r(k) = \frac{n - 2 - ikD}{n - ikd}$$

and

$$t(k) = \frac{-2}{n - ikD}.$$

3.13.2   Lasso Graph

Exner (1997) considered a charged spinless quantum particle confined to a graph consisting of a loop of radius R attached to a half line where the loop is in a homogeneous magnetic field perpendicular to the loop plane. The Hilbert space of this system is $\mathcal{H} = L^2(\Gamma) = L^2(0, L) \oplus L^2(\mathbf{R}_+)$ where L is the loop perimeter. On the loop

$$H_{loop}(B) = (-i\partial_x + A)^2$$

where $A = \frac{1}{2}BR = \Phi/L$. Here Φ is the magnetic flux through the loop and $H_{halfline} = -\partial_x^2$. The wavefunction

$$\psi = \begin{pmatrix} u \\ f \end{pmatrix}$$

with Dirichlet boundary conditions $u(0) = u(L) = f(0) = 0$. The operator H_{loop} has simple discrete spectrum $(n/2R)^2$ with eigenfunctions

$$\chi_n(x) = \frac{1}{\sqrt{\pi R}} e^{-iAx} sin(nx/2R) \tag{3.1}$$

which are embedded in the continuous spectrum of the operator $H_{halfline}$ is $[0, \infty)$.

Exner derives the $S-$matrix for the scattering problem on Γ, viz. the reflection of a particle traveling along the halfline from the magnetic loop end. He finds an infinite ladder of resonances half of which turn into true embedded eigenvalues; that is bound states are found to exist only at integer or halfinteger values of magnetic flux with eigenfunctions given by (3.1).

One should also consult Kostrykin and Schrader (1998) and Avron and Sudan (1991) for more details on this example.

3.14   T-Shaped or Appendix Graph

Exner and Seresova (1994) have treated the scattering problem for a simple graph consisting of an infinite line with a finite length appendix. This model extends earlier work of a line with a stub by Porod, Shao and Lent (1992) and Tekman and Bagwell (1993). We will see that this elementary example demonstrates the concept of dissolving eigenvalues, to which we return in Chapter eight and nine. The Hilbert space of the problem is $\mathbf{H} = L^2(\mathbf{R}) \oplus L^2(0, \ell)$. Assume the appendix is coupled at the point $x = 0$. The Hamiltonian is given by

$$(H\psi)_1(x) = -f''(x)$$

and

$$(H\psi)_2(x) = (-u'' + Vu)(x)$$

where $\psi = \begin{pmatrix} f \\ u \end{pmatrix}$. The boundary conditions are:

$$f(0+) = f(0-) = f(0)$$

$$u(0) = bf(0) + cu'(0)$$
$$f'(0+) - f'(0-) = df(0) - bu'(0)$$
$$u(\ell) = 0.$$

In this model the line supports a potential, which could be an external potential such as an electric field perpendicular to the line, $V(x) = Ex$, or it could be due to the geometry of the appendix. We assume that V is a real-valued, measurable function in $L_{loc}^1(0, \ell)$ with finite limits at both endpoints of the appendix.

The disconnected appendix has the Hamiltonian

$$h_c = -\frac{d^2}{dx^2} + V$$

with boundary condition $u(0) + cu'(0) = 0$ at the junction.

Let $u(x) = \beta u_\ell(x)$ where u_ℓ is a solution to $-u'' + Vu = k^2 u$ corresponding to the boundary condition $u_\ell(\ell) = 0$. The reflection and transmission terms are found to be

$$t(k) = \frac{-2ik(cu'_\ell - u_\ell)(0)}{b^2 u'_\ell(0) + (d - 2ik)(cu'_\ell - u_l)(0)}$$

and

$$r(k) = \frac{b^2 u'_\ell(0) + d(cu'_\ell - u_l)(0)}{b^2 u'_\ell(0) + (d - 2ik)(cu'_\ell - u_\ell)(0)}.$$

This system may have a bound state. E.g. if $b = 0$, an isolated eigenvalue exists if $d < 0$, given by $\kappa^2 = -\frac{1}{4}d^2$, determined by the the denominator being zero (with $k = i\kappa$). And it remains isolated for $|b|$ small.

If $b = 0$ the appendix has simple purely discrete spectrum whose eigenvalues are positive. Thus, they are embedded in the continuous spectrum of the Hamiltonian. As the coupling is turned on, one expects these eigenvalues to turn into resonances, which Exner and Seresova demonstrate. The authors use Krein's formula to determine explicitly the resolvent of the Hamiltonian. In this expression, they find that the singularities of the resolvent are given by the zeros of

$$D(k) = -\frac{ib^2}{2k} u'_\ell(0) - (1 + \frac{id}{2k})(cu'_\ell(0) - u_\ell(0)).$$

For the case, $b = 0$, the zeros correspond to the eigenvalues of h_c which are embedded in the continuous spectrum of the line Hamiltonian. The weak coupling expansion is given by:

Theorem 38 (*Exner and Seresova*) *Let k_n denote the nth eigenvalue of h_c with corresponding normalized eigenfunction χ_n. Then the condition $D(k) = 0$ has for sufficiently small $|b|$ just one solution in the vicinity of k_n given by*

$$k_n(b) = k_n - \frac{ib^2 \chi'_n(0)^2}{2k_n(2k_n + id)} + \mathcal{O}(b^4)$$

and resonance pole positions

$$z_n(b) = k_n^2 - \frac{ib^2 \chi'_n(0)^2}{2k_n + id} + \mathcal{O}(b^4).$$

In the non-weak coupling case, particular examples can still be treated. In the case $V = 0$ and $c = d = 0$, then the embedded eigenvalues are k_n^2 where $k_n = n\pi/\ell$, with the nth Dirichlet eigenfunctions $\chi_n(x) = \sqrt{2/\ell}cos(k_nx)$. The equation $D(k) = 0$ in this case reduces to $tan(kl) = -\frac{1}{2}i\beta^2$, which solutions are given by:

$$k_n(b) = \frac{n\pi}{\ell} + \frac{i}{2\ell}ln\frac{2-b^2}{2+b^2}$$

for $|b| < \sqrt{2}$ and

$$k_n(b) = \frac{(2n-1)\pi}{2\ell} + \frac{i}{2\ell}ln\frac{b^2-2}{b^2+2}$$

for $|b| > \sqrt{2}$. If $|b| = \sqrt{2}$ the resolvent has no poles. For small $|b|$, the first expression gives $k_n(b) = \frac{n\pi}{\ell} - \frac{ib^2}{2\ell} + \mathcal{O}(b^4)$ which is consistent with the above theorem. For large $|b|$ the poles move towards the real axis with the limit $|b| \to \infty$ of $\frac{\pi}{\ell}(n + \frac{1}{2})$, which correspond to h_∞, the appendix Hamiltonian with Neumann boundary condition at $x = 0$.

Exner and Seresova have numerically examined pole trajectories in the case $V(x) = Ex$. They evaluated the transmission coefficient and find for weak coupling, the transmission is one except for small neighborhoods of the embedded eigenvalues. For large b, the transmission is possible only in the vicinity of the "Neumann" eigenvalues.

3.15 Branching Graphs

Extending the model of the appendix graph, Exner (1996) has treated the case of a Schrödinger operator on a graph, which consists of N links joined at a single point. Each link supports a real, locally integrable potential V_j. Each link can be a finite or semi-infinite interval. The Hilbert space in this case is $\mathcal{H} = \bigoplus_{j=1}^{N} L^2(0, \ell_j)$. Here $V_j \in L^1_{loc}(0, \ell_j)$ and for $\ell_j < \infty$, $\psi(\ell_j)cos\omega_j + \psi'(\ell_j)sin\omega_j = 0$. The Hamiltonian is

$$H_\alpha(V)(\psi_j) = (\psi_j + V_j\psi_j)$$

with the boundary conditions $\psi_1 = ... = \psi_N = \psi, \sum_{j=1}^{N} \psi'_j = \alpha\psi$ at the vertex where $\psi_j = lim_{x\to 0+}\psi_j(x)$ and $\psi'_j = lim_{x\to 0+}\psi'_j(x)$. For $\alpha = \infty$, the boundary condition is the Dirichlet condition $\psi_j = 0, j = 1, ..., N$ and the operator is decoupled

$$H_\infty(V) = \bigoplus_{j=1}^{N} h_j(V_j).$$

Extending the result for the case $V = 0$ by Exner and Seba (1989), Exner proved:

Theorem 39 *(Exner) Under the stated conditions, the operator $H_\alpha(V)$ is self-adjoint for any $\alpha \in \mathbb{R} \cup \{\infty\}$.*

Exner has shown that if the coupling is ideal $(\alpha = 0)$ and all the links are semi-infinite, an arbitrarily weak potential which is not repulsive in the mean and decays fast enough produces a single bound state. Viz., assume $\ell_j = \infty$ and $V_j \in L^2(\mathbf{R}_+, (1 + |x|)dx)$. Then

Theorem 40 *(Exner) The operator $H_0(\lambda V)$ for sufficiently small $\lambda > 0$ has a single negative eigenvalue $\epsilon(\lambda) = -\kappa(\lambda)$ if and only if*

$$\sum_{j=1}^{N} \int_0^\infty V_j(x)dx \leq 0.$$

The proof of this theorem is essentially an application of the Birman-Schwinger principle and Krein's resolvent formula. Viz., let $H_0(\lambda V) = A_0 + \lambda V$ where $A_0 = H_0(0)$. Then the possible negative eigenvalues of $H_0(\lambda V)$ are given by the Birman-Schwinger method; i.e. such an eigenvalue exists if and only if λK has the eigenvalue -1, where

$$K = |V|^{1/2}(A_0 - k^2)^{-1}V^{1/2}.$$

Here $V^{1/2} = |V|^{1/2}sgnV$. And Krein's formula allows for an explicit evaluation of the kernel of this operator. The remainder of the proof follows that of Klaus (1977) and Blanckenbecler, Goldberger and Simon (1977). In particular for the case $N = 2$, the result reduces to the formula for the Schrödinger operator on the line, derived in these cited references.

Exner has determined the asymptotic expansion, which to first order is

$$\kappa(\lambda) = -\frac{\lambda}{N}\sum_{j=1}^{N}\int_0^\infty V_j(x)dx.$$

The essential spectrum of the Dirichlet link operators $h_j(V_j)$ is $[0, \infty)$. Exner shows that $H_\alpha(V)$ and $H_\infty(V)$ differ by a rank one perturbation in the resolvent, hence $\sigma_{ess}(H_\alpha(V)) = [0, \infty)$. The operators $h_j(\lambda V_j)$ have no discrete spectrum for small λ, so $H_\alpha(\lambda V)$ has at most one negative eigenvalue. More generally, Exner also estimates the number of negative eigenvalues, i.e. the number of bound states, of $H_0(V)$.

3.16   Quantum Wire Superlattice Models

Avishai and Luck (1992), Gratus et al. (1994) and Exner and coworkers have studied graph models of quantum wire superlattices. Consider a rectangular lattice with spacings l_1, l_2 in the x and y directions. Assume the lattice links are coupled by δ_s' interactions. Set

$$f_m(x) = e^{im\theta_2 l_2}(a_n e^{ikx} + b_n e^{-ikx})$$

for $x \in (nl_1, (n + 1)l_1)$ and

$$g_n(y) = e^{in\theta_1 l_1}(c_m e^{iky} + d_m e^{-iky})$$

for $y \in (ml_2, (m+1)l_2)$. The boundary conditions in this case are

$$f'_m(nl_1 + 0) = -f'_m(nl_1 - 0) = g'_n(ml_2 + 0) = -g'_n(nl_2 - 0) = G_{nm}$$

and

$$f_m(nl_1 + 0) + f_m(nl_1 - 0) + g_n(ml_2 + 0) + g_n(nl_2 - 0) = DG_{nm}.$$

This gives rise to the condition

$$\frac{cos\theta_1 l_1 + coskl_1}{sinkl_1} + \frac{cos\theta_2 l_2 + coskl_2}{sinkl_2} - \frac{Dk}{2} = 0.$$

Set $\theta = l_1/l_2$ and let $\sigma(l, \theta, D)$ denote the spectrum of the corresponding δ'_s Hamiltonian.

Exner (1997) has shown that for this model there are always infinitely many gaps for $D \neq 0$:

Theorem 41 *(Exner) The spectrum of a δ'_s Hamiltonian has a band structure which can be described as follows. For any nonzero D, the number of gaps is infinite:*

$$\sigma(l, \theta, D) = \cup_{r=1}^{\infty} [\alpha_r, \beta_r]$$

with $\alpha_r < \beta_r < \alpha_{r+1}$. If $D = 0$, then $\sigma(l, \theta, 0) = [0, \infty)$

If $D > 0$, then each $\alpha_r = (\pi n/\lambda)^2$ for $\lambda = l_1$ or l_2 and $n \in \mathbf{Z}$; and $\beta_r = (\pi m/\lambda)^2$ with $m \in \mathbf{Z}$ for $D < 0$ and $r \geq 2$.

$\alpha_1 = 0$ for $D \geq 0$ and $\alpha_1 < 0$ for $D < 0$.

For $-l_1 - l_2 < D < 0$, then $\beta_1 < 0$ and $\alpha_2 = 0$.

Finally, $\sigma(l, \theta, D') \cap \mathbf{R}_+ \subset \sigma(l, \theta, D) \cap \mathbf{R}_+$ for $|D'| > |D|$.

The bandwidths of the δ'_s lattice have the following asymptotic behavior:

Theorem 42 *(Exner) The bandwidths of the δ'_s lattice with $D \neq 0$ satisfy the asymptotic bounds*

$$\frac{8}{DL} + O(r^{-1}) < \beta_r - \alpha_r < \frac{8}{D}(l_1^{-1} + l_2^{-1})e(\theta) + O(r^{-1})$$

where $e(\theta) < 4/3$ and $L = max(l_1, l_2)$.

Exner also studied the δ lattice model, which has interesting number theoretic properties. We refer the reader to his papers for more details.

3.17   Periodic Chain

Albeverio, Geyler and Kostrov (1999) have considered the problem of a three dimensional Schrödinger operator with a uniform magnetic field in a periodic chain of point scatterers, using the restriction-extension methods of Pavlov. Let H_0

denote the free Hamiltonian in a uniform magnetic field and let Λ denote the one dimensional crystal lattice. The Green's function is given by Krein's formula:

$$G(r, r', \zeta) = G_0(r, r', \zeta) - \sum_{\lambda, \mu \in \Lambda} [Q(\zeta) + A]^{-1}_{\lambda\mu} G_0(r, \lambda, \zeta) G_0(\mu, r', \zeta)$$

where G_0 is the Green's function for H_0 and $Q(\zeta)$ for $\zeta \in \mathbf{C} \backslash \sigma(H_0)$ is a bounded operator

$$Q_{\lambda\mu}(\zeta) = \begin{cases} G_0(\lambda, \mu, \zeta) & \text{if } \lambda \neq \mu \\ \frac{1}{4}(\frac{|\xi|}{\pi})^{1/2} Z(1/2, 1/2 - \zeta/\omega) & \text{otherwise} \end{cases}$$

where $Z(u, \zeta)$ is the Hurwitz zeta function, $\omega = 4\pi|\xi|$ and ξ is the density of the magnetic flux through the $x - y$ plane. There is a one-one correspondence between all point perturbations H of H_0 supported by Λ and all self adjoint operators A in $l^2(\Lambda)$. Let H be denoted by H_A. Albeverio and Geyler (1999) have shown that

Theorem 43 *(Albeverio and Geyler) For every Cantor set $C, C \subset (-\infty, \omega/2)$ there is a bounded self-adjoint operator A in $l^2(\Lambda)$ such that $\sigma(H_A) \cap (-\infty, \omega/2) = \sigma_{sc}(H_A) \cap (-\infty, \omega/2) = C$.*

Albeverio, Geyler and Kostrov have shown that the spectrum of H_A has a band structure and contains the semi-axis $[\omega/2, \infty)$. And this spectrum contains K bands $B_1, ..., B_K$ where B_k is the closure of $\{E_{k0}(p), p \in X\}$; here $E_{kn}(p), 1 \leq k \leq K, n \geq 0$ is the discrete spectrum of H_A. These bands may be mutually intersecting and may intersect the semi-axis $[\omega/2, \infty)$. In particular,

Theorem 44 *(Albeverio, Geyler and Kostrov) If $K = 1$, then the spectrum of $H_A(p)$ is purely absolutely continuous.*

Chapter 4

Mathematical Modeling and Quantum Devices

4.1 Introduction

In this chapter we review the mathematical modelling which has been performed on quantum wires and quantum devices related to the precise results presented in Chapter 3. Not all of the modelling results have been captured in the precise results to date, and thus remain as open problems. Several of these problems will be specifically spelled out below.

This modelling work has used a variety of numerical methods. These include the finite element method, the transfer matrix method, the coupled-channel method, the mode-matching method and the method of recursive calculation of the discrete Green function.

4.2 Mesoscopic Systems

Consider the case of a quantum wire which is connected to two 2DEG reservoirs. Due to z-confinement, the model Hamiltonian is

$$H = -\frac{\hbar^2}{2m^*}(\frac{\partial^2}{\partial x^2} + \frac{\partial^2}{\partial y^2}) + V(x,y)$$

where m^* is the effective mass of the electron in the semiconductor material (e.g., $m^* = 0.067m_o$ for GaAs). The detailed shape of the potential which confines the electron to the channels has been shown not to be important in determining the presence of bound states; e.g., infinite square well potentials and harmonic potentials have been utilized.

In the presence of a magnetic field the Hamiltonian becomes

$$H = \frac{1}{2m^*}(-i\hbar\nabla + \frac{e}{c}A)^2 + V.$$

E.g., in a perpendicular magnetic field with $A = (-\frac{By}{2}, \frac{Bx}{2}, 0)$, then the Hamiltonian is

$$H = \frac{1}{2m^*}[(-i\hbar\frac{\partial}{\partial x} - \frac{eBy}{2})^2 + (-i\hbar\frac{\partial}{\partial y} + \frac{eBx}{2})^2] + V.$$

Assume there is a weak potential difference between two 2DEG reservoirs which causes the flow of electrons. In ballistic transport the total current density is obtained by integrating over all the incoming electrons:

$$J(x, y) = 2\int_{k_0}^{k_F} kdk \int_{-\pi/2}^{\pi/2} d\phi j_{\mathbf{k}}(x, y)/(2\pi)^2.$$

Here $\mathbf{k} = k(cos\phi, sin\phi)$. Let E_F denote the Fermi energy and eV the drop in potential. By the Pauli principle, only electrons in the range $(E_F - eV, E_F)$ are allowed to flow from the emitter to the collector. So $k_F = (2m^*E_F)^{1/2}/\hbar$ and $k_0 = (2m^*(E_F - eV))^{1/2}/\hbar$ and if the applied voltage difference is small

$$J(x, y) = \frac{2m^*eV}{h^2}\int_{-\pi/2}^{\pi 2} d\phi j_{\mathbf{k}}(x, y)$$

where $|\mathbf{k}| = k_F$.

Associated with the current given by all states $\psi_K(x, y)$ in the channel in the energy window $(E_F - eV, E_F)$, the electron probability density is given by

$$\rho(x, y) = \frac{2m^*|eV|}{h^2}\int_{-\pi/2}^{\pi/2} |\psi_K(x, y)|^2.$$

The conductance is given by

$$G = \frac{-2em^*}{h^2}\int_{-\pi/2}^{\pi 2} d\phi j_{\mathbf{k}}|_{|\mathbf{k}|=k_F}.$$

4.3   Wire-Resonator-Wire Model

Bulla, Gesztesy, Renger and Simon (1997) have discussed the problem of bound states in quantum wires with bumps. Vacek, Okiji and Kasai (1992) considered a simple model of a protrusion in a straight quantum wire, viz. a straight well of length L and potential $V < 0$ in this region, attached to quantum wires on either side. They modelled this with and without a magnetic field. For $B = 0$, the transmission probability is simply

$$T_{mn} = \delta_{mn}\frac{4p_m^2 q_m^2}{4p_m^2 q_m^2 cos^2(q_m L) + (p_m^2 + q_m^2)^2 sin^2(q_m L)}$$

where p_m and q_m are the wavenumbers of the mth channel of the leads and well respectively.

The conductance of this model shows the standard steplike structure as a function of E_F. For nonzero magnetic fields the wavefunction in the well and in the leads are no longer orthogonal and coupling occurs between various channels in the well and leads. Off diagonal terms appear in T_{mn} and sharp dip structures are observed in the conductance. The energies of the dips correspond to the energy of quasibound states in the well and the width of the dips is determined by the coupling strength.

Vacek, Okiji and Kasai (1992) similarly model the conductance of a quantum wire with a circular bend, with and without a magnetic field.

4.4 One Dimensional Mesoscopic Crystal

A standard numerical method used in studying mesoscopic systems is the mode matching method. Wu et al. (1991) in particular studied the one dimensional mesoscopic crystal introduced by Ulloa, Castano and Kirczenow (1990). This problem allows a convenient formulation of the mode matching method in terms of the transfer matrix. The problem is to solve the Schrödinger equation

$$\Delta\phi + (E - V)\phi = 0.$$

For a quantum wire of uniform width $2d$ and where the potential depends only on x, one can separate variables to write

$$\phi_n(x,y) = \psi(x)sin(n\pi y/2d)$$

where

$$\psi''(x) + [E_{1d} - V(x)]\psi(x) = 0$$

with $E_{1d} = (E - (n\pi/2d)^2)$. We assume that the potential $V = V_0$ on the interior of the mesoscopic crystal composed of $n+1$ segments attached to two infinite leads where $V = 0$. In the leads ψ has the form

$$\psi_k(x) = c_k exp(i\alpha(x - x_k)) + \bar{c}_k exp(-i\alpha(x - x_k))$$

where $\alpha = \sqrt{E_{1d}}$. On the interior of the crystal,

$$\psi(x) = f exp(i\gamma(x - x_1)) + \bar{f} exp(-i\gamma(x - x_1))$$

with $\gamma = \sqrt{E_{1d} - V_0}$. By matching the wavefunction and its derivative at each end, one finds:

$$\left(\begin{array}{c} c_1^+ \\ c_1^- \end{array} \right) = M \left(\begin{array}{c} c_2^+ \\ c_2^- \end{array} \right)$$

where

$$M = \left(\begin{array}{cc} cos(\gamma d) & -i\alpha/\gamma sin(\gamma d) \\ -i\alpha/\gamma sin(\gamma d) & cos(\gamma d) \end{array} \right).$$

For $V_0 = 0$ and $d = l$, the transfer matrix is

$$P = \left(\begin{array}{cc} cos(\alpha l) & -i sin(\alpha d) \\ -i sin(\alpha l) & cos(\alpha l) \end{array} \right)$$

and the total transfer matrix for a device with $n + 1$ cells is

$$M^{total} = (MP)^n M.$$

The conductance for this one dimensional system is

$$g = \frac{2e^2}{h} T$$

where

$$T = | \sum_{i,j=1}^{2} M_{i,j}^{total} |^2$$

and for the two dimensional system

$$G(E) = \sum_{n \in O} g(E - (n\pi/2)^2)$$

where the sum is over the open or propagating channels where $E_{1d} > 0$. For this system $E_0 = (\pi/2d)^2$ is the threshold of the first propagating channel.

Using these equations one finds numerically:

(1) there is a steplike repeating structure for the conductance in a 1D mesoscopic crystal;

(2) there are resonant peaks below each step, in particular below the first threshold;

(3) the number of peaks is equal to the number of potential wells or cells in the mesoscopic crystal.

4.4.1 Single Mode Approximation

$L-$shaped quantum wires were studied by Weisshaar et al. using mode matching methods. They show strong resonance effects. The model of a periodic quantum wire was considered by Wu et al. (1993) based on mode matching. In the case of a single channel or the single mode approximation, the model is defined by the equations

$$\begin{pmatrix} 1 \\ r_1 \end{pmatrix} = \begin{pmatrix} M_{11} & M_{12} \\ M_{21} & M_{22} \end{pmatrix} \begin{pmatrix} t_1 \\ 0 \end{pmatrix}$$

for a single cell. And a bound state occurs if

$$\begin{pmatrix} 0 \\ r_1 \end{pmatrix} = \begin{pmatrix} M_{11} & M_{12} \\ M_{21} & M_{22} \end{pmatrix} \begin{pmatrix} t_1 \\ 0 \end{pmatrix}$$

i.e. $M_{11} = 0$. The condition for a bound state in the N-cell case is

$$(M^N)_{11} = 0.$$

One can show that

$$\frac{1}{T_N} = 1 + \frac{|sinN\phi|^2}{|sin\phi|^2}(\frac{1}{T_1} - 1)$$

and $cos(\phi) = Tr(M)/2$ where ϕ is the Bloch phase for a single cell.

Theorem 1 *(Wu et al.) In the single mode approximation, an $N-$cell periodic quantum wire will produce N tunneling resonances.*

Wu et al. also examined by single mode approximation the periodic multi-stub device studied by Brum (1991). In addition, Wu et al. examined the sub-threshold tunneling resonances which occur in this quantum dot superlattice. A similar quantum dot superlattice was examined by Ji and Berggren (1992).

4.5 L-Shaped Quantum Wires: Mode Matching

The proof we present follows that of Martorell, Klarsfeld, Sprung and Wu (1991). The quantum mechanical problem is given by

$$\frac{\hbar^2}{2m^*}(\frac{\partial^2}{\partial x^2} + \frac{\partial^2}{\partial y^2})\psi + E\psi = 0.$$

Consider the symmetric and antisymmetric functions

$$\chi_p^\pm(x,y) = sin(\frac{p\pi y}{d})sin(k_p x) \pm sin(\frac{p\pi x}{d})sin(k_p y)/sin(k_p d)$$

where $k_p^2 + p^2\pi^2/d^2 = 2m^*E/\hbar = \kappa^2$ for $p = 1, 2, ...$ In the arm parallel to the x-axis

$$\phi_p^\pm(x,y) = \sum_q A_{pq}^+ sin(k_q x) + B_{pq}^\pm cos(k_q x)sin(q\pi y/d).$$

The wavenumber k_q will be real for open channels and purely imaginary for closed channels. The matching conditions at $x = d$ require:

$$\phi_p^\pm(d,y) = \chi_p^\pm(d,y)$$

and

$$\frac{\partial \phi_p^\pm}{\partial x}(d,y) = \frac{\partial \chi_p^\pm}{\partial x}$$

which gives:

$$A_{pq}^\pm = \delta_{pq}sin(k_p d) + g_{pq}^\pm cos(k_q d)$$

$$B_{pq}^\pm = \delta_{pq}cos(k_q d) - g_{pq}^\pm sin(k_q d)$$

and

$$g_{pq}^\pm = \delta_{pq}cot(k_p d) \pm \frac{(-1)^{p+q}2pq\pi^2}{k_p d(k_p^2 d^2 - q^2\pi^2)}.$$

Thus, the problem is to find a solution of the form

$$\psi = \sum(\alpha_p^+ \phi_p^+ + \alpha_p^- \phi_p^-)$$

which satisfies the boundary conditions at infinity. For a bound state this results in the equation:

$$\sum \alpha_p^\pm(B_{pq}^\pm + iA_{pq}^\pm) = 0$$

in which case the energy of the bound state can be determined as the zero of a determinant similar to the results of Evans and coworkers discussed in Chapter 2. Truncating this system to first order gives:

$$B_{11}^{\pm} + iA_{11}^{\pm} = 0$$

or

$$\pi\nu(1 + \cosh(\pi\nu)) = \pm 2/(1 + \nu^2)$$

where $k_1 d = i\pi\nu$ and the $+(-)$ term refers to the symmetric (resp. antisymmetric) case. In the first case there is one real root, $\nu_0 \approx .232$.

4.5.1 Reflection and Transmission

The reflection and transmission amplitudes are given by

$$r_{nq} = \sqrt{k_q} e^{-ik_q d}(\alpha_q^+ - \alpha_q^-) - \delta e^{-2ik_q d}$$

$$t_{nq} = \sqrt{k_q} e^{-ik_q d}(\alpha_q^+ + \alpha_q^-)$$

and the reflection and transmission coefficients are:

$$R_{nq} = |r_{nq}|^2$$

$$T_{nq} = |t_{nq}|^2$$

and $R_n = \sum_q' R_{nq}, T_n = \sum_q' T_{nq}$ where the sum is over open channels only. Conservation of probability flux means that $R_n + T_n = 1$.

4.6 L-Shaped Quantum Wires: Single Mode Approximation

For the case of a single open channel $n = 1$ and truncating the equations to $N = 1$ Wu et al. (1991) showed that

$$T_1(E) = [1 + (u - 1/u)^2/4\sin^2\pi\eta]^{-1}$$

where

$$u = \frac{2}{1 - \eta^2} \frac{\sin\pi\eta}{\pi\eta}$$

and

$$\pi\eta = k_1 b = \sqrt{\kappa^2 b^2 - \pi^2}$$

gives a very good approximation to $T(E)$.

The bound states for an L-shaped for the case $N = 1$ reduce to

$$B_{11}^{\pm} + iA_{11}^{\pm} = 0$$

or

$$\pi\nu(1 + \coth\pi\nu) = \pm 2/(1 + \nu^2)$$

where $k_1 b = i\pi\nu$. Numerically one finds a single real root ν_0 to this equation with $\nu_0 \sim .232$ or $\xi_0 \sim .973$. Taking higher order approximations Wu et al. find $\xi_0 = .964$ which agrees with Schult et al. and Exner et al.

4.7 T-Shaped Electron Waveguides

T-shaped election waveguides have been studied by Sols, Macucci, Ravioli and Hess (1989), Avishai and Band (1990), Martorell et al. (1991), Berggren and Ji (1990), Ji and Berggren (1992) among others. Sols et al. and Martorell et al. noted that the transmission coefficient or conductance has a periodic structure as a function of the stub length c. In using mode matching methods, Wu et al. (1991) noted that even in the fundamental mode, $T = T(c)$ is a periodic function with period $\delta c = \pi/\sqrt{E - (\pi/d)}$ where d is the width of the stub.

Sols et al., Martorell et al. and Berggren and Ji (1990) showed that there is a single bound state for a T-shaped stub below the first threshold $E_t = \pi^2 \hbar^2/2m^* d^2$, where the bound state is found by the same technique as in the last section. The energy of the bound state was estimated by Wu et al. At this energy the conductance has a peak value $G = 2e^2/h$.

As a function of the length of the sidearm for these structures, Berggren and Ji found that G is relatively constant although it has an oscillatory structure. Sols et al. (1989) noted strong interference terms which give rise to vanishing of the conductance in T-shaped waveguides. T-shaped circuits have also been studied by D'Amata et al. (1989) using the tight binding model and experimentally by Haug, Lee and Hong (1990). Tight binding models will be discussed in Chapter 6 in greater detail.

For T-shaped structures, Berggren and Ji show that there are peaks near the second plateau due to quasibound states.

4.8 Electron Stub Tuner (EST)

The connection of a quantum wire to a resonant cavity has been described above. In the literature it is also referred to as an electron stub tuner (EST) where the length of the stub laterally attached to the quantum wire can be controlled by an independent gate.

It has been found that if the width of the quantum wire and the stub are such that only a single propagating mode exists for a given energy, then the conductance G is a periodic function of the stub length, with G oscillating between 0 and 1 in units of $2e^2/h$. Viz., for an EST with a stub length c and width d a periodic conductance is obtained as a function of c for infinite square-wall confinement where the period is

$$\delta c = \pi/\sqrt{2m^* E_F/\hbar^2 - (\pi/d)^2} = \lambda_s/2.$$

Applications of the EST for quantum transistors have been mentioned above. The interest here lies in the possibility of femtosecond order tuning and very low power dissipation.

A related device is the double electron stub tuner (or DEST), which can produce a nearly square-wave conductance output. One recalls the results of Popov discussed earlier. Applications of this device in analog-to-digital converters have been considered. Recent work by Debray and coworkers on DESTs have shown:

(1) for a symmetric DEST, the period is doubled, so that $\delta c = \lambda_s$;

(2) there are minima in G as a function of a weak perpendicular magnetic field B; here conductance minima or antiresonances correspond to bound states in the stubs; weak magnetic field will break the symmetry and allow states in the cavity and wire that were previously orthogonal to couple giving rise to new dips in the conductance;

(3) edge states do not occur in these conditions;

(4) new quasibound states are created when $B \neq 0$ that are not present in a zero field due to couplings between the wire and the stub wavefunctions.

Numerical work of Akis et al. show that reflection resonance peaks occur. These resonances correspond to conductance minima. These are quasi-bound states which satisfy $det(M) = 0$. For a symmetric DEST, even quantum wire states couple only with even DEST states to produce resonances. However, if an offset is introduced the symmetry in the middle of the quantum wire is broken. This enables even quantum wire states to couple with odd DEST states.

4.9 The Double-Bend Quantum Wire

The double-bend quantum wire has been studied by Weisshaar et al. (1989), Bar-Touv and Avishai (1990), Wu et al. (1991), Wu, Sprung and Martorell (1992), and Wang, Berggren and Ji (1995). Wu et al. (1991) have a very interesting data set. Their experiment shows the conductance plateaus, two sharp peaks in the conductance below the first conductance plateau, a dip (or anti-resonance) in the conductance just prior to the second propagation threshold and five resonance peaks superimposed on the conductance plateaus. The suggestion by Wu et al. that the sharp peaks in the conductance are due to impurities has resulted in several modeling papers, including Wang, Berggren and Ji (1995), Carini et al. (1997), and Wu, Sprung and Martorell (1992).

The results of Exner and Seba (1989) and Goldstone and Jaffe say that in a quantum channel with a double bend discontinuity, there should exist two local bound states which are localized at the bends. The relationship of these bound states to the conductance peaks was studied by Wang et al.

Wu, Sprung and Martorell (1992) found that even in the single mode approximation to the double bend quantum wire there are two peaks in the conductance below the first conductance plateau. They identified the first peak at $.908(\pi/d)^2$ as an even symmetry bound state and the next peak at $.973(\pi/d)^2$ as an odd bound state.

Let t_0 denote the width of each segment in the double bend; let $t_x + t_0$ denote the length of the first segment and let $W = t_0 + t_y$ denote the length of the second segment. In each of the segments let

$$\psi^j(x, y) = \sum (a_n^j e^{q_a^j x} + b_n^j e^{-q_a^j x}) \phi_n^j(y)$$

where $\phi_n^j(y)$ are the transverse eigenstates with eigenvalues

$$E_n^j = (\frac{\hbar^2}{2m^*})(\frac{n^2\pi^2}{w_j^2})$$

and $(\frac{n^2\pi^2}{w_j^2}) + (q_n^j)^2 = (\frac{2m^*}{\hbar^2})E$.

Wang et al. (1995) have performed a numerical study using the mode matching method. In the course of their work they evaluate the transmission and reflection matrices as well as the electric current. The conductance is given by

$$G = -\frac{2em^*}{h^2} \int_{-\pi/2}^{\pi/2} d\phi j(K)|_{K=K_F}$$

at $T = 0K$, and the current density

$$j(K) = -\frac{e\hbar}{m^*}[\overset{Re}{\sum_n} q_n^i(a_n^i a_n^{i*} - b_n^i b_n^{i*}) + \overset{Im}{\sum_n} q_n(a_n^i b_n^{i*} - b_n^i a_n^{i*})]$$

where $i = I, II, III$ denotes the region.

In addition these authors examine the finite temperature conductance which is given by

$$G(\mu, T_B) = \int_0^\infty dE(-\frac{\partial f}{\partial E})G(E, T_B = 0)$$

where

$$f = (e^{(E-\mu)/k_B T_B} + 1)^{-1}$$

is the Fermi function and μ is the chemical potential. Here k_B is Boltzmann's constant and T_B is the temperature.

Wang et al. show:

(1) the first transverse mode energy of the first conductance plateau is given by $\hbar^2\pi^2/2m^*(V - t_y)^2$;

(2) if $t_y < zW/2$ there is one resonant peak below the first plateau;

(3) for $t_y \geq W/2$ there exist two resonant peaks below the first plateau;

(4) as t_x increases the peaks become narrower but their height remains the same;

(5) as the separation in the right angles increases, the two peaks decrease.

(6) as T increases, the quasi-bound states are broadened and their amplitudes are lowered;

(7) peaks of Wu et al. are determined to be due to resonant tunneling via the two quasibound states and not due to impurities;

(8) peaks superimposed on the conductance plateau are due to resonant tunneling via longitudinal resonant states;

(9) peaks can be interpreted as symmetric and antisymmetric superpositions of two local bound states, localized at the right angles;

(10) temperature effects broaden the sharp peaks and lower their heights.

4.9.1 Numerical Methods: Mode-Matching

Carini, Londergan, Murdock, Trinkle and Yung (1997) rewrite the equations in dimensionless coordinates $x = x'/W, y = y'/W$ and $R = H/W$ so that the Schrödinger equation is rewritten as the Helmholtz equation

$$(\Delta + k^2)\psi = 0$$

where $k^2 = 2m^* E W^2 / \hbar$, and W and H are the width and heigth of the channel.

In region I,

$$\psi_I(x,y) = \sum_{n=1}^{\infty} A_n sin(n\pi y) e^{\alpha_n x}$$

where $\alpha_n = \sqrt{n^2 \pi^2 - k^2}$. In region II, viz. $0 \le x \le 1, 0 \le y \le 1$, the boundary conditions are

$$\psi_{II}(x,y)|_{y=0} = \psi_{II}(x,y)|_{x=0} = \psi_{II}(x,y)|_{x=y=1} = 0$$

and the wave equation has the form

$$\psi_{II}(x,y) = \sum_{n=0}^{\infty} (B_n sin(n\pi y) sinh(\alpha_n x) + C_n sin(n\pi x) sinh(\alpha_n y)).$$

In region III, viz. $0 \le x \le 1, 1 \le y \le R - 1$

$$\psi_{III}(x,y) = \sum_{n=1}^{\infty} D_n sin(n\pi x) F_n(\alpha_n(y - R/2))$$

where

$$F_n^s(t) = \begin{cases} cosh(t) & \text{for } n \text{ odd} \\ sinh(t) & \text{for } n \text{ even} \end{cases}$$

and

$$F_n^a(t) = \begin{cases} sinh(t) & \text{for } n \text{ odd} \\ cosh(t) & \text{for } n \text{ even} \end{cases}$$

for the symmetric and antisymmetric cases. Truncating these equations to order N, the bound states occur when

$$det(Z(k^2)) = 0$$

where

$$Z = TS - I$$

with

$$T_{nm} = \frac{m\pi(-1)^m d_{nm}}{\alpha_n \{sinh(\alpha_n) F_{n+1}(\alpha_n[1 - R/2])/F_n(\alpha_n[1 - R/2]) - cosh\alpha_n\}}$$

$$S_{nm} = \frac{(-1)^{m+1} m\pi exp(-\alpha_n) d_{nm}}{\alpha_n^2}$$

and

$$d_{nm} = \frac{(-1)^{n+1} 2n\pi sinh(\alpha_m)}{\alpha_m^2 + (n\pi)^2}.$$

In these coordinates, the continuum begins at $k^2 = \pi^2$.

Carini et al. have observed:

(1) for $R < 2.5$, only the symmetric state is bound; the antisymmetric state is unbound for smaller values of R;

(2) the bound state wave function has a single peak centered in the middle of the double bend;

(3) for $R > 2.5$, both the symmetric and antisymmetric bound states appear; both symmetric and antisymmetric wave functions have a peak at each bend in the quantum wire;

(4) for $R > 7$, both symmetric and antisymmetric bound states approach the same energy $\epsilon = .930$, corresponding to the bound state energy for an infinitely long wire of unit width and a single right angle bend;

(5) the number of bound states and their locations are completely determined by the geometry of the bent quantum wire.

Carini et al. have verified their numerical work by measurements of TE modes in a microwave waveguide of width 1.905 cm. Bound states then correspond to localized states below the cutoff frequency of the waveguide, which is nominally $c/2W = 7.87$ GHz for their model. The field distribution was measured following the method of Sridhar (1991) of moving a steel ball in the structure which allows one to map out the electric and magnetic fields.

Carini and coworkers also calculated the conductance, which in terms of their expansion is given by

$$G(E_F, T_B = 0) = \frac{2e^2}{h} \int_{-k_F}^{k_F} \sum_{n \in Open} [B_n^* B_n - C_N^* C_n| +$$

$$\sum_{n \in Closed} [C_n^* B_n - B_n^* C_n] dK / 2\pi k$$

where $q_N = \sqrt{k_F^2 - (n\pi)^2}$ is real for open channels (i.e., $k_F > n\pi$), otherwise $q_n \to i|q_n|$. As with Berggren and coworkers, they predict conductance peaks above the conductance threshold due to bound states in the quantum wire.

4.10 Intersecting Quantum Wires

Schult et al. as we mentioned earlier found two bound states that reside at the intersection of the two infinite quantum wires of constant width; (v., also Peeters (1989)). Tunneling through the discrete states in two intersecting wires was modelled by Berggren and Ji (1991), Berggren, Besev and Ji (1992). In Berggren and Wang (1997) the authors argued that the lowest bound state would remain singly occupied while a second electron would be pushed into the continuum because of spin polarization. Therefore, the conductance associated with bound state resonance would be reduced from $2e^2/h$ to e^2/h because of interaction.

4.10.1 Quantum Cross Bar

This case was studied by Berggren, Besev and Ji (1992) and Berggren and Ji (1991). The structure consists of two perpendicular channels one of which connects two reservoirs of 2DEG. When a weak potential difference is applied to the 2DEG regions, ballistic transport of electrons occurs within subbands. In addition there are sharp resonances associated with bound states at the intersection. For long channels the conductance G is quantized as $G = 2e^2/h$.

Berggren and coworkers have found:

(1) resonances are due to bound states and antiresonances are due to interference effects

(2) there are resonant peaks at $E_a = .66E_1$ where $E_1 = \hbar(\pi/d)^2/(2m^*)$ and $E_b = .93E_2$ where $E_2 = 4E_1$. The conductance at each of these energies is $G = 2e^2/h$ and $G = 4e^2/h$ if the stubs are sufficiently extended;

(3) strong interference which gives rise to vanishing of the conductance;

(4) the energies E_a, E_b are insensitive to extension of the sidearms indicating the wave function are well localized at the center of the cross bar;

(5) other peaks appear which are sensitive to variations in d_1 and d_2 where the wavefunction is mainly localized in the stubs; the first peak appears at

$$E_c \sim \hbar^2/2m^*[(\pi/d_1)^2 + (\pi/W)^2];$$

the determinant does not vanish for shorter stubs; in other words E_c is pushed into the second subband where it becomes a quasibound state;

(6) resonances are due to standing waves associated with the stubs and their position depends on the length of the stubs;

(7) conductance vanishes due to antiresonances or interference blockades;

(8) resonance peaks correspond to bound states discovered by Schult, Ravenhall and Wyld (1989).

Berggren and coworkers have calculated the relative probability distribution associated with resonant tunneling current via bound states, resonances of interference energies as well as quantum mechanical velocity distributions, where they have observed a transition from "laminar" to "vortex" flow. Furthermore, small changes in energy or geometry can result in dramatic changes in the flow.

4.11 Periodic Quantum Wire and Dot Structures

In this section we examine periodic quantum wire structures and quantum dot superlattices. Periodic structures on quantum wires have been examined numerically by Ji and Berggren (1992), Wu et al. (1993) and Brum (1991), among others. One recalls the theoretical work of Popov discussed earlier.

Wu et al. examined the problem in terms of mode matching for the one dimensional quantum crystal and a multi-stub quantum wire.

4.11.1 N-Crosses

The N-cross periodic structure was modelled by Ji and Berggren (1992) following experimental work of Haug et al. Ji and Berggren found:

(1) resonance peak below the lowest subband split into N peaks if there are N intersecting crosses;

(2) one can view transport as tunneling via coupled quantum dots;

(3) if the stubs are sufficiently extended, the conductance is equal to $2e^2/h$ at the resonance energy;

(4) N−periodic T-structures will also have N peaks;

(5) the spatial distribution of currents can be complex with a transition from "laminar" to "vortex" flow.

4.12 Quantum Structures with Circular Bends

The model of Sols and Macucci (1990) considers a bent electron waveguide with a circular bend model. Let d denote the wire width, R_0 the internal bend radius and Θ the bending angle. In the two leads the equations are

$$\psi_I(\rho) = \sum \xi_n(y)(a_n e^{k_n x} + e_n e^{-k_n x})$$

and

$$\psi_{III}(\rho) = \sum \xi_n(y')(b_n e^{k_n x'} + f_n e^{-k_n x'}),$$

where $\chi_n(y) = (2/\pi)sin(n\pi y/d)$ and $k_n^2 = (n\pi/d)^2 - 2mE/\hbar^2$. For evanescent waves $k_n^2 > 0$, which requires that $e_n = f_n = 0$ and for propagation modes $k_n^2 < 0$ in which case one writes $k_n = -i|k_n|$.

In the bend region the Schrödinger equation can be written in polar coordinate as $\phi(\rho) = P(\rho)\Phi(\phi)$ where

$$\frac{d^2\Phi}{d\phi^2} = \nu^2 \Phi$$

and

$$\rho^2 \frac{d^2 P}{d\rho^2} + \rho \frac{dP}{d\rho} + \rho^2 k^2 P = -\nu^2 P.$$

The general solution in this region has the form

$$\psi_{II} = \sum P_n(\rho)(c_n e^{\nu_n \phi} + d_n e^{-\nu_n \phi})$$

with boundary conditions $P_n(\rho_0) = P_n(\rho_0 + d) = 0$. Matching conditions at the interfaces of regions I-II and II-III gives a system of coupled homogeneous equations for a, b, c, d, e, f. For a bound state all modes are evanescent so that $e_n = f_n = 0$ and the bound state satisfies an equation of the form $det(M) = 0$.

The energy threshold in the first angular mode (where $\nu_1^2 = 0$) for a large radius of curvature is $E_1 \sim \hbar^2\pi^2/2md^2$. Numerically Sols and Macucci show that there is one or more bound state in a circular bend and they show that the binding energy $E_1 - E_{bs}$ decreases with increasing radius and decreasing angle. They show that the circular case is more weakly bound than the L−shaped bend; viz. $E_{bs} - L = .92E_1$ while $E_{bs} - cir = .97E_1$ in the case $\rho = .15W$ and $\Theta = 180$ degrees.

Resonance structure for the case of a circular bend was noted by Sols and Macucci. They showed that the transmission matrix $T_{nm}(E)$ undergoes rapid variation just below the threshold energies, which is a signature of resonances. They viewed these as quasibound states of the subband whose threshold lies just above and the finite width comes from coupling to states in the lower subbands; e.g., they found that $T_{11}(E)$ has a dip of width $.005E_1$ which is at a distance $.006E_1$ below E_2.

Berggren and Ji (1993) have examined double bend structures with circular bends; they show once again a transition from "laminar" to "vortex" current flow and attribute the weak oscillations on the plateaus in G to standing waves in the middle section. That is, the current flow is laminar in regions of energy where the conductance is well quantized; however at subthreshold energy where interference blockage occurs, the current becomes vortical and quite complex in texture with rapid shift in the direction of the vortices as the resonance is passed.

In narrow energy regions just below the thresholds there is a strong interference between localized and propagating states causing an interference blockade where the current flow becomes "vortical". This region is unstable where minor changes in energy cause drastic changes in the flow, whereas the current flow is laminar in regions of energy in which conductance is well quantized.

4.12.1   Circular Bends in a Magnetic Field

The case of a circular bent wire in a magnetic field was considered by Vacek, Kasai and Okiji. Let w denote the width of the quantum wire, R_0 the inner bending radius, Θ the bending angle and L the distance between bends. Here the Hamiltonian has the form

$$H = \frac{1}{2m}(-i\hbar\nabla + \frac{e}{c}A)^2 + V.$$

In their paper they consider a hard wall potential, viz. $V = 0$ in the wire and $V = \infty$ outside. If region I is parallel to the x_1-axis through the center of the wire, the Landau gauge is selected to be $A = (-y_1B, 0, 0)$; and similarly in region III. In the region of the bend with axes x_2, y_2, the Landau gauge is $A = (-y_2B/2, x_2B/2, 0)$. Since ψ_I, ψ_{II} and ψ_{III} have been obtained in different gauges, as a global condition Vacek, Kasai and Okiji use the symmetrical gauge condition

$$\psi_I^s(x_1, y_1) = exp\{-\frac{ieB}{2\hbar c}[(R_0 + \frac{d}{2})x_1 + x_1 y_1]\}\psi_I(x_1, y_1)$$

$$\psi^s(r, \phi) = \psi_{II}(r, \phi)$$

and

$$\psi_{III}^s(x_3, y_3) = exp\{-\frac{ieB}{2\hbar c}[(R_0 + \frac{d}{2})x_3 + x_3 y_3]\}\psi_{III}(x_3, y_3).$$

Again one requires continuity of the wave function and its normal derivatives at the interfaces I,II and II,III which gives rise to a system of linear equations for e_l, b_l, c_l, d_l.

Taking the initial conditions $a_l = \delta_{lm}$, $f_l = 0$ one has $t_{mn} = b_n$ and $r_{mn} = e_n$; and $T_{mn} = |t_{mn}|^2$, $R_{mn} = |r_{mn}|^2$ and

$$G = \frac{2e^2}{h} \sum_{l,n=1}^{N} T_{ln}.$$

In their numerical work they find that G has the standard step like structure; however, their plot of conductance for $R_0 = .5d$ and $\Theta = \pi/2$ shows no bound state peak below the first plateau. In terms of resonance structure, they find that the presence of the magnetic field effects the coupling between the channels and it differs depending on whether the bend is a right turn or a left turn. They find a sharp dip in the conductance just below the thresholds of the next channels. E.g., at the step from $G = 2e^2/h$ to $G = 4e^2/h$, the width of the first dip is $.0021E_1$ for zero magnetic field and $.0011E_1$ for a magnetic field of $B = \pm 5B_0$ for $R_0 = .5d$ and $\Theta = \pi/2$. The the Fermi energy of the center of the first dip is $E_F \sim 3.9965E_1$ for $B = 0$ and $E_F \sim 4.1738E_1$ for $B = \pm B_0$. Here $B_0 = \hbar c/ed^2$.

Vacek, Okiji and Kasai (1993) have studied both S-shaped and U-shaped bent quantum wires, with and without a magnetic field. In the S-shaped case if $L \leq 2w$ then there exists one quasibound state; but if $L > 2w$, they found two quasibound states which merge as $L >> 2w$; the lower state has even symmetry and the higher odd. The resonance widths of the quasibound states were related to the coupling of the quasibound states with the propagating modes in the straight sections. Adding a magnetic field can narrow or widen the conductance dips or make them vanish.

They also examined the conductance in a bent channel with and without a magnetic field. Breaking the symmetry by bending the wire or by applying a magnetic field leads to coupling between bound states in the created potential well and the propagating states in the asymptotic leads. This leads to dips of finite width which appear in the two terminal conductance plots.

4.12.2 Approximations

Sprung, Wu and Martorell (1992) have shown that the circular bend in a quantum wire can be approximated with a square well potential where the longitudinal motion in the bend is described by

$$\frac{\partial^2 \Phi_n}{\partial x^2} + (p_n^2 + \frac{1}{4\bar{R}^2})\Phi_n = 0$$

where $a = \bar{R}\Theta$. This leads to an approximation for the energy of the bound state given by

$$E/E_0 = (1 - (d/2\pi\bar{R})^2(1 - 4\nu^2)$$

where ν is relate to the bending angle by

$$\Theta = (\frac{1}{\nu})\tan^{-1}[\sqrt{1/4\nu^2) - 1}]$$

The authors suggested the use of $\bar{R} = R + d/2$. They also calculated the transmission coefficient, T, and compared it to Lent's (1990) results. Here

$$T^{-1} = 1 + \lambda^4 sin^2(\epsilon q a)/4p^2(p^2 + \lambda^2)$$

where $\lambda = d/2\pi\bar{R}$, $2qa = (\Theta/\lambda)\sqrt{p^2 + \lambda}$, $P = pd/\pi$ and p is the wavenumber in the lead, $p^2 = (2m^8E/\hbar^2) - (\pi/d)^2$.

For higher modes Sprung et al. predict a bound state for each transverse mode n which becomes a quasibound state or subthreshold resonance with an energy shift ΔE_n with respect to the threshold given by

$$\Delta E_n/E_0 = -(d/2\pi\bar{R})^2(1 - 4\nu^2).$$

This approximation agrees well with the results of Sols and Macucci for the case of a 90 degree bend and $\bar{R}/d = .7$.

4.13 Arbitrary Sharp Bending Angle

This case has been considered by Wu, Sprung and Martorell (1992) and Carini, Londergan, Mullen and Murdock (1993), which extends the work of Exner and co-workers in the case of an L-shaped wire. For this model $R = d/sin(\alpha/2)$. For bound states they show that if α is not too small there is a unique bound state and when $\alpha \to \pi$, i.e. the bending angle approaches zero, the energy goes to $E_1 = (\pi/d)^2$. When $\alpha \to 0$, one can approximate the model as a circular sector of radius $2d/\alpha$, where

$$\Psi = sin(\theta n\pi/\alpha)J_{\pi n/\alpha}(\sqrt{e}r)$$

for $0 \leq \theta \leq \alpha$. The quantization condition is then

$$J_{\pi n/\alpha}(\sqrt{E}2a/\alpha) = 0$$

in which case the eigenenergies are

$$E/E_1 = x_{n_i}^2\alpha^2/4\pi^2$$

where x_{n_i} is the ith root of $J_{\pi n/\alpha}(x) = 0$. They found numerically that the second, third and fourth bound state appear at $\alpha = 28.8, 16.6$ and 11.7 degrees respectively and all these states become degenerate as $\alpha \to 0$ with energy $(\pi/2d)^2$. These bound states all have even symmetry and no odd symmetry states were found.

4.14 Serpentine Quantum Dot Wires and Arrays

Multibend quantum wires have been examined by Wu et al. (1992), Wu and Sprung (1993), Xu, Ji and Berggren (1991), and Vacek, Okiji and Kasai (1993). Wu and coworkers have examined the problem from the single mode approximation.

Xu, Ji and Berggren have studied serpentine quantum dot arrays, including the now famous double bend case, using the transfer matrix method. For the serpentine structures they find $N - 1$ resonance states below the first conductance plateau if there are N bends.

These authors also examined an array of square-like potential barriers in the center of a straight channel. Again N barriers were found to produce $N - 1$ peaks below the first plateau.

Their model was somewhat idealized using rectangular barriers of height V_b in a straight channel of width w with hard walls. They evaluated the conductance G. If $V_b = 0$, they find the standard plateaus increasing by one unit in $2e^2/h$ for each unit increase in $\xi = (2/\hbar\pi)(2m^* E_F)^{1/2}$ where E_F is the Fermi energy in the 2DEGs.

For a double-bend case as V_b is increased a single sharp peak occurs below the lowest subband threshold E_t. For a double-double-bend curve, they observe that the single peak splits into two peaks below E_t, and for a triple-double-bend they find a splitting into three sharp peaks. These conductance peaks are associated with resonant tunneling via quasi-bound states in their paper.

Xu et al. also modeled multiple square-like potential repulsive barriers implanted in the center of an otherwise straight quantum channel delimited by hard-wall boundaries. For two barriers, they observed for $V_b = 5E_1$, a single conductance peak below the first threshold. For three barriers they observe a splitting of the peak into two peaks. For four barriers, they observe a splitting into three peaks and for six barriers they observe a splitting into five peaks. In the last case, the splitting peaks can be easily seen also below the first, second, third and fourth conductance thresholds.

Vacek, Okiji and Kasai (1993) have studied the serpentine quantum wire, i.e. a periodic circular bend structure, as well as one with a stair-step serpentine structure. They showed for multiple bend ballistic transport:

(1) the number of states in the miniband structure is equal to the number of segments in the system;

(2) gap structures are related to interference of intrachannel coupling.

They found resonance peaks as well as anti- resonance gaps appearing on each plateau. For the serpentine case with $R_0 = .5w, \Theta = \pi$ and $L = 0$, the distance between each segment, they found that the number of resonance peaks on the first plateau, before the first gap is equal to the number of segments in the serpentine structure. The location of the nth gap in the first plateau was found to be $E_{l=1}^{n=1} = 1.084E_1, E_{l=1}^{n=2} = 1.414E_1, E_{l=1}^{n=3} = 1.942E_1, E_{l=1}^{n=4} = 2.639E_1, E_{l=1}^{n=5} = 3.483E_1$ and in the second plateau $E_{l=2}^{n=1} = 4.095E_1, E_{l=2}^{n=2} = 4.478E_1$ in the region $E_F < 4.5E_1$, where $E_1 = \hbar^2\pi^2/2m^* w^2$. The five gap structures in the first conductance plateau are related to the angular momentum values $\hbar\nu_1 \sim n\hbar$, for $n = 1, .., 5$ and in the second plateau for $\hbar\nu_2 \sim n\hbar$ for $n = 1, 2$, i.e. where the phase shift $\nu_l\Theta$ is equal to $n\pi$; here $\hbar\nu_l$ corresponds to the angular momentum of an electron in the subband l in each bend.

Similar resonance and antiresonance structures appear in the stairstep serpentine quantum wire studied by these authors. A wide gap with almost perfect vanishing of the conductance appears even for a small number of segments for the case $R_0 = .5w, \Theta = \pi/2$ and $L = 0$ just below the onset of the second plateau, in the region $E_F = (3.968E_1, 4.027E_1)$; in fact this extends beyond the energy threshold $4.0E_1$ for the second conductance plateau. This wide gap arises from the overlap of two vanishing channels (v., Vacek et al. for the details).

No peaks below the first plateau appear in the work of Vacek et al.

4.15 Quantum Wires with Bulges

Okiji, Kasai and Nakamura (1991) have considered quantum wires with step and smooth bulge like structures. In their paper they consider a parabolic confining potential, both with and without a magnetic field. The Hamiltonian has the form

$$H = -\frac{\hbar^2}{2m}(\frac{\partial^2}{\partial x^2} + \frac{\partial^2}{\partial y^2}) + \frac{1}{2}m^*\omega^2(y)x^2.$$

In this paper the authors note that the anti-resonance or dip structures are due to the reduction of G as a result of the enhancement of the reflection probability for an electron initially in $n = 0$ channel which occurs if the energy of the incident electron is close to the energy of the bound state. Thus, the number of channels contributing to the conductance is reduced by one whereas if the Fermi energy differs from that, the electrons can transfer without reflection and the conductance recovers its magnitude. No peak structures are seen in their work but resonance dips occur. The number of dips increases as the length L of the bulge increases; viz., two dips appear when $L = 2d$ and one for $L = d/2$ to $3d/2$ for $B = .2B_0$ and with width of the bulge $W = .6d$.

4.16 Quantum Wire with Ring Geometry

A quantum wire with a ring geometry in the presence of a magnetic field was modelled by Okiji, Kasai and Mitsuka (1992). Their Hamiltonian had the form

$$H = \frac{1}{2m^*}(-i\hbar\partial/\partial x)^2 + (-i\hbar\partial/\partial y + eBx)^2 + V_c(x,y) + V_s(x,y)$$

where $V_c(x,y) = \frac{1}{2}m^*\omega^2(y)x^2$ is the confining potential and

$$V_s(x,y) = V_0 exp(-(x^2 + y^2)/R_0^2)$$

is the potential barrier at the center of the ring, where R_0 is the radius of the ring.

They calculated the conductance for the case $V_0 = 0$ and showed that the conductance has sharp dips due to bound states in the intersection. They find that the conductance takes the value $2e^2/h$ for $B > B_0$ where only the $n = 0$ channel is open in the asymptotic regions ($y \to \pm\infty$). For $B_0 < B < 2B_0$, they found that G has dip structures which shows a period of order h/e, which they attribute to intrachannel coupling of $N = 0$ channel. For $V_0 > 0$, the conductance show oscillations as a function of the magnetic field for $B > .7B_0$, similar to A-B oscillations. They also found a peak structure in the conductance at $E_F \sim 1.7\hbar\omega_0$ and $2.5\hbar\omega_0$ which they attribute to coupling between the electronic states in $n = 0$ and $n = 2$ channels.

4.17 Electron-Electron Interaction

Schult, Ravenhall and Wyld (1989) examined the question of electron-electron interaction for cross quantum wires. One should also compare the work of Toyama and Nogami (1994) and Berggren and Wang (1997).

4.18 Diffraction and Mesoscopic Devices

Diffraction in the semiclassical description of mesoscopic devices is developed in Vattay et al. (1997) where they apply it to impurity and wedge diffraction. See also Bogomolny, Pavloff and Schmit (1999).

4.19 Electromagnetic Waveguides

In electromagnetic waveguides there is transverse-electric (TE) and transverse-magnetic (TM) modes. However, in a very flat rectangular electromagnetic waveguide, the high cutoff frequency of the TM modes and the $TE_{mn}, n \neq 0$ leaves only the TE_{m0} modes as active in propagation and scattering. For the TE modes if

$$(\Delta + k^2)\psi(x,y) = 0$$

with Dirichlet boundary conditions, then

$$\mathbf{E} = ik\hat{z}\psi(x,y)$$

and

$$\mathbf{B} = -\hat{z} \times \nabla\psi(x,y)$$

satisfy Maxwell's equations. The frequency and k are related by

$$k = \frac{2\pi\sqrt{\mu\epsilon}}{c}f.$$

The primary experiments in the area of trapped modes in electromagnetic waveguides are those of Carini et al. (1997).

Using the method of Sridhar (1991) to map the field distribution for the bound states, Carini et al. (1993) showed that the perturbation in the resonant frequency produced by the presence of a metal ball of radius r is given by

$$\frac{\Delta f(x,y)}{f_0} = -\frac{4\pi r^3}{2Dw^2}(C|\psi(x,y)|^2 - \frac{1}{2k^2}(|\frac{\partial\psi(x,y)}{\partial x}|^2 + |\frac{\partial\psi(x,y)}{\partial y}|^2)).$$

Here ψ is the normalized wave function associated with the bound state of interest and w is the width of the channel. C is a dimensionless constant with depends on the size of the ball and the depth, D, of the waveguide, and is greater than 2.4 in general. In this technique the small steel ball is moved to known positions in the waveguide and the changes in the resonant frequency of mode are mapped as a function of these positions. Using the steel ball one is able to map out both the

electric and magnetic fields, the positions of the ball that give rise to the minima
in the resonant frequency correspond to the antinodes of the electric field and the
maxima correspond to the antinodes of the transverse magnetic field.

In the theoretical and numerical modelling area there are several references. As
noted earlier, Popov considered the impact of bulges in electromagnetic waveguide
propagation. Mehran (1978) calculated the reflection and transmission coefficient
of TM waves in Y junctions versus the angle of the Y. Other papers include
Kirilenko et al. (1979, 1985), Cambell and Jones (1974) and Bostrom (1983). In
optical waveguides, see Tsao and Gambling (1989) and Weder(1988).

4.20   Problems with Wire Growth

There are two characteristic types of disorder which arise in the manufacture of
GaAs/AlAs quantum wires using molecular-beam epitaxy (MBE), viz., (1) surface
roughness, the interface between the GaAs and the AlAs region is not smooth;
and (2) impurities, within the GaAs there will be islands of AlAs. Monte-Carlo
simulations of vicinal surface grown quantum well wires has been performed by
Hugill et al. (1989). For a discussion see Nikolić and MacKinnon (1994).

4.21   Tight Binding Model Quantum Wire

Nikolić and MacKinnon (1994) have numerically modelled the conductance and
conductance fluctuations of narrow disordered quantum wires with two types of
disorder, boundary roughness and islands of strong scattering impurities within
the bulk of the wire. Their model is consists of two infinite perfect leads with
a section of imperfect wire; i.e., a lead-sample-lead model. They used the tight
binding Hamiltonian to describe the quantum wire and they used the recursive
Green's function technique to solve their equations.

The conductance is given by the two term Landauer formula for spin degenerate
states:

$$G = 2\frac{e^2}{h} \sum_{n=1}^{N_L} \sum_{n=1}^{N_R} |t_{mn}|^2$$

where the summation is over the N_L open channels in the left lead and the N_R open
channels in the right lead. Here t_{mn} are the amplitude transmission coefficients.
Conductance fluctuations are measured by the square root of the variance:

$$rms(G) = (<G^2> - <G>^2)^{1/2}.$$

The standard three regimes are covered in their modelling:

- the quasi-ballistic regime where the wire length, L, is comparable to the
  mean free path length, l;

- the mesoscopic regime where $l < L < \lambda$; here λ is the localization length;

- the strong localization regime where $L > \lambda$.

4.21.1 Boundary Roughness

Nikolić and MacKinnon (1994) have modelled boundary roughness in their quantum wires. For average conductance, they found that higher subbands are more affected than lower ones. In the quasi-ballistic regime, they found:

- edge roughness destroys the conductance quantization steps, first near the band center for very small disorder and then spreading towards the band edge as the disorder and length of the wire increases;

- conductance fluctuation decreases as l increases and vice versa;

- conductance fluctuations increase as disorder or wire length increases;

- Nikolic and MacKinnon conjecture a relationship of l and the density of states and thus to conductance fluctuation.

In the mesoscopic regime, $l < L < \lambda$, they found:

- the conductance curve showed sample specific fluctuations, but the amplitude of the fluctuations is on the order of e^2/h, independent of energy or length of wire; that is, conductance fluctuation is close to the universal value for a metallic quasi-one dimensional systems, $rms(G) = .729e^2/h$, for energies in the first subband; this quantum interference effect is similar to universal conductance fluctuations (UCF);

- scale of sensitivity to energy differs with length of wire or disorder, which agrees with Takagaki and Ferry's (1992a) suggestion that it is due to phase modulation of the electron wavefunction due to multiple elastic scattering in the wire and not scattering from the rough edges;

- typical spacing between peaks and valleys in the conductance as a function of energy depends on wire length as $E_c \sim 1/L$, versus UCF in the metallic regime where $E_c \sim 1/L^2$, (v. Lee and Stone (1985));

- as length of wire increases, there is rapid decrease of average conductance $< G >$ near the band center.

In the strong localization regime, one finds that the conductance is reduced to set of peaks of different amplitudes, with maximum values of $2e^2/h$; in this case, electron transport through the wire is via resonant tunnelling; the average conductance decays exponentially with length (v., Johnston and Kunz (1983))

$$< G(E, L) > \sim exp(-2L/\lambda(E)).$$

One notes that if $< ln\, G >$ is fitted to a straight line, then $\lambda(E)$ can be estimated from its slope; and similarly, $rms(G(E))$ follows $\lambda(E)$. The conductance fluctuations decrease as the wire becomes longer due to decrease in conductance. Nikolić and MacKinnon argue that $rms(G)$ and $< G >$ are proportional; in particular they assert that conductance statistics of quasi-1D samples in this regime are dominated by a single channel; the data supports log-normal distribution for conductance distribution (v., Johnston and Kunz (1983)).

Makarov and Yurkevich (1989) considered the transport properties of narrow 2D conductors with a finite rough boundary section of length L. The lower boundary at $x = d$ is ideally flat but the other boundary is rough; viz., $x = \zeta(y)$ in the interval $-L \leq y \leq 0$. The width of the strip d is chosen such that electrons fill only the lowest transverse quantization level: $1 < k_F d/\pi < 2$. In addition, one assumes that the fluctuations of the boundary are small compared with the width of the strip: $\zeta(y) << d$. The conductance is given by $G = (e^2/\pi\hbar)|T_L|^2$ where T_L is the transmission coefficient of the rough part of the boundary. Let $k_y = [k_F^2 - (\pi/d)^2]^{1/2}$ denote the longitudinal wave number. Makarov and Yurkevich reduce the 2D problem to the similar problem for a 1D system in a random potential.

Theorem 2 *(Makarov and Yurkevich) Assume that $\zeta(y)$ is a Gaussian random process with mean zero and a correlation length of ℓ, viz. $< \zeta(y) >= 0$ and $< \zeta(y)\zeta(y') >= \zeta^2 w(|y - y'|/\ell)$. Let $W(2k_y\ell)$ denote the Fourier transform of the correlation function $w(s)$. The localization length is given by*

$$L_0 = 2\pi d (\frac{k_d d}{\pi})^3 (\frac{d}{\pi\zeta})^2 [2k_y\ell W(2k_y\ell)]^{-1}.$$

The average conductance and resistance are given by

$$< G >= \frac{e^2}{\pi\hbar} \frac{4}{\sqrt{\pi}} (\frac{L_0}{L})^{3/2} exp(-\frac{L}{4L_0}) \int_0^\infty \frac{x^2 dx}{chx} exp(-\frac{L_0}{L}x^2)$$

and

$$< R >= \frac{\pi\hbar}{e^2} \frac{1}{2} [exp(2L/L_0) + 1].$$

In particular, there is full localization in a distance L_0 for electron states in a 2D system with an inhomogeneous boundary.

For additional work in this area, see Takagaki and Ferry (1992a, 1992b, 1992c), Kumar and Bagwell (1992), Masek and Kramer (1989), and Timp and Howard (1991). We return to the problem of localization in rough quantum wires in Chapter 6 and the work of Kleespies and Stollmann (1999).

4.21.2 Impurity Islands

Nikolić and MacKinnon modelled the effects of strong scattering centers or islands in the bulk of a quantum wire. They found that even a small concentration of islands can cause an almost complete suppression of conductance quantization. They found that island disorders have a similar effect on each subband, unlike edge roughness. They found that the average conductance as a function of energy has local maxima near the energies of the subband edges of a perfect wire.

In the quasi-ballistic regime, they found:

- conductance fluctuations increase with length of wire.

In the mesoscopic regime, they found:

- conductance fluctuates as a function of energy with fluctuations on the order of e^2/h;

- they estimate that $E_c \sim 1/L^2$, similar to UCF;

- conductance fluctuations achieve a maximum in this regime as a function of the length of the wire.

In the strong localization regime, they found:

- conductance is reduced to set of peaks of maximum amplitude of $2e^2/h$ where each peak corresponds to occurrence of resonant tunnelling through the wire;

- peaks are higher toward the center of the band;

- average conductance decrease exponentially with length of the wire and slope of the line $< ln(G) >$ gives an estimate for $\lambda(E)$;

- conductance fluctuations decrease with length of wire.

Nikolić and MacKinnon compare their work with the work of Masek and Kramer (1989) who modelled a disordered quantum wire using the Anderson model with a uniform distribution for the site energies of the wire. They also discuss the relationship to the conjectures of Ando and Tamura (1992) and Tamura and Ando (1991), who predicted a broader region of universal conductance fluctuation. We refer the readers to Nikolić and MacKinnon (1994) for the details.

4.21.3 Islands and Edge Roughness

In real wires, one expects to find both types of disorder, edge roughness and scattering impurities. In this case Nikolić and MacKinnon have found that the presence of both types of disorder causes further decrease in conductance; the influence of islands is larger for energies near the band edge and edge roughness is dominant towards the band center; in particular for quasi-ballistic regime, conductance fluctuations are similar to edge roughness alone except for energies near the band edge at lower energies. They note that at higher energies conductance fluctuations are similar to rough wires without islands. The conductance fluctuations decrease faster with presence of islands as length of wire increases and the localization length of wires with islands and edge roughness is usually about half of the localization length of rough wires without islands.

4.22 Conductance and Symplectic Matrices

The result of Johnston and Kunz (1983) was mentioned above. In this section we consider their model in more detail. They examined a wire with a disordered section and derived the asymptotic properties of the conductance as the length of the disordered section tends to infinity. In particular they showed that the conductance decreases exponentially with the length of the disordered section.

They considered a tight binding Hamiltonian lattice which is infinite along the axis of the wire. Let $\mathbf{x} = (x, \mathbf{r})$, where $x \in \mathbf{Z}$ and $1 \leq r_j \leq L_j < \infty$ where $j = 1, 2, ...\nu - 1$ where ν is the dimension of the host lattice. The potential is assumed to be zero outside a section Λ of finite length with N sites. Within the section Λ the random potential is described by a set of identically distributed, independent random variables with probability distribution ρ.

Within the disordered section, the stationary Schrödinger equation can be transformed into a transfer matrix equation of the form

$$T_x = \begin{pmatrix} (eI - H_x) & -I \\ I & 0 \end{pmatrix}$$

where H_x is the Hamiltonian of the $\nu - 1$ dimensional system associated with the cross section at point x. Here I is of rank d and T_x is a real symplectic matrix in $Sp(d, \mathbf{R})$. That is, $T_x^t J T_x = J$ where $J = \begin{pmatrix} 0 & -I \\ I & 0 \end{pmatrix}$.

Outside of the region Λ the system is described by the scattering matrix. The analysis can be reduced to a study of the product of symplectic matrices, say $T(N)$, which is described by d distinct, strictly positive numbers $\gamma_1 < \gamma_2 ... < \gamma_d$, the Lyapunov exponents associated with $T(N)$.

Theorem 3 *(Johnston and Kunz) If the distribution of the potential satisfies*

$$\int_{\mathbf{R}} d\omega \rho(\omega)[log(1 + \omega^2)]^2 < \infty,$$

then if $t_N^e(q', q)$ is any element of the transmission matrix,

$$\frac{1}{N} log|t_N^e(q', q)| \to^P -\gamma_1,$$

and the central limit theorem holds

$$lim_{N \to \infty} Pr\{N^{-1/2}[(log|t_N^e(p', p)|) + N\gamma_1] \leq x\} = \frac{1}{(2\pi\lambda)^{1/2}} \int_{-\infty}^{x} e^{-y^2/2\lambda} dy.$$

Similarly, for the conductance $G(N)$:

$$\frac{1}{N} log|G(N)| \to^P -2\gamma_1.$$

Here $\to^P$ *denotes convergence in probability. So the conductance decreases exponentially with the length of Λ, the exponent is the smallest Lyapunov characteristic exponent and the asymptotic distribution is log normal.*

4.23 Random Waveguides with Absorption

Brouwer (1997) has computed the statistical distribution of the transmittance of a random waveguide with absorption in the limit of many propagating channels.

He was motivated in part by the measurements of Stoytchev and Genack (1997) of microwave transmission through a copper tube with randomly placed polystyrene scatters, making up a random waveguide. Let N denote the number of transverse propagating channels in the waveguide, let L denote the length of the waveguide and let ℓ denote the elastic mean free path. Assume that the random waveguide is quasi one dimensional, i.e. the width W is much smaller that L. Let ξ_a denote the exponential absorption length and let $\xi = N\ell$ denote the localization length. Brouwer assumes that $\ell \ll \xi_a \ll \xi$, i.e. absorption is weak on the scale of a single scattering event. Let ℓ_a denote the ballistic absorption length, where $\ell_a = 2\xi_a^2/\ell$.

Transmission through the waveguide is given by the $N \times N$ transmission matrix t, which determines the transmittance $T = \sum_{a,b=1}^{N} |t_{ab}|^2 = tr\, t^\dagger t$ or $T_{ab} = |t_{ab}|^2$. The scattering matrix S is given by

$$S = \begin{pmatrix} r & t' \\ t & r' \end{pmatrix}$$

where r is the reflection matrix. The reflectance is given by $R = tr\, r^\dagger r$.

Brouwer considers both the diffusive regime where the transmittance distribution is Gaussian and the localized regime where the transmittance distribution is log-normal.

In the diffusive regime $\ell \ll L \ll \xi$ Brouwer computes $< T >$ to leading order in N. Assume the initial conditions $< tr t^\dagger t >= N$ and $< tr r^\dagger r >= 0$ at $L = 0$, then:

Theorem 4 *(Brouwer) For $\ell_a \gg \ell$:*

$$< T >=< tr\, t^\dagger t >= \frac{\xi}{\xi_a sinh\, s}$$

and

$$< R >=< tr\, r^\dagger r >= N = \frac{\xi}{\xi_a} coth\, s$$

where $s = L/\xi_a$ and $\xi_a = [\ell\ell_a/2]^{1/2}$. In the weak absorption regime, $L \ll \xi_a$, then this result reduces to Ohm's law

$$< T >= N- < R >= \xi/L.$$

In the strong absorption regime, $L \gg \xi_a$, the reflectance saturates at $N - \xi/\xi_a$ and T decays exponentially with decay length ξ_a:

$$< T >= \frac{2\xi}{\xi_a}e^{-L/\xi_a}.$$

Brouwer also computes the weak localization correction :

Theorem 5 *(Brouwer) The weak localization corrections to average reflectance and transmittance are:*

$$\delta T = \delta_{\beta,1}\left(\frac{coths - 2s}{4sinhs} - \frac{2}{4sinh^3 s}\right)$$

and

$$\delta R = \delta_{\beta,1}(\frac{1}{4} + \frac{scoths - 1}{4sinh^2 s}).$$

In the weak absorption regime one gets the result of Mello and Stone (1991)

$$\delta R = \delta T = \frac{1}{3}\delta_{\beta,1}$$

and in the strong absorption regime one gets

$$\delta T = -\delta_{\beta,1}\frac{L}{\xi_a}e^{-L/\xi_a}$$

and

$$\delta R = \frac{1}{4}\delta_{\beta,1}.$$

Brouwer also calculates $var(T), var(R)$ and $cov(R,T)$. In particular for weak absorption $L << \xi_a$, he shows that one recovers the universal conductance fluctuation results:

$$var(T) = -cov(R,T) = var(R) = 2/15\beta.$$

For the derivation of results on the localized regime, $L >> \xi$ we refer the reader to Brouwer's paper. We summarize the results:

Theorem 6 *(Brouwer) In the localized regime with weak absorption*

$$< log T >= -2L/\beta\xi$$

and

$$var\, log T = 4L/\beta\xi;$$

and in the strong absorption regime

$$< log T >= -L/\xi_a + \mathcal{O}(L/\xi)$$

and

$$var\, log T = L/2\xi.$$

For additional reading, see Beenakker's (1999) review and Beenakker (1998) where he discusses random lasers and chaotic cavities. We briefly review one topic in this area in the next section.

4.24 Excess Noise

Based on random matrix theory, Patra and Beenakker (1999) (see also Beenakker and Patra (1999)) have calculated the excess noise for coherent radiation propagating through an amplifying random medium; in particular they treat the case of an amplifying disordered waveguide and an amplifying disordered cavity. To obtain these results they used the work of Brouwer discussed in the last section.

Let N denote the number of propagating modes, let $\bar{I}$ denote the mean photocurrent and let P denote the noise power. Finally, let $(...)_{m_0 m_0}$ denote the m_0, m_0−element of the matrix.

For electrons the Büttiker formula states that

$$P/P_{Poisson} = 1 - \frac{tr(tt^\dagger)^2}{tr(tt^\dagger)}$$

in terms of the transmission matrix t at the Fermi energy. In the absence of scattering, e.g. at a ballistic point contact, all eigenvalues of $tt^\dagger$ are equal to unity and hence $P = 0$. If all the transmission eigenvalues are $<< 1$, then $P = P_{Poisson}$.

The optical analogue of Büttiker's formula is given by the result of Patra and Beenakker. Here one considers a monochromatic laser beam with frequency ω_0 incident in a single mode m_0 on a waveguide containing a disordered medium at temperature T. Then one can show

$$P/P_{Poisson} = 1 + 2f(\omega_0, T)\frac{[t^\dagger(1 - rr^\dagger - tt^\dagger)t]_{m_0 m_0}}{[tt^\dagger]_{m_0 m_0}}$$

where f is the Bose-Einstein function

$$f(\omega, T) = [exp(\hbar\omega/kT) - 1]^{-1}.$$

As an example, if the scattering matrix is unitary, then $rr^\dagger + tt^\dagger = 1$ and we have $P = P_{Poisson}$.

Theorem 7 *(Patra and Beenakker) The optical shot noise is given by*

$$\bar{I} = I_0[tt^\dagger]_{m_0 m_0}$$

and

$$P = \bar{I} + 2I_0 f(\omega_0, t)[t^\dagger(1 - rr^\dagger - tt^\dagger)t]_{m_0 m_0}.$$

Consider the case of a weakly amplifying, strongly disordered waveguide of length L. Brouwer (1998) has calculated the averages of the moments of $rr^\dagger$ and $tt^\dagger$ for this system. Let P_{exc} denote the excess noise, viz. $P_{exc} = P - \bar{I}$. In an amplifying medium P increases more than $\bar{I}$ due to excess noise resulting from spontaneous emission. The noise figure is defined by $\mathcal{F} = (P_{exc} + \bar{I})I_0/\bar{I}^2$ where I_0^2/P_0 is the squared signal-to-noise ratio at the input. For coherent radiation at the input $P_0 = I_0$. Let α denote the detection efficiency. They assume that all N outgoing modes are detected with the same efficiency. Assume that the illumination is in a single propagating mode, m_0.

For measurement in transmission and reflection, Patra and Beenakker have shown:

Theorem 8 *(Patra and Beenakker) The mean photocurrent in transmission is*

$$\bar{I} = \alpha I_0 (t^\dagger t)_{m_0 m_0}$$

and the excess noise is

$$P_{exc} = 2\alpha^2 f I_0[t^\dagger(1 - rr^\dagger - tt^\dagger)t]_{m_0 m_0}.$$

And for measurement in reflection, replace r by t' and t by r'.

Let ℓ denote the mean free path, τ_a is the absorption time, $\tau_s = \ell/c$ is the scattering time, and let $\xi_a = \sqrt{D\tau_a}$ denote the amplification length, where $1/\tau_a$ is the amplification rate and $D = c\ell/3$ is the diffusion constant. Set $s = L/\xi_a$. Consider the regime $1/N << \ell/\xi_a << 1$. Patra and Beenakker show that for a measurement in transmission, the mean photocurrent $\bar{I}$, is

$$\bar{I} = \frac{4\alpha\ell}{3L} I_0 \frac{s}{\sin s}$$

and for a measurement in reflection one finds

$$\bar{I} = \alpha I_0 [1 - \frac{4\ell}{3L} s\cot an\, s].$$

They also calculate P_{exc} in these cases. It is noted that the noise figure $\mathcal{F}$ increases sharply on approaching the laser threshold at $s = \pi$; i.e., $P/\bar{I}$ increases without bound as $L \to \pi\xi_a$, the laser threshold. In terms of the scattering matrix $S(\omega)$, in the absence of amplification all poles of S are in the lower half of the complex plane. As Beenakker and Patra note, amplification shifts the poles upwards by an amount $1/2\tau_a$ and the laser threshold is reached when the first pole hits the real axis. For ω near the threshold frequency Ω, the scattering matrix has the form

$$S_{nm} = \frac{\sigma_n \sigma_m}{\omega - \Omega + \frac{1}{2}\Gamma - i/2\tau_a}$$

where σ_n is the complex coupling constant of the resonance to the nth mode in the waveguide and Γ is the decay rate. The laser threshold is at $\Gamma\tau_a = 1$. Patra and Beenakker have shown that while P and $\bar{I}$ diverge at the laser threshold, the signal to noise ratio $\bar{I}^2/P$ has a finite limit, independent of σ_n, Γ and τ_a.

For absorbing random media, Beenakker and Patra note that the optical analogue of the universal limit $P/P_{Poisson} \to 1/3$ for $L >> \ell$ in electronic shot noise is given by

$$P/P_{Poisson} \to 1 + \frac{3}{2} f(\omega_0, T)$$

for $L >> \xi_a$ which follows from the above stated results.

Since the laser threshold is found to be $1/\tau_a = \pi^2 D/L^2$ in the large N limit, $\Gamma = \pi^2 D/L^2$ appears to be the minimal decay rate. Thus, Beenakker and Patra conjecture that the density of the poles of the S-matrix for a disordered waveguide without amplification should vanish for $Im(\omega) > -\pi^2 D/2L^2$ in the limit $N \to \infty$.

Problem 1 *Verify the Beenakker-Patra pole density conjecture.*

One notes a similar gap in the density of poles for the scattering matrix of a chaotic cavity in the work of Fyodorov and Sommers (1997).

Chapter 5

Problems in Open Quantum Devices

5.1 Introduction

Quantum chaos studies quantum ballistic structures such as squares, circles and stadia, e.g. see Ji and Berggren (1995) where they examine the energy spectrum of a stadium shaped quantum dot with and without a magnetic field. As closed structures, the theoretical results appear to agree fairly accurately with the measured data, e.g. the statistical properties of the spectra. Building on work on disordered systems, semiclassical theories were being developed which provided explicit predictions regarding open quantum structures. However, as we discuss below, these predictions are not agreeing with measured data at least for instances of low number of modes entering the quantum dot. The work of Ferry, Bird and coworkers and Berggren's group in this area will be reviewed below. Briefly, the work of Ferry, Bird, Akis and others have found that for quantum mechanical transport in an open quantum dot where the leads support only a small number of propagating modes:

(1) conductance fluctuations as a function of magnetic field show quasi periodic oscillation at low magnetic fields as compared to random aperiodic variations for ergodic systems;

(2) the conductance fluctuations show several resonance features which occur periodically and the wave functions corresponding to these resonances appear to be the same "scarred" feature;

(3) the scar effect occurs at magnetic field levels in which classical scattering is expected to be chaotic;

(4) only a few regular orbits appear to dominate quantum transport at low magnetic fields, even through the structure is classically chaotic;

(5) the amplitude of the associated wavefunction is concentrated along the underlying classical trajectory;

(6) the scars are roughly independent of the size of the dot;

(7) collimation and lead placement play a crucial role in nonuniform excitation of orbits;

(8) the scarring appears for different lead configurations and it appears to be universal;

(9) position of the peaks scale inversely with the dot's linear dimension;

(10) the orbits are highly stable, invariant to temperature and gate voltage variations;

(11) interference occurs when the electrons returns to the injection point contact which in a weak magnetic field happens after several rotations around the cavity before orbit closure is achieved; thus the crucial requirement is that the electrons remain coherently trapped in the quantum dot over a very long time; this is reflected in the sensitivity of the fluctuations to increased temperature.

5.2   Quantum Transport

5.2.1   Mesoscopic Parameters

A few terms from mesoscopic physics will be defined in this section. The average level spacing of a quantum dot of area A is

$$\Delta = h^2/2\pi m^* A.$$

Let L denote the length of the side of a square quantum dot. The Fermi wavevector k_F is related to the carrier density n_s by

$$k_F = \sqrt{2\pi n_s}.$$

The cyclotron radius r_c is given by

$$r_c = \phi_0/\lambda_F B$$

where $\lambda_F = 2\pi k_F$ is the Fermi wavelength and $\phi_0 = h/e$ is the unit of quantum flux. If μ is the mobility, then the elastic mean free path l_o is

$$l_o = \mu h/e\lambda_F.$$

Typical values for these parameters are $L = .3 - 1.8\mu m$, $n_s = 4.5 \times 10^{15} m^{-2}$, $\mu = 30 - 80 m^2/Vs$, $l_o = 4 - 9\mu m$, which is greater than the lateral dimension of the dot so that transport is ballistic; $\lambda_F \sim 35 - 40 nm$, and $\Delta/k_B = .03 - .92 K$. In the devices studied by Bird and coworkers, the gate size varied from .4 to 2.0 μm and the number of modes in the point contact leads ranged from one to five; v. Bird, Ishibashi et al. (1997).

5.2.2   Universal Conductance Fluctuations

In disordered systems electrons are diffusively scattered and the resulting interference gives rise to the phenomenon of universal conductance fluctuations (UCF). In this model conductance fluctuations are aperiodic and are conveniently described by random matrix theory (RMT); v. Beenakker (1997). For ballistic quantum dots it was believed that one could extend RMT to this case by assuming a chaos induced uniformity of transport in the dot. In the related semiclassical model the conductance is calculated as a sum over all classical paths connecting the input and the output (v., Jalabert, Baranger and Stone (1990)). Here one assumes ergodic scattering, or a uniform sampling of phase space by the electrons, which follows from the distribution of swept trajectory areas being exponential.

Jalabert, Pichard and Beenakker (1994) computed the statistical properties of conductance of open ballistic quantum dots assuming the scattering matrix belongs to Dyson's circular ensemble. In this case for large N

$$< T >= \frac{1}{2}N + \delta T$$

where the weak localization correction is

$$\delta T = \frac{\beta - 2}{4\beta}.$$

Recall $\beta = 1$ when $B = 0$ and $\beta = 2$ when $B \neq 0$. In particular when $\beta = 1$, then $\delta T = -1/4$. By Landauer's formula T is just the conductance g measured in units of $2e^2/h$. More precisely

$$< T >= \frac{N}{2} - \delta_{\beta 1} \frac{N}{4N + 2}$$

and

$$var(T) = \begin{cases} \frac{N(N+1)^2}{(2N+1)^2(2N+3)} \to \frac{1}{8} & \text{COE} \\ \frac{N^2}{4(4N^2-1)} \to \frac{1}{16} & \text{CUE} \end{cases}$$

where the limit is as $N \to \infty$. The weak localization correction is given by

$$\delta G = < G >_{B \neq 0} - < G >_{B=0} \to \frac{1}{4} \frac{2e^2}{h}.$$

In the case $N = 1$, Jalabert, Pichard and Beenakker showed that the conductance probability distribution is

$$P(g) = \frac{1}{2}\beta g^{-1+\beta/2}$$

for the case of ideal point contacts and no dephasing.

Brouwer and Beenakker (1997) have recently shown that the $N = 1$ case has certain corrections. Viz.,

$$P(G) \to \frac{\gamma\beta}{2}(1 + |x| - \delta_{\beta 1}x)e^{-|x|}$$

for $\gamma \gg 1$; here $\gamma = 2\pi\hbar/\tau_\phi\Delta$ is the dimensionless dephasing rate and Δ is the mean level spacing in the quantum dot; $x = 2\gamma\beta(G - 1/2)$. Assuming ideal point contacts , i.e. $\Gamma_1 = \Gamma_2 = 1$, where Γ_n is the transmission probability of a mode, the mean and variance in this case are:

$$< G >= \frac{1}{2} - \frac{1}{2}\delta_{\beta 1}\gamma^{-1} + O(\gamma^{-2})$$

and

$$Var(G) = \frac{1}{4}(1 + 2\delta_{\beta 1})\gamma^{-2} + O(\gamma^{-3}).$$

Jalabert, Stone and Alhassid (1992) have used RMT to derive the universal peak height probability distributions:

$$P_{B=0}(\alpha) = \sqrt{2/\pi\alpha}e^{-2\alpha}$$

and

$$P_{B\neq 0}(\alpha) = 4\alpha(K_0(2\alpha) + K_1(2\alpha))e^{-2\alpha}$$

in the case the leads are statistically identical and independent. In particular

$$\int \alpha P_{B=0}(\alpha)d\alpha = 1/4$$

and

$$\int \alpha P_{B\neq 0}(\alpha)d\alpha = 1/3.$$

A primary object of study in conductance fluctuations is the correlation function

$$F(\Delta B) = < \delta g(B)\delta g(B + \Delta B) >$$

where $g(B)$ is the conductance in units of e^2/h at magnetic field B, $\delta g(B) = g(B) - < g(B) >$, and $< g(B) >$ is the ensemble average. In chaotic geometries the classical trajectories are exponentially distributed according to

$$N(A) = N_0 exp(-2\pi\alpha|A|)$$

where A is the trajectory area and $\sqrt{2\pi}\alpha$ is the rms coherent area enclosed by the electrons. This is related to the correlation function by

$$F(\Delta B) \sim | \int_{-\infty}^{\infty} dA e^{2\pi i \Delta B A/\alpha} N(A)|$$

$$\sim [1 + (\frac{2\pi\Delta B}{\alpha\phi_0})^2]^{-2}.$$

viz., $F(\Delta B)$ has the form

$$F(\Delta B) = F(0)/[1 + (\frac{2\pi\Delta B}{\alpha\phi_0})^2]^{-2}$$

and the Fourier spectrum is

$$S(f) = S(0)[1 + 2\pi\alpha\phi_0 f]exp(-2\pi\alpha\phi_0 f).$$

Here f is the magnetic frequency which corresponds to the path area A by $A = \phi_0 f$, (v. Marcus et al. (1993)).

At large magnetic fields where the cyclotron radius r_c is less than the dot radius, skipping orbits or edge states will dominate the conductance features.

At sufficiently large fields, a fraction of the total number of edge states may be confined within the quantum dot, thus defining an effective ring geometry. While the edge states propagating through the quantum dot generally do so with very little scattering between them, the strong curvature of the confining profile thus gives rise to electron tunneling via the trapped edge states. This resonant tunneling process may either enhance the reflection or the transmission of edge states through the dot and is reflected as the observation of Aharonov-Bohm magneto-resistance oscillation with a well defined period. At small magnetic fields aperiodic behavior arises.

For quantum dots, low field magnetoresistance measurements are found to exhibit ballistic weak localization peaks. The standard theory states that the weak localization effect in ballistic dots is sensitive to the shape of the cavity, i.e. whether the cavity is chaotic or regular. The semiclassical theory of line shape is developed in Baranger, Jalabert and Stone (1993), where they predict different line shapes for chaotic and regular geometries. In particular, this theory predicts a different line shape of the averaged magnetoresistance peak near $B = 0$ for chaotic and regular cavities; viz., it is Lorentzian for chaotic cavities:

$$< R(B) >= R_0 + \frac{\Delta R}{[1 + (2B/\alpha\phi_0)^2]}$$

and linear for regular cavities:

$$R_0 - < R(B) > \sim |B|.$$

Here R_0 is the average reflection coefficient in the absence of weak localization effect, $\Delta R = R(B = 0) - R_0$, and α is related to the inverse of the type area S enclosed by the classical path, $S = (2\pi\alpha)^{-1}$.

A second characteristic is that the conductance fluctuations for RMT should have certain properties; they are aperiodic; the related correlation function is always positive definite and should decay rapidly to zero (v., Jalabert, Baranger and Stone (1990)). In addition the fluctuation amplitude should change if the dot lead opening changes, v. Baranger and Mello (1995).

Early work based on "larger" dots showed that conductance fluctuations are highly aperiodic, which is consistent with a broad distribution of classical trajectories contributing to the transport properties, v. Marcus, Rimberg et al. (1992) and Marcus, Westervelt et al. (1993).

However, problems arose almost immediately. Chan et al. (1995) reported Lorentzian shape of weak localization for square billiards which should be regular. More dramatically, Bird, Olatona et al. (1997) observed a transition from a Lorentz peak to a linear peak for a single dot as the width of the quantum point contact (i.e., the leads connecting the dot to the reservoir) were narrowed.

Even in the now standard experimental work of Marcus et al. (1992) for stadium shaped dot which showed conductance fluctuation consistent with ergodicity, if one looks at the data prior to their ensemble averaging, the spectra of the fluctuations for low magnetic field. $B < .07T$ revealed the presence of strong peaks at a few discrete frequencies. As one increases the magnetic field the stadium will

undergo a transition from a chaotic to a regular system. However, for the region
identified the system is chaotic; Marcus et al. identified the transition to take place
at $B = .45T$. Ferry et al. (1997) in their modelling of this system found strong
scarring of the wave function, which they attribute to the dominance of a stable
orbit for this supposedly chaotic system, which we will discuss further below.

5.2.3   Transport in an Open Square

Transport in an open square has been modelled by Akis, Ferry and Bird (1997)
where they performed spectral analysis on the correlation function

$$F(\Delta B) = < \delta g(B) \delta g(B + \Delta B) >$$

where $\delta g(B) = g(B) - < g(B) >$. Here $g(B)$ is the dimensionless conductance at
magnetic field B and $< g(B) >$ is the ensemble average.

Bird et al. define the correlation field B_c as the halfwidth

$$F(B_c) = F(0)/2.$$

The rms amplitude of fluctuation, δg_{rms} is defined to be

$$\delta g_{rms} = \sqrt{F(0)}.$$

For the RMT case, $\delta g_{rms} = .25$ (v. Baranger and Mello (1994)).

In contrast to the semiclassical results mentioned above, Bird, Akis, Ferry et
al. (1997) found negative excursions and quasi-periodic oscillations in the correlation function. They found a periodicity in the power spectra of the conductance
fluctuations, viz. strong peaks and their harmonics which are consistent with the
interference being dominated by a single classical orbit.

Bird et al. have found B_c to be insensitive to the dot lead opening width and
temperature whereas the semiclassical theory predicts otherwise (v., Marcus et al.
(1993) and Clarke, Chan et al. (1995)).

In addition, Bird et al. have found that the periodic nature of the fluctuation
was increasingly resolved as the cavity size is reduced. One notes that the root
mean square amplitude of the fluctuation does note really change much as the dot
size is varied. However, the periodic nature of the fluctuations does increase.

5.2.4   Quantum Point Contact

5.2.5   Leads

Akis, Ferry and Bird (1996) and Bird, Ferry, Akis et al, (1996) have demonstrated
that for the QPC the quantization of the transverse momentum of the leads causes
the electrons to enter the cavity in a collimated beam at an angle. In the case
of many modes, Beenakker and van Houten (1989) have shown that the electrons
form a beam centered on the transport axis of the contact. However, when only a
few modes are present the quantization of the transverse modes causes a diffraction
effect whereby the electrons exit the QPC at an angle to the transport axis (v.,

Jacoboni et al. (1995)). For a single QPC a state with $k_\perp$ inside the QPC is mostly coupled to the outgoing state with the same transverse wavevector (v., Szafer and Stone (1989)).

Numerical simulations of Bird et al. show that for a quantum point contact (QPC) with only a few modes, the electrons are injected into the cavity in a highly collimated beam (v. Akis, Ferry and Bird (1996) and Akis, Bird and Ferry (1996)) at an angle to the transport axis. The electrons from the external reservoirs when only a few modes are excited are able to enter the cavity by matching their transverse momentum with the quantized values in the contact. More specifically, the energy of the electron in the jth eigenstate in the QPC is $E = (\hbar^2/2m^*)k_F^2$ where $k_F^2 = k_\parallel^{j2} + k_\perp^{j2}$ where $k_\perp^j = \pi j/w$ where w is the width of the QPC. From energy conservation principles, the longitudinal wave numbers of the outgoing states are close to $k_\parallel^j$. Thus at the exit of the QPC the angular distribution of the electrons is roughly confined to the cone with opening angle $\beta \sim tan^{-1}[k_\perp/k_\parallel]$ from which one sees that smaller j implies smaller ratio which in turn implies a narrower cone. Thus, the semiclassical condition of wide angular distribution is achieved only in the case of many occupied subbands. Zozoulenko and Berggren (1996) note that this can be modified by rounding the QPC opening.

5.2.6 Phase Breaking Time

Bird, Isibashi, Ferry et al. (1996) have shown that the correlation field B_c is related to the phase breaking time τ_ϕ by

$$B_c(B) = [8\pi^2 m^* B/hk_F^2\tau_\phi]$$

over a fairly wide range of parameters. At high magnetic field B_c is found to be a linear function of the magnetic field. In this regime, τ_ϕ is then independent of the magnetic field. Using this relationship they can estimate τ_ϕ, which they find to be approximately several hundred picoseconds for the mesoscopic systems they were studying. In particular, it is much larger than the ballistic transit time. In this situation, an electron may undergo up to 100 collisions with a confining wall before coherence is totally lost.

As a function of temperature they find

$$\tau_\phi(ns) = .03/T(K).$$

However, they find an unexpected low temperature saturation which is still being studied. One possibility is that it is related to a transition from two dimensional to zero dimensional phase breaking which occurs when the discrete levels of the dot become resolved. For other work on phase breaking time, see Linke, Bird, Cooper et al. (1997), Bird, Linke, Cooper et al. (1997) and Bird, Micolich, Linke et al. (1998).

5.2.7 Dot Spectroscopy

For the devices which they were studying, Bird et al. found that the discrete spectrum of the dot can be resolved since the total broadening of the levels is

$\sim 100mK$. This is smaller than the average level spacing for device temperatures less than $50mK$ and for device sizes less than $1\mu m$.

5.2.8 Line Shape Analysis

As noted above, Bird, Olatona et al. (1995) observed a transition from a Lorentz peak to a linear peak for a single dot as the width of the quantum point contact (i.e., the leads connecting the dot to the reservoir) were narrowed. Thus, Bird, Ishibashi et al. (1997) and Bird Akis et al. (1997) have concluded that line shape is not a reliable indicator of chaos as suggested by the semiclassical analysis. In one case they found a structure which can not support chaos but the line shape is Lorentzian and in another case they demonstrated a transition from Lorentzian to linear shape for a single dot. They argue that line shape does not reflect a transition from chaos to regular behavior but a change in the coupling of the cavity as the number of modes in the leads is varied.

Simulations of Bird et al. show "chaotic" like behavior even for a perfect square. This has supported their conclusion that the observed line shape is not due to a variation in the shape of the cavity caused by variation in the QPC but is related to the collimation effect.

The reader is also directed to papers of Zozoulenko and Berggren (1996, 1997) which discuss the possibility of alternative explanations of this behavior.

5.2.9 Fluctuation Amplitude

Bird and coworkers have shown that the fluctuation amplitude, δg_{rms}, is invariant with respect to changes in the dot lead opening. This is in contrast to RMT predictions, v. Marcus et al. (1993). In fact, Bird et al. found that the amplitude is independent of the cavity size. They also showed that the amplitude of fluctuation decays exponentially with increasing temperature, whereas the correlation field is largely unchanged. In Akis, Bird and Ferry, numerical results are presented which show that the exponential decay of the fluctuations is consistent with a similar sensitivity of the wavefunction scarring to phase breaking.

5.2.10 Reproducible Fluctuations

Bird, Ferry, Akis et al. (1996) and Ferry, Bird, Akis et al. (1997) have shown that the quasiperiodic conductance fluctuations are reproducible over a wide range of magnetic fields. The quasiperiodic structure is independent of the details of the sample material; they used both AlGaAs and InGaAs. And it is independent of the gate technology, where both surface Schottky barrier gates and in-plane gates were tested.

5.2.11 Wave Function Scarring

Ferry, Bird, Akis et al. (1997) have modelled open quantum dots with attached ideal quantum wires using the iterative matrix technique of Usuki et al. (1995) to calculate the transmission coefficient and quantum probability density function of the cavity. The conductance is then calculated using the Landauer-Büttiker formula. They have modelled a variety of systems including open square dots, open stadia and open circles. They have found wave function scarring which reoccurs periodically over a significant range in magnetic field. E.g. for the open square dot of side $0.3\mu m$ with $0.04\mu m$ lead openings which allowed two modes to enter the cavity, they observed identical diamond shaped scar features for $B = 0.069, 0.173, 0.283$ and $0.397T$.

The scars were found to be associated with sharp "resonances" in the conductance fluctuations $\delta g(B)$. The Fourier transform of the correlation function associated with $\delta g(B)$ showed harmonically related peaks at $8, 16, 24T^{-1}$ and $13, 26T^{-1}$. Assuming $\Delta B = .11T$ as above, $(\Delta B)^{-1} = 9T^{-1}$, which roughly agrees with this result. However, a resonance does not necessarily yield a well defined scar.

The scars were found to be highly stable and independent of the rounding of the corners of the dot but sensitive to lead opening or rounding to the potential profile of the leads.

If the QPCs were replaced by uniform tunnel barriers, the scarring was completely suppressed; instead the wavefunction exhibited a uniform sampling of the cavity.

Scarring here is a weak field effect which is observed prior to edge state formation, which as mentioned above gives rise to the Aharonov-Bohm-like oscillations.

The amplitudes of the wavefunctions were found to be highly concentrated along underlying single classical closed orbits which are strongly overlapped with the leads.

Thus, Bird et al. have concluded that collimation by the QPC selects a set of trajectories of the quantum dot. The recent work of Okubo, Ochiai, Vasileska et al. (1997) supports the view that the periodic conductance fluctuations and associated scarring in square dots does not appear to be a consequence of regular, square geometry.

5.2.12 Open Ballistic Square

Transport properties of a regular square dot has been shown to be very different from chaotic quantum dots, viz. the conductance is characterized by periodic conductance oscillations rather than aperiodic fluctuations which can be described by random matrix theory (RMT). Recall that the eigenstates of a closed quantum square of side L are

$$\psi_{mn} = (2/L)sin(\pi mx/L)sin(\pi ny/L)$$

with eigenvalues

$$e_{mn} = \frac{\hbar^2}{2m^*}(k_m^2 + k_n^2)$$

where $k_m = \pi m/L$. For an open square, even as the number of modes in the openings exceeds $N = 10$, transport through the open structure is still mediated by just a few eigenstates of the correspond isolated dot – i.e. transport in the open structure is effectively mediated by a few eigenstates of the corresponding closed system with eigen energies lying near the Fermi energy, say

$$e_{m,n} = \hbar^2/2m^*(k_m^2 + k_n^2) \sim E_F$$

where $k_m = \pi m/L$.

5.2.13 Density of States Theory

Berggren and coworkers, (v. Zozoulenko and Berggren (1997)) have proposed the density of state analysis of wave function scarring and the related spectral analysis. In this approach it is argued that the scarred features are not related to classical trajectories but should be thought of as a superposition and interference of several regular eigenstates corresponding to the relate closed structure. On occasion they argue these may resemble a classical trajectory, but rarely. They argue that the electron is essentially delocalized in the whole dot and one can not speak of an orbit.

Zozoulenko, Schuster, Berggren and Ensslen (ZSBE) (1997) have measured the longitudinal resistivity R_L and they have noted that it is periodic as a function of k_F.

ZSBE also have modelled transport in an open square quantum dot. They find that the eigenstates that contribute are those excited states on the circle $\sqrt{m^2 + n^2} = k_F L/\pi$ where $k_F = (2\pi n_s)^{1/2}$; but that it is primarily the eigenstates with $m \sim n$ i.e. $k_m \sim k_n \sim k_F/\sqrt{2}$. For a collimated beam $k_\perp \sim 0$ and $k_\parallel \sim k_F$; thus $< k_m > \sim < k_n > \sim k_F/\sqrt{2}$. Thus, it is the injection properties of the leads which define the selected states.

Increasing k_F by $\Delta k_F^2 = 2\Delta k_F^1$ corresponds to excitation of state $\{m+1, n+1\}$ where both quantum numbers are changed by one. These authors have calculated the longitudinal resistance $\Delta R_L(\Omega)$ as a function of k_F and found a periodicity corresponding to $\Delta k_F L = 2.22$, although their measured data showed 4.44; we refer the reader to their paper for a discussion of this discrepancy.

The density of state approach arrives at a convenient explanation of the spectral features seen in the resistivity measurements as a function of k_F. Based on the shell structure of the density of states, they show that one should expect a features like $\Delta k_F^1 \sim \Delta k L/\sqrt{2}$ where $\Delta k = \pi/L$ for a dot with side L, caused by starting in a state $\{m, n\}$ in the dot and transitioning to the state $\{m + 1, n\}$. In their analysis they found peaks in the Fourier transform of the density of states at $1/\Delta k_F \sim 32 \times 10^{-8} m$ and $1/\Delta k_F \sim 44 \times 10^{-8} m$ which correspond to periodicities of $\Delta_F L = \pi$ and $\pi\sqrt{2}$, respectively.

5.2.14 Triangle Model

In an interesting paper Berggren's group modelled conductance in a triangle, v. Christensson, Linke et al. (1997); they found what appears to be a fairly close correspondence of the periodicity of the conductance oscillations and the density of states of the corresponding isolated structure. At high temperatures, Linke et al. (1997) have compared the magnetoresistance with simulation of classical electron trajectories; they showed that maxima of resistance peaks could be correlated with specific commensurate trajectories, and they showed that the three peaks could be correlated with specific trajectories (i.e., in the case of no quantum interference). At lower temperatures considered by Christensson et al., they examined the local fluctuations about a maximum at $B = B_c \sim 50mT$; here the rms amplitude conductance fluctuations were $\sim .06e^2/h$; the Fourier transform showed a dominant frequency $f \sim 130T^{-1}$ with corresponding area $f/\phi_0 = .54\mu m^2$, which corresponded to the reflected trajectory causing the peak at $B = B_c$.

In the density of state theory, the detailed structures described in the experiments and simulations of Bird's group are not really addressed.

5.2.15   Spectral Periods and Peaks

In the experiments the positions of the peaks in the Fourier spectra of the fluctuations were found to be insensitive to changes in gate voltage. The period of the scars was shown by Ferry, Bird et al. (1997) to be determined by the length of the orbit, i.e. $1/\Delta B \sim A^{1/2}$. For the example above where $\Delta B \sim .11T$, this corresponds to an effective area of $.04\mu m^2$ which agrees fairly closely with the area enclosed by the diamond shaped wave function. The authors note that it is probably the symplectic area swept out as the electrons make several circuits around the cavity and not the area inside the scar which is relevant, v. Berry (1989).

The spectral peaks in the power spectra of the conductance fluctuations were shown by Bird, Ferry, Akis et al. (1996) to increase by two orders of magnitude as the size of the dot is reduced. They also noted a non-monotonic decay of the successive harmonics.

5.2.16   Interference and the Magnetic Field

Under the coherence conditions discussed above, constructive interference between electrons is predicted. However, an applied magnetic field and the associated Lorentz curvature of the ballistic orbits cause the electrons to undergo a precessing motion within the cavity, undertaking several tours of the cavity before returning to the injection point contact and causing interference. This repetitive motion gives rise to the scarring orbits.

5.2.17   Wavefunction Scarring: Open Stadium Case

Ferry et al. have modelled transport in the open stadium, which is a classically chaotic system. They have calculated the wavefunction in this system and have found it to be scarred. The electron orbits in this case are only exploring a limited area of the entire stadium due to collimation of the input beam by the QPC.

5.2.18   Open Square

Zozoulenko and Berggren (1997) have modelled the conductance of a square dot of side L as function of the sheet electron density ($n_s = E_F m^*/\pi\hbar^2$). They have calculated the wave function inside the dot and expanded it in terms of the eigenstates of the closed dot, $\psi_{mn} = (2/L)sin(\pi mx/L)sin(\pi ny/L)$ with energies

$$e_{mn} = \hbar^2/2m^*(k_m^2 + k_n^2)$$

with $k_m = (\pi n/L)$; viz.,

$$\Psi(x,y,E) = \frac{2}{L}\sum_{m,n} c_{mn}sin\frac{\pi mx}{L}sin\frac{\pi ny}{L}.$$

They find that at a given k_F only those coefficients c_{mn} associated to the circle of radius R in k space, where $R = k_F = \sqrt{2m^*e_{mn}}/\hbar = \pi\sqrt{m^2 + n^2}/L$, contribute to the wavefunction. This lead Zozoulenko and Berggren to conclude that the transport through the open system is still effectively mediated by a few eigenstates of the corresponding closed dot with eigenenergies lying close to the Fermi energy $e_{mn} \sim E_F$.

The "scarred" features of $|\psi(x,y)|^2$ resembling classical billiard ball trajectories in the square should be thought of as an interference of several particle in the box eigenstates; viz., Zozoulubo and Berggren (1997) propose that the scars are not classical trajectories but result from interference of several eigenstates of the corresponding closed structure.

5.2.19 Spectroscopy

Consistent with Bird et al.'s statement above, the mean energy level spacing is $\Delta = 2\pi\hbar^2/m^*l^2 \sim 70mK$ for a dot size $l = 1\,\mu m$. A broadening of the resonant energy levels is typically less than Δ so transport measurements at a very low temperature in a single-mode regime in the leads may probe a single resonant energy level of the dot. However, Zozoulenko and Berggren (1997) show that there is not a simple correspondence between e_{mn} and resonance or antiresonance structure in G.

As noted above, Zozoulenko and Berggren argue that the scarred features seen in the wavefunction density distribution in a square dot is not related to classical trajectories but is the result of a superposition and mutual interference of several regular eigenstates of the corresponding close system, which states are contributing to the transport properties.

In the model of Zozoulenko and Berggren the conductance oscillations for a square dot are found to have a periods corresponding to $\delta kL = \pi$ and $\delta kL = \pi/\sqrt{2}$. They discuss these results in relationship to their theory of the formation of a global shell structure, where is shell is ascribed to a certain family of periodic orbits in the square. Based on the arrangement of leads, only certain families of periodic orbits mediate transport through the open dot. We direct the reader to their paper for more details.

5.2.20 Are Level Statistics Relevant?

The statistics of the spectra for open dots have been shown to be exactly the same as that for closed dots by Wang, Wang and Guo (1996). However, work of Ishio

(1996), Seba (1996) and Berggren and Ji (1996) suggest that leads attached to the dot may change the level statistics. Ishio studied the distribution of energies at resonances of conductance fluctuations in a chaotic stadium and in a regular circular dot. In both billiards he found the statistics to follow the Wigner distribution.

Zozoulenko and Berggren (1997) showed that in the many mode regime the concept of statistics of the spectra is blurred since it is not possible to introduce any reasonable definition of spacings between resonances due to the irregular character of broadening of the resonance states. This has led Zozoulenko and Berggren to conclude that eigenlevel spacing statistics of the corresponding closed system is not relevant to average transport properties.

5.2.21 Multiple Dots and Superlattice Structures

Ferry, Bird, Akis et al. (1997) have observed oscillations at low magnetoresistance for multiple quantum dots and superlattice structures. Work is still in progress to see whether this is related to the results already mentioned in this section. See also Okubo, Ochiai, Vasileska et al. (1997).

5.3 Resonant Tunneling Diodes

Consider a GaAs quantum well enclosed between AlGaAs tunnel barriers. In an applied magnetic field, say $B = (B_x, 0, B_z)$, the component $B_x = B\cos\theta$ perpendicular to the plane of the 2DEG quantizes the 2D in-plane motion into Landau levels. For large magnetic field for most θ values, only the lowest Landau level is occupied. Resonant tunneling occurs when the energy level of the emitter coincides with the energy of a subband of the quantum well. As the bias voltage V is changed, the tunneling electrons scan the energy level spectrum of the quantum well.

Recent work of Eaves and co-workers has showed distinct sets of periodic resonances in the current-voltage (I-V) characteristics of resonant tunneling diodes (RTDs). Their work suggests that for certain tilt angles and voltage ranges the tunneling rates into individual "scarred" states in quantum wells whose wave function are localized along paths of particular closed classical orbits are much higher than for the transition into adjacent unscarred states.

They examined an RTD with 60 nm quantum well. At $\theta = 20°$ the electrons in the quantum well exhibit predominantly strong classical chaos. The calculated I-V characteristics of this well showed two series of resonant peaks which persisted over a wide range of magnetic field values, viz. pulsed magnetic fields up to 37 T.

Recall that under a bias voltage, a 2DEG is formed in the undoped layer adjacent to one of the tunneling barriers and the current flows as electrons travel from the 2DEG into the quantized nth energy levels of the quantum well. This current is proportional to the transition rate $W_n \sim M_n^2$, where M_n depends on the overlap between the emitter and quantum well states. Eaves group calculated these using the Bardeen transfer Hamiltonian formalism. They found that M_n^2 is large for a subset of almost equally space energy levels.

In Muller et al. (1995) they examined the I-V peaks in the entire parameter space of magnetic field and voltage, varying the tilt angle from 0 to 45 degrees. In this way they were able to analyze the transition from chaos to regular or integrable regimes.

In their earlier work, Eaves et al. examined the density of energy levels $D(e)$ and associated successive minima of $D(e)$ with Gutzwiller fluctuations produced by unstable periodic orbits in the quantum wells. These orbits were associated with the strong scars in the wavefunctions corresponding to a subset of energy levels. The more recent work has shown that the narrow quantum wells allow one to perform resonant tunneling spectroscopy, i.e. individual energy levels can be resolved showing a regularly space subset of individual levels whose eigenfunctions are scarred by particular closed orbits. They have found that this effect is clearly distinguishable from that due to Gutzwiller fluctuations.

The reader is also directed to the recent papers of Stone and coworkers on RTDs in this area, v. Narimanov and Stone (1997) and Narimanov, Stone and Boebinger (1997), and the work of Saraga and Monteiro (1999).

5.4   Cleaved Edge Overgrowth

Growth techniques such as molecular beam epitaxy (MBE) or chemical vapor deposition (CVD) have been used to fabricate planar devices. The MBE technique of cleaved edge overgrowth (CEO), which uses regrowth on the side wall of a previously prepared multilayer, has been utilized to construct quantum wire structures including quantum dots and T-shaped quantum wires. Using CEO Pfeiffer's group has constructed quantum wires with nominal cross section of 250A by 250A; in these devices the mean free path exceeded 10 microns. They found the quantized conductance in step sizes of e^2/h and quantum wire subband separations in excess of 20 meV. Using the two orthogonal sets of cleavage planes in GaAs single crystals, T-shaped quantum wires with dimensions of 7 nm $\times$ 7 nm have been constructed by Wegscheider's group (v., Wegscheider et al. (1997)).

In their T-shaped intersection of two quantum wires they have found evidence of bound states for holes, electrons and excitons. They have measured the photoluminesce and photoluminesce excitation spectra for this geometry.

Using this approach Wegscheider, Pfeiffer et al. (1993) showed lasing from excitons in one dimension. For a recent review of work in this area on quantum wire exciton lasers see Pfeiffer (1998). For other work on CEO, see Grundmann and Bimberg (1999). For a discussion of the electronic and optical properties of both T-shaped and V-shaped quantum wires, see Goldoni, Rossi, Molinari and Fasolino (1996).

Based on CEO T-shaped quantum wires, Goldoni, Rossi and Molinari (1997) have calculated the conductance for ballistic transport in these structures. Unlike the proposal to control the conductance based on the length of the lateral closed arm or stub, the model of Goldoni et al. uses an open lateral arm and the conductance is controlled by modulating the chemical potential or injection energy. When the width of the side arm matches an integer number of semiperiods of the incoming wave, they found strong reflection resonances at the energies of the 1D states localized at the intersection. If the matching condition does not hold, they found sudden changes in the conductance as new propagating channels open in the arms.

5.5 T-shaped Quantum Dot Transistors

Liang, Frost, Pepper, Ritchie and Jones (1997) have measured the low temperature properties of a T-shaped quantum dot transistor. In this model they found two conductance dips superimposed on two 1D conductance steps. In addition they found conductance oscillations superimposed on the 1D conductance steps. These oscillations disappeared gradually as a magnetic field was applied. Also, as the temperature was increased the conductance oscillations became weaker. They interpreted the oscillations as electron phase-coherent length resonance effects from the ballistic channel.

5.6 Quantum Wire Transistors

One of the goals of quantum device engineering is to build a quantum wire transistor. To do this one needs to form a scattering free one dimensional electron waveguide with gate control. The first steps in this direction have had to work with devices running in the mK temperature range. A second problem is that the waveguides have been limited to 600 nm. One effort to extend these limits has been the work of Okada, Hashizume and Hasegawa (1995).

Okada, Hashizume and Hasegawa (1995) have studied the low temperature transport properties of an AlGaAs/GaAs Schottky in-plane gate (IPG) quantum wire transistor. In split-gate devices the 2DEG is controlled by an electric field perpendicular to the 2DEG layer. In this case quantum waveguide lengths have been limited to 600 nm. For in-plane gate devices the electric fields are perpendicular to the edge of the 2DEG layer and stronger electron confinement has been demonstrated. In the Schottky gates of these authors, quantum conductance, with steps of $2e^2/h$, were shown at a temperature of 4 K; clear conductance quantization is seen up to 10 K and the first plateau remained visible up to a temperature of 40 K, with a device length of 1600 nm.

The device fabrication produced quantum well wires with a geometrical width of 400 to 700 nm and length of 1600 nm, with good field-effect transistor characteristics.

The effect of the magnetic field on the quantized conductance levels was investigated by these authors who showed that the width of the conductance plateaus increased, as compared to the case $B = 0$ T, since the magnetic field enhances the subband energy due to magnetic confinement; thus a larger gate voltage is required for the occupation of the subband. The number of occupied 1D subbands in the presence of a magnetic field is given by

$$n = int[\frac{E_F - eV_0}{\hbar \omega} + \frac{1}{2}]$$

where $\omega = \sqrt{\omega_0^2 + \omega_c^2}$, ω_c is the cyclotron frequency, ω_0 is the harmonic frequency of the electrostatic confinement potential, E_F is the Fermi energy, and eV_0 is the electrostatic energy of the wire. This equation is used to estimate the confinement strength, $\hbar\omega_0$, of the Schottky IPG quantum well wire. They found a value of 4-5 meV, which is contrasted to 1 meV for a split-gate device. The higher temperature observation of the quantized conductance is attributed to this stronger confinement potential of the electrons in the Schottky IPG.

Oscillatory behavior was observed on the first plateau, in addition to fine structure and dips on the conductance curves. One of the devices in this study showed no clear conductance plateaus in no magnetic field, although normal quantized conductance steps were apparent with a magnetic field of 5 T. The authors were not able to determine the origins of fine structure or oscillations; possible causes due to impurities and resonance tunneling were mentioned.

5.7   Fractal Conductance Fluctuations

Ketzmerick (1996) has argued that conductance fluctuations due to phase coherent ballistic transport through a chaotic cavity is generically fractal, i.e. $< (\Delta G)^2 >\sim (\Delta B)^\beta$ for $\beta \leq 2$, due to the mixed phase space, including both chaotic and regular regions. This is sometimes referred to as the hierarchical phase space structure of mixed chaotic systems. Sachrajda, Ketzmerick et al. (1997) have noted that the magnetoconductance fluctuations in a soft wall stadium and a Sinai billiard are fractal. Hegger et al. (1999) have looked at the magnetoconductance fluctuations of weakly disordered quasiballistic gold-nanowires of various lengths $L < L_\phi$, the phase coherence length. They found that the variance

$$< (\Delta G)^2 >=< (G(B) - G(B + \Delta B))^2 >,$$

when analyzed for ΔB much smaller than the correlation field B_c, varies according to $< (\Delta G)^2 >\sim \Delta B^\beta$ for $\beta < 2$, which indicates that the graph of G versus B is fractal. This behavior is attributed to long lived states in the mesoscopic wire with a dwell time probability $P(t)$ decaying much slower than exponential, $P(t) \sim t^{-\beta}$. The wires had a cross section of $30\,nm^2$ with lengths between $400\,nm$ and $1000\,nm$. Measurements were taken at $T \approx 60\,mK$ with fields $|B| \leq 6\,T$.

The samples here are in the quantum chaos regime. And the long lived states are a generic feature of systems with mixed (chaotic and regular) phase of the corresponding classical system.

Huckestein, Ketzmerick and Lewenkopf (1999) have modeled the transport through a two dimensional billiard attached to two infinite leads. The billiard was a cosine billiard defined by two hard walls at $y = 0$ and $y(x) = W + (M/2)(1 - cos(2\pi x/L))$ for $0 \leq x \leq L$. There are two semi-infinite leads of width W attached to the openings of the billiard at $x = 0$ and $x = L$. In this case they find no indication of fractal behavior of g versus E. Instead, the conductance is characterized by narrow isolated resonances. They examined both the conductance and the Wigner-Smith time delay

$$\tau = -i\hbar Tr(S^\dagger dS/dE)/2N$$

for N modes. And $2N$ is the dimension of the S matrix. Their results were found to follow

$$\tau(E) = \sum_i \tau_i \frac{\Gamma_i^2/4}{(E - E_i)^2 + \Gamma_i^2/4} + \tau_{smooth}(E)$$

where $\tau_{smooth}(E) \sim E^{-1/2}$. Here the resonance at energy E_i is characterized by a width Γ_i and a height τ_i. The authors found a power law dependence of conductance increments and a power law distribution of resonance widths, governed by the classical dwell time exponent coming from the phase space structure. For the fully chaotic case they found the statistical properties of the S-matrix to agree with the predictions from semiclassical theory and random matrix theory. We direct the reader to this paper for the details.

Micolich et al. (1998) have extended the work of Hegger et al. on fractal magneto-conductance fluctuations (MCF) in the quasi-ballistic regime in gold nanowires to the fully ballistic regime of mesoscopic semiconductor billiards. In this work the measurements used AlGaAs/GaAs heterostructures. Surface gate geometry was used to construct a $1\ \mu m$ square billiard in the 2DEG located 71 nm below the heterostructure surface. The billiard width is smaller than the electron mean free path which ensures ballistic transport. The entrance and exit to the billiard used quantum point contacts with lithographic width of $0.1\ \mu m$. Measurements were taken over a temperature range of $0.03 - 4\,K$. A magnetic field was applied perpendicular to the device.

As discussed above, the soft-walled potentials which define the semiconductor billiards produce a mixed phase space. Electron escape is then blocked by the infinite hierarchy of Cantori found at the chaotic-regular boundary of the phase space, which results in a MCF with

$$< (\Delta G)^2 > \sim (\Delta B)^\beta,$$

with $\beta < 2$. The magneto-conductance trace is predicted to have a fractal dimension of

$$D_F = 2 - \frac{\beta}{2}$$

with $1 < D_F < 2$.

For $B > B_{cyc}$, the cyclotron field, electron transport is dominated by skipping orbits. Thus, the focus was on the region $B < B_{cyc} = 0.2\,T$. In this range, fractal behavior is predicted to be limited to

$$\Delta B < B_c \sim 10\,mT$$

where B_c is the correlation field determined by the half-width of the autocorrelation function

$$F(\Delta B) = < \delta G(B)\delta G(B + \Delta B) >$$

where $\delta G = G(B) - < G(B) >$. Viz., B_c is given by $F(B_c) = F(0)/2$. One notes that

$$F(\Delta B) = F(0) - (1/2) < (\Delta G)^2 > .$$

The measurements in these papers show $1 < D_F < 2$ for the temperature range indicated. Limitations which impact these measurements include the phase coherence length l_ϕ and $T < T_{semiclass} = T_Q = \Delta E/k = 2\pi\hbar^2/m^* k_B A_B$, at temperatures where the discrete energy level structure of the billiard becomes resolved. Here A_B is the area of the billiard. Other work discussed earlier in this chapter has shown a phase breaking time of $\tau_\phi \sim 90\,ps$ at $30\,mK$, which gives rise to $l_\phi = v_F \tau_\phi \sim 24\,\mu m$. For a diamond shaped periodic orbit in the $1\,\mu m$ square billiard, this gives rise to 8 orbits before the electron loses phase coherence. As the temperature is increased, thermally induced scattering will decrease τ_ϕ. Measurements confirmed a reduction in D_F with increasing temperature. For $T < T_Q$, a saturation value of τ_ϕ has been observed in other work, with value approaching the Heisenberg time $\tau_H = \hbar/k_B T_Q$, which for the billiard in the experiment is $\tau_H \sim 95\,ps$. Thus, it is expected that D_F will approach a fixed value for $T < T_Q$.

We note that the fractal behavior of MCF results in a statistical self-similarity. Taylor, Newbury et al. (1997) and Taylor, Micolich et al. (1997, 1998) have discussed the exact self-similarity in the magneto-conductance of semiconductor Sinai billiard. We direct the reader to the original papers for the details.

Chapter 6

Anderson Localization

6.1 Introduction

Anderson localization is the general phenomena where disorder can cause localization of electron states. This in turn will impact transport properties such as conductivity and Hall currents as well as the statistics of energy level spacings. Anderson (1958) using tight binding models predicted that there is a phase transition for electrons in condensed matter from the classical diffusion regime or Ohm's law regime to a localized state where the material behaves as an insulator in the presence of disorder. This prediction was generalized by Mott and Twose (1961) and Landauer (1970) who conjectured that all states are localized for any strength of disorder. In one dimension this conjecture was proven by Abou-Chacra, Anderson and Thouless (1973) and Kunz and Souillard (1980) among others. The study of Anderson localization now extends beyond solid state physics. E.g., Anderson localization has been examined in acoustic, microwave and light propagation. Diffusion of light through a disordered material also has an Ohm's law regime where conductance decreases linearly with length. Localization of light is expected to occur for $kl \leq 1$, the modified Ioffe-Regel regime. However, for light the experimental problem is to realize a medium with sufficiently strong scattering. Precursors to Anderson localization of light have been reported including weak localization effects, i.e. enhanced backscattering of light from a disordered media. For microwave systems Anderson localization has been reported for a two dimensional system with rods and a tubular system with metallic and dielectric spheres, although for these systems the absorption component is very strong. Most recently, Anderson localization has been noted in a deformed three dimensional microwave resonator, consisting of twenty cells connected together.

Anderson localization is in fact a study of the spectral theory of random Schrödinger operators, H_ω. If H_ω has pure point spectrum in a certain energy region, then the corresponding electron wave function essentially remains restricted to a bounded region, i.e. is localized for all time. The system is then an insulator. Extended states on the other hand arise when H_ω has absolutely continuous spectrum and the system is then a conductor. For the quantum Hall effect we

105

will see that there is a requirement for a system to have both localized states and extended states.

In this chapter we will review a variety of topics related to recent work on Anderson localization in periodic systems and in deformed quantum wires; in the following chapter Anderson localization and the quantum Hall effect are developed. The proofs will be minimized with an attempt to survey the type of results which have been developed related to spectral theory of Schrödinger operators for these systems. To a large extent the proofs are variations on the multiscale analysis approach of Fröhlich and Spencer (1983) and von Dreifus and Klein (1989). The elements of multiscale analysis are reviewed in Section 13. There are excellent compendiums on Anderson localization largely devoted to treating one dimensional systems, v. Carmona and Lacroix (1990), Pastur and Figotin (1992), Kirsch (1989) and Cycon et al. (1987). Stollmann's (1999c) upcoming monograph will treat in detail the more recent literature which we are reviewing and we direct the reader to these volumes for more information.

6.2   Ergodic Operators

Let $(\Omega, \mathcal{F}, \mathbf{P})$ be a mesure space with probability measure $\mathbf{P}$. Let $L(\mathcal{H})$ denote the space of bounded linear operators on a Hilbert space $\mathcal{H}$. The mapping $B : \Omega \to L(\mathcal{H})$ is called measurable if for all x, y in $\mathcal{H}$, the mapping $\Omega \ni \omega \to (B(\omega)x, y) \in \mathbf{C}$ is measurable. Let $S(\mathcal{H})$ denote the space of self-adjoint operators on $\mathcal{H}$. If H is measurable, then $H : \Omega \to S(\mathcal{H})$ is called ergodic if there is an ergodic family T_i on $(\Omega, \mathcal{F}, \mathbf{P})$ and a family of unitary operators U_i on $\mathcal{H}$ such that

$$H_{T_i(\omega)} = U_i^* H_\omega U_i.$$

6.3   Deterministic Systems

Measurable, ergodic operators are said to have deterministic spectra; viz. one can show:

Theorem 1 *Let $H : \Omega \to S(\mathcal{H})$ be measurable and ergodic. Then there exists a set $\Sigma \subset \mathbf{R}$ and a set $\Omega_1 \in \mathcal{F}$ such that $\mathbf{P}(\Omega_1) = 1$ and*

$$\sigma(H_\omega) = \Sigma$$

for all $\omega \in \Omega_1$.

The spectral components of H_ω, the pure point spectrum, $\sigma_{pp}(H_\omega)$, absolutely continuous spectrum, $\sigma_{ac}(H_\omega)$, and singular continuous spectrum, $\sigma_{sc}(H_\omega)$ where

$$\sigma(H_\omega) = \sigma_{ac}(H_\omega) \cup \sigma_{sc}(H_\omega) \cup \sigma_{pp}(H_\omega)$$

are almost surely non-random sets. That is, ergodicity implies that there are sets $\Sigma_{pp}, \Sigma_{ac}, \Sigma_{sc} \subset \mathbf{R}$ such that $\sigma_{pp}(H_\omega) = \Sigma_{pp}, \sigma_{ac}(H_\omega) = \Sigma_{ac}$, and $\sigma_{sc}(H_\omega) = \Sigma_{sc}$ with probability one.

6.4 Integrated Density of States

Consider a random Schrödinger operator

$$H_\omega = -\Delta + V_\omega.$$

Let Λ represent a cube centered at 0 in $\mathbf{R}^d$. Let $H_{\omega,\Lambda}^D$ denote the Dirichlet restriction of H_ω to Λ. For $E \in \mathbf{R}$ define

$$N_{\omega,\Lambda}(E) = \frac{1}{Vol(\Lambda)} |\{ \text{ eigenvalues of } H_{\omega,\Lambda}^D \text{ smaller than } E\}|.$$

Then for a large class of random potentials one has:

Theorem 2 *There exists a non random, non decreasing, non negative, right continuous function $N(E)$ such that almost surely for all $E \in \mathbf{R}$, E a continuity point of $N(E)$, $N_{\omega,\Lambda}(E)$ converges to $N(E)$ as Λ exhausts $\mathbf{R}^d$.*

For a proof, see Kirsch (1989).

The function $N(E)$ is called the integrated density of states (IDOS) of H_ω. One sees that the IDOS is really the probability of finding an eigenvalue of $H_{\omega,\Lambda}$ lying below a given energy E.

6.5 Anderson Spectra – Discrete Model

Figotin and Klein (1994) have studied the localization in gaps of the spectrum of random lattice operators on $l^2(\mathbf{Z}^d)$ of the form

$$H_\omega = H_{per} + gv$$

where $H_{per} = -\Delta + w$; here Δ is the lattice Laplacian, $w(x)$ is a real valued, periodic function, g is a positive coupling constant and $v(x), x \in \mathbf{Z}^d$, are real valued, independent, identically distributed random variables. Define the norm $|x|_\infty$ for $x \in \mathbf{Z}^d$ by $|x|_\infty = max_{1 \le j \le d}|x_j|$. Let H_{per} denote a local periodic operator, i.e. H_{per} acts on $l^2(\mathbf{Z}^d)$ and there is a number ρ, called the range, such that if $|x - y|_\infty > \rho$, then $H_{per}(x,y) = 0$ and there is a vector $q \in \mathbf{Z}^d$ with positive components such that $H_{per}(x,y) = H_{per}(x+q',y+q')$ for all $x,y \in \mathbf{Z}^d$ and for all $q' \in q_1\mathbf{Z} \times ... \times q_d\mathbf{Z}$.

Floquet-Bloch theory still applies to the discrete operator H_{per}. Viz., the spectrum of H_{per} has band structure:

Theorem 3 *(Floquet-Bloch) If H_{per} is a periodic operator on $l^2(\mathbf{Z}^d)$, then its spectrum consists of a finite number J of intervals*

$$\sigma(H_{per}) = \cup_{1 \le i \le J}[\mu_i^{(0)}, \lambda_i^{(0)}]$$

where $0 \le \mu_i^{(0)} \le \lambda_i^{(0)}$ for $1 \le i \le J$ and $\lambda_i^{(0)} < \mu_{i+1}^{(0)}$ for $1 \le i \le J - 1$.

For a proof, see Figotin and Klein (1994).

The assumptions for the discrete periodic model are:

(D1) H_{per} is a local, q-periodic self-adjoint operator on $l^2(\mathbf{Z}^d)$ with $J - 1 > 0$ gaps $(\lambda_i^{(0)}, \mu_{i+1}^{(0)})$ for $1 \leq i \leq J - 1$

(D2) the random potential v is an operator on $l^2(\mathbf{Z}^d)$ given by multiplication by $v(x)$ where $v(x)$ are real valued, independent, identically distributed random variables on a probability space with probability measure $\mathbf{P}$. The probability distribution μ of $v(0)$ has a bounded density h with $\|h\|_\infty \leq D_0$ and v has support, $supp(v(x)) = [-1, 1]$.

Figotin and Klein have shown that this discrete Anderson model has deterministic spectra since H_ω is metrically transitive and if the coupling constant g is sufficiently small, H has a band gap structure which is naturally associated with the band gap structure of H_{per}:

Theorem 4 *(Figotin and Klein) Let* $\xi(x) = \xi_\omega(x), x \in \mathbf{Z}^d$, *be a set of real-valued, independent, identically distributed random variables on the probability space* $(\Omega, \mathcal{F}, \mathbf{P})$ *such that for some finite constants* ξ_1, ξ_2 *the support is given by* $supp(\xi(x)) = [\xi_1, \xi_2]$. *Assume* $H_\omega = H_{per} + \xi$ *where* H_{per} *satisfies (D1) and* ξ *is the operator multiplication by* $\xi(x)$. *Then with probability one, the spectrum* $\sigma(H_\omega)$ *is deterministic,* $\sigma(H_\omega) = \Sigma$ *and*

$$\Sigma = \sigma(H_\omega) = \sigma(H_{per}) + [\xi_1, \xi_2].$$

If $\xi(x) = gv(x)$ *where* v *satisfies (D2), then for any*

$$0 \leq g \leq g_i = (\mu_{i+1}^{(0)} - \lambda_i^{(0)})/2$$

for $1 \leq i \leq J - 1$, *with probability one the spectrum* $\sigma(H)$ *has a nonempty gap*

$$(\lambda_i, \mu_{i+1})$$

where $\lambda_i = \lambda_i^{(0)} + g < \mu_{i+1} = \mu_{i+1}^{(0)} - g$ *which is naturally associated with the gap* $(\lambda_i^{(0)}, \mu_{i+1}^{(0)})$ *of the unperturbed periodic operator.*

It was known that operators of this form have pure point spectrum with exponentially decaying eigenfunctions for low energies and near the endpoints of the spectrum. Figotin and Klein showed that the operator H_ω develops pure point spectrum with exponentially decaying eigenfunctions in a vicinity of the gaps:

Theorem 5 *(Figotin and Klein) Let* $H_\omega = H_{per} + gv$ *and assume the*

$$\mu\{|v(0) \pm 1| \leq \epsilon\} \leq C\epsilon^\eta$$

for a finite constant C *and* $\eta > d$. *Then, if* $0 \leq g \leq g_i$, *there is a* $\delta_\pm = \delta_\pm(d, H_{per}, D_0, g, C, \eta)$ *such that* H_ω *is exponentially localized in the intervals* $(\lambda_i - \delta_+, \lambda_i)$ *and* $(\mu_i, \mu_i + \delta_-)$ *with probability one.*

The distance to the spectrum of a self adjoint operator A acting on the Hilbert space $\mathcal{H}$ is denoted $dist(\sigma(A), \lambda)$. One can show

Theorem 6 *If λ is a real number, then*

$$dist(\sigma(A), \lambda) = min_{\psi \in \mathcal{H}, \|\psi\|=1} \|(A - \lambda)\psi\|.$$

For exponential estimate for the resolvent of the local operator H, a version of the Combes-Thomas (1973) estimate is:

Theorem 7 *(Figotin and Klein) If A is a local operator with range ρ_A acting on $L^2(\mathbf{Z}^d)$ and Green's function*

$$G(x, y, z) = <\delta_x|(A - z)^{-1}|\delta_y >$$

for $x, y \in \mathbf{Z}^d$ and $z \notin \sigma(A)$, and if $|A(x, y)| \leq c$ and $dist\{z, \sigma(A)\} = \delta > 0$, then there is a $b > 0, b = b(c, \rho_A)$ such that

$$|G(x, y, z)| \leq 2\delta^{-1}e^{-b\delta|x-y|}$$

where $|x| = \sum_{1 \leq j \leq d} |x_j|$.

6.6 Crystal with Randomly Distributed Impurities

Let $|y| = max_i|y_i|$, for $y \in \mathbf{R}^d$. Let $\Lambda_l(x) = \{y \in \mathbf{R}^d | |x - y| < l/2\}$ denote the cube with center $x \in \mathbf{Z}^d$ and sidelength $l \in \mathbf{N}$. The Anderson model is given by the Hamiltonian

$$H_\omega = H_{per} + V_\omega$$

with the assumptions:

- $H_{per} = -\Delta + V_{per}$ describes the electron in a crystal without impurities, where $V_{per} \in V_{loc}^p(\mathbf{R}^d)$; here V_{per} is real valued and periodic with respect to $\mathbf{Z}^d$; here $p = 2$ for $d \leq 3$ and $p > d/2$ if $d \geq 4$;

- the impurities are alloy like, i.e. they are fixed to the period lattice and described by an Anderson like random potential of the form

$$V_\omega(x) = \sum_{k \in \mathbf{Z}^d} q_k(\omega)f(x - k);$$

- the single site potential f satisfies: $f \in L^p$, the support of f is contained in $\Lambda_1(0)$, $f \geq 0$, and for some $0 < s \leq 1$, the potential $f \geq \chi_{\Lambda_s(0)}$;

- the coupling constants q_k are independent, identically distributed, real valued random variables with a common, absolutely continuous distribution μ with density $h \in L^\infty$ where $supp(h) = [q_-, q_+]$

In this case the probability space is given by $\Omega = [q_-, q_+]^{\mathbf{Z}^d}$ and

$$\mathbf{P} = \otimes_{\mathbf{Z}^d} \mu = \otimes_{k \in \mathbf{Z}^d} h(q_k) dq_k$$

on Ω. For $\omega \in \Omega$, $q_k(\omega) = \omega_k$.

Let $T_k \Omega \to \Omega : \omega \to \omega(\cdot - k)$ and define $U_k : L^2(\mathbf{R}^d) \to L^2(\mathbf{R}^d)$ be given by $U_k f = f(\cdot - k)$. Then one can show:

Theorem 8 *(Kirsch, Stollmann and Stolz) For the Anderson model H_ω is measurable and ergodic with respect to T_k. In particular H_ω is deterministic.*

6.7 Kotani and Simon Model

The case of the Anderson model described above with $V_{per} = 0$ was treated by Kotani and Simon (1987). They showed that the almost sure spectrum of H_ω is a half line $[E_0, \infty)$ and for the case $f = \chi_{\Lambda_1(0)}$ there is localization at low energies; i.e., there is a $\delta > 0$ such that H_ω almost surely has dense pure point spectrum in $[E_0, E_0 + \delta]$ with exponentially decaying eigenfunctions.

6.8 Anderson Spectra – Continuous Model

Returning now to the general Anderson model of electron motion in a crystal, under minor conditions on V_{per}, H_{per} has Floquet-Bloch structure, i.e. it has purely absolutely continuous spectra consisting of closed intervals which are called bands. The addition of the random potential V_ω changes the spectral properties, viz. it produces band edge localization, as will be described below.

The Anderson model spectra is given by

Theorem 9 *(Kirsch, Stollmann and Stolz) For the Anderson model*

$$\Sigma = \sigma(H_\omega) = \cup_{q \in [q_-, q_+]} \sigma(H_{per} + q f^{per})$$

where $f^{per} = \sum_{k \in \mathbf{Z}^d} f(\cdot - k)$.

As we see, Σ is the union of closed disjoint intervals which are called bands with open intervals in between, which are called gaps. The points in the boundary $\partial \Sigma$ of Σ are given by band edges of either $H_{per} + q_+ \sum_k f(\cdot - k)$ or $H_{per} + q_- \sum_k f(\cdot - k)$. Since energies at the boundary points are rare events (requiring all coupling constants $q_k(\omega)$ to be near q_+ or q_-), $\partial \Sigma$ is called the fluctuation boundary. It was the work of Lifshitz (1963, 1965, 1968) who showed that Anderson models in solid state physics should exhibit pure point spectra near fluctuation boundaries.

Extending the result of Figotin and Klein for the discrete case discussed above, Kirsch, Stollmann and Stolz (1999a) have shown that for Anderson type random perturbations of a periodic Schrödinger operator on $\mathbf{R}^d$, there is exponential localization near the band edges.

Theorem 10 *(Kirsch, Stollmann and Stolz) Assume there is a $\tau > d/2$ such that the coupling constant density g satisfies:*

$$\int_{q_-}^{q_-+h} h(s)ds \leq \varepsilon^\tau$$

and

$$\int_{q_+-h}^{q_+} h(s)ds \leq \varepsilon^\tau$$

for small $\varepsilon > 0$. Then in a neighborhood of $\partial\Sigma$ the spectrum of H_ω is pure point **P***-almost surely with exponentially decaying eigenfunctions.*

This result is an improvement over the work of Figotin and Klein (1996, 1997) which work excludes unbounded f and f's with small support. The work of Kirsch, Stollmann and Stolz also extends that of Barbaroux, Combes and Hislop (1997) to be discussed below, in particular their condition (H9), v.i. The advantage of Barbaroux, Combes and Hislop is that they derive a linear version of the Wegner estimate versus the $|\Lambda|^2$ term in Kirsch, Stollmann and Stolz. The linear term allows Barbaroux, Combes and Hislop to show Lipshitz continuity for the integrated density of states (IDOS).

Near the bottom of the spectrum Kirsch, Stollmann and Stolz are able to show:

Theorem 11 *(Kirsch, Stollmann, and Stolz) If V_{per} is reflection symmetric with respect to all coordinate axes, then in the neighborhood of $\inf(\Sigma)$, the spectrum of H_ω is pure point* **P***-almost surely with exponentially decaying eigenfunctions.*

The approach to the proof of these results follows the standard approach of multiscale analysis, which we discuss in Section 13. Briefly the steps involve: (1) the Wegner estimate, (2) initial scale estimate and (3) multiscale analysis. Their proof also requires extra steps involving spectral averaging of Kotani and Simon (1987) and expansions in generic eigenfunctions.

The Wegner estimate deals with box Hamitonians

$$H_\omega^\Lambda = H_{per} + V_\omega$$

on $L^2(\Lambda)$ where $\Lambda = \Lambda_l(i)$ with $l \in \mathbf{N}$ and $i \in \mathbf{Z}^d$. Here also one imposes periodic boundary conditions. The Wegner estimate of Kirsch, Stollmann and Stolz states that it is unlikely to hit an eigenvalue of H_ω^Λ for fixed Λ and varying ω:

Theorem 12 *(Wegner Estimate) Let $a \in \partial\Sigma$. Then there is an open interval I containing a and a constant C such that for all intervals $J \subset I$ and all $\Lambda = \Lambda_l(i)$ with $l \in \mathbf{N}$ and $i \in \mathbf{Z}^d$, we have*

$$\mathbf{P}\{\sigma(H_\omega^\Lambda) \cap J \neq \emptyset\} \leq C|J||\Lambda|^2.$$

A second fundamental step in their proof is to take finite box resolvent estimates and turn them into basic estimates for the resolvent

$$R(z) = (H_\omega - z)^{-1},$$

which are needed to prove localization and exponential decay for the eigenfunctions.

6.9   Combes-Thomas Estimate

Let H be a self-adjoint operator with resolvent $R_H = (H - z)^{-1}$. Set

$$d(E) = dist(\sigma(H), E).$$

Let χ_x denote a bounded function with compact support localized near $x \in \mathbf{R}^2$. Then the Combes-Thomas (1973) estimate states that for $x, y \in \mathbf{R}^2$ with $\|x - y\|$ large enough, there are finite, positive constants C_1, C_2 such that

$$\|\chi_x R_H(E) \chi_y\| \le C_1 d(E)^{-1} e^{-C_2 d(E) \|x - y\|};$$

that is, the resolvent decays exponentially in the distance, $d(E)$, when it is localized between two functions with separated supports.

Barbaroux, Combes and Hislop (1997a) have improved the Combes-Thomas estimate for the case of a spectral gap (B_-, B_+). They show that decay is proportional to $\sqrt{\Delta_+(E)\Delta_-(E)}$ where $\Delta_+(E) = dist(E, \lambda \in \sigma(H))$ where $\lambda \ge B_+$ and $\Delta_-(E) = dist(E, \lambda \in \sigma(H))$ where $\lambda \le B_-$.

6.10   Simon-Wolff Criterion and Localization

Let H_ω satisfy the conditions stated above for the Anderson model. Let $Q : \mathbf{R}^d \to \mathbf{R}$ be a bounded, measurable compact supported function and let U be a neighborhood of $\partial\Sigma$ or $inf(\Sigma)$. Define the conditions:

(SW1) with probability one for almost every $E \in U$

$$sup_{\epsilon \neq 0} \|R(E + i\epsilon)Q\| < \infty$$

(SW2) with probability one for almost every $E \in U$ there is a C, γ such that

$$sup_{\epsilon \neq 0} \|\chi_{\Lambda_1(x)} R(E + i\epsilon)Q\| \le Ce^{-\gamma|x|}$$

for all $x \in \mathbf{Z}^d$.

Theorem 13 *(Kirsch, Stollmann and Stolz) For the Anderson model, H_ω conditions (SW1) and (SW2) hold.*

The Simon-Wolff (1986) criterion states that (SW1) implies that H_ω has pure point spectrum in U for almost every ω.

Let $H_s^1 = \{f| < x >^s f \in H^1\}$ where H^1 is the first order Sobolev space on $\mathbf{R}^d$ and $< x >= (1 + |x|^2)^{1/2}$. One can show that generalized eigenfunctions exist for H:

Theorem 14 *Let $H = -\Delta + V$ on $L^2(\mathbf{R}^d)$ where V is uniformly local in L^p with p as above. Let μ be the spectral measure for H and $s > d/2$. Then for μ-almost every $E \in \mathbf{R}$ there is a nontrivial weak solution ϕ of $H\phi = E\phi$ with $\phi \in H_{-s}^1$.*

Condition (SW2) implies that eigenfunctions decay exponentially for $E \in U$ for almost every ω. Viz., one can show:

Theorem 15 *(Kirsch, Stollmann and Stolz) For every $\omega \in \Omega$ there is a subset $S_\omega \subset U$ such that $|U \backslash S_\omega| = 0$ and if $E \in S_\omega$ and ϕ is a weak solution of $(H_\omega + \lambda f)\phi = E\phi$ in H^1_s, for some $s \in \mathbf{R}$, then there is a C and γ such that*

$$|\phi(x)| \leq Ce^{-\gamma|x|}$$

for all $x \in \mathbf{R}^d$.

Specifically, one takes $\tilde{S}_\omega$ to be the set of $E \in U$ for all $y \in \mathbf{Z}^d$ such that there is a C, γ with

$$sup_{\epsilon \neq 0}\|\chi_x R(E + i\epsilon, \omega)\chi_y\| \leq Ce^{-\gamma|x|}$$

for all $x \in \mathbf{Z}^d$ and $\chi_x = \chi_{\Lambda_1(x)}$. Define

$$S_\omega = \tilde{S}_\omega \backslash \{ \text{ eigenvalues of } H_\omega \}.$$

One finds from the conditions (SW1) and (SW2) that $|U \backslash S_\omega| = 0$ with probability one, i.e. there is a $\Omega_0 \subset \Omega$ such that $P(\Omega_0) = 1$ and $|U \backslash S_\omega| = 0$ for all $\omega \in \Omega_0$. Let $\omega \in \Omega_0$ be fixed and set $H(g) = H_\omega + gf$. It follows that $H(g)$ has pure point spectrum in U and all its eigenfunctions decay exponentially from which it can be concluded that H_ω has pure point spectrum in U (v., Kirsch, Stollmann and Stolz for the details).

6.11 Long Range Interactions

Kirsch, Stollmann and Stolz (1999b) have studied the problem of Anderson models with long range interactions, where V_ω has the form

$$V_\omega(x) = \sum q_i(\omega)f(x - i).$$

The long range interaction f behaves at infinity like $|f(x)| \leq |x|^{-m}$ for $|x|$ large. The coupling constants q_i are bounded, independent identically distributed random variables with a bounded density h, with $supp(h) = [q_-, q_+]$. In addition we assume $f \geq 0$, $f \geq c > 0$ on an open set, and f belongs to $l^1(L^p)$ with $p = 2$ for $d \leq 3$ and $p > d/2$ or $d \geq 4$.

Again one can show that H_ω is deterministic and $\sigma(H_\omega)$ has the form described above; viz.

Theorem 16 *(Kirsch, Stollmann and Stolz) H_ω is deterministic, $\Sigma = \sigma(H_\omega)$ and*

$$\Sigma = \cup_{q \in [q_-, q_+]} \sigma(H_{per} + q \sum_{k \in \mathbf{Z}^d} f(\cdot - k)).$$

For this model Kirsch, Stollmann and Stolz have shown the following Wegner estimate; define the event

$$A(E, l, \epsilon) = \{\omega | dist(E, \sigma(H_\Lambda)) < \epsilon\}$$

where $\Lambda = \Lambda_l(x)$. Then in a neighborhood of $\partial\Sigma$ this event has very small probability:

Theorem 17 *(Kirsch, Stollmann, and Stolz) Let H_ω be as above; for $a \in \partial\Sigma$ there is a neighborhood U of a and a constant $C > 0$ such for $E \in U$*

$$\mathbf{P}(A(E, l, \epsilon)) \le C\epsilon |\Lambda_l|^2.$$

In terms of localization, they have shown:

Theorem 18 *(Kirsch, Stollmann, and Stolz) For the Anderson model with long range interactions if $m > 4d$, then for almost every ω the spectrum of H_ω is pure point spectrum in a neighborhood of $inf(\Sigma)$ with exponentially decaying eigenfunctions.*

Theorem 19 *(Kirsch, Stollmann, and Stolz) For the Anderson model with long range interactions, if $m > 4d$, and if there is a $\tau > d/2$ such that the density g satisfies:*

$$\int_{q_-}^{q_-+h} h(s)ds + \int_{q_+-h}^{q_+} h(s)ds \le \varepsilon^\tau$$

for small ε, then for almost every ω the spectrum of H_ω is pure point spectrum in a neighborhood of $\partial\Sigma$ with exponentially decaying eigenfunctions.

Their result also applies to a closing gap, i.e. where the upper band edge of $H_{per} + q_+ \sum f(\cdot - k)$ coincides with a lower band edge of $H_{per} + q_- \sum f(\cdot - k)$. However, one can show that Σ will have nontrivial gaps if $\sigma(H_{per})$ has gaps and q_+ is close to q_-.

6.12 Lifshitz Asymptotics

Let $H_0 = -\Delta + V_0$ on $L^2(\mathbf{R}^d)$ and set $E_0 = inf\,\sigma(H_0)$; then Weyl's law states that the integrated density of states behaves like

$$N(H_0, E) \sim C(E - E_0)^{d/2}$$

as $E \to E_0$. The Lifshitz asymptotics states that for the standard Anderson model H_ω the integrated density of states behaves like

$$N(H_\omega, E) \sim C exp(-\gamma(E - E_0)^{d/2})$$

as $E \to E_0$ in the case that E_0 is a fluctuational edge. In particular Lifshitz considered that case that $V_\omega(x) = \sum_j f(x - X_j(\omega))$ where $\{X_j(\omega)\}$ is the Poisson point field of density μ in $\mathbf{R}^d$ and $f(x)$ is a nonnegative single-site potential with compact support. In this case $E_0 = 0$ is a fluctuational edge.

Another asymptotic regime arises in the case $inf\,f(x) < 0$, so that the lower edge of the spectrum is $E = -\infty$. E.g., let $f(x) = -g_0\delta(x)$ where $g_0 > 0$. Lifshitz asymptotics in this case are

$$log N(E) = -2\sqrt{\frac{E}{E_0}} log\sqrt{\frac{E}{E_0}}(1 + o(1))$$

as $E \to -\infty$ where $E_0 = -g_0^2/4$. In the more general case, Klopp and Pastur (1999) have examined asymptotics of the form

$$log N(E) \sim -g(E) log g(E)$$

as $E \to -\infty$.

6.12.1 Poisson Model

The Poisson model considers impurities randomly distributed in a crystal. Viz., let $X_i(\omega), i \in \mathbf{Z}$ represent the points of a Poisson process in $\mathbf{R}^d$, then the Poisson potential is

$$V_\omega(x) = \sum_{i \in \mathbf{Z}} f(x - X_i(\omega)).$$

Very little is known in this case. In one dimension Stolz (1997) has shown that all states are localized. Tip (1994) has considered the IDOS for a class of repulsive potentials, showing that it is absolutely continuous at high energies. Klopp (1999a) has considered a low concentration asymptotic expansion for the density of states for the Poisson model, where concentration is the expected number of impurities found in a set of Lebesgue measure one. He shows that if $N_\mu(d\lambda)$ denotes the density of states for H_ω then

$$N_\mu(d\lambda) \sim_{\mu \to 0, \mu > 0} \sum_{k \geq 0} \mu^k n_k$$

where $\{n_k\}$ is a sequence of distributions. In particular, n_0 is the density of states of the free Laplace operator and $n_1(\lambda)$ is given in terms of the spectral shift function $\theta(\lambda; \Lambda)$, viz.

$$n_1(\lambda) = -\theta'(\lambda; 0)$$

where Λ is a set of k sites – i.e. k distinct points in $\mathbf{R}^d$. Here $\theta'(\lambda, 0)$ is the derivative of the spectral shift of the pair of operators $(-\Delta + V, -\Delta)$.

Let A be a finite subset of $\mathbf{R}^d$ and set $H(A) = -\Delta + \sum_{y \in A} V_y$ where $V_y(x) = V(x - y)$. So $H(A)$ is a relatively compact perturbation of $-\Delta$. In this case if $q > d/2$ and $z \notin \sigma(H(A))$, the operator $(z - H(A))^{-q} - (z + \Delta)^{-q}$ is trace class. Then we can define the spectral shift function for a pair of operators H_A and $H = -\Delta$ as the distribution in the Schwartz space $\mathcal{S}'(\mathbf{R})$ given by

$$Tr(\phi(H_A) - \phi(H)) = \int_0^\infty \theta(\lambda, A)\phi'(\lambda)d\lambda$$

for $\phi \in \mathcal{S}(\mathbf{R})$.

Broderix et al. (1999) have considered the Poisson random potential for the case of a Schrödinger operator in the Euclidean plane with a perpendicular magnetic field. Viz.,

$$H_\omega = \frac{1}{2}(i\nabla + A)^2 + V_\omega$$

where

$$V_\omega(x) = \sum_j f(x - X_j(\omega))$$

where $X_j(\omega)$ are random points in the Euclidean plane $\mathbf{R}^2$ with Poisson distribution (and density ρ). The vector potential A is assumed to be given in the symmetric gauge as $A(x) = \frac{B}{2}(x_2, -x_1)$. Let $E_n(B)$ denote the Landau levels $E_n(B) = (n + 1/2)B$.

Assume the single-site potential f is nonnegative and decays algebraically:

$$lim_{|x|\to\infty}|x|^\alpha f(x) = \mu$$

where $0 < \mu < \infty$ and $\alpha > 2$. In addition assume $f \in L^2_{loc}(\mathbf{R}^2)$.

Theorem 20 *(Broderix et al.) For the Poisson model above, for all $B > 0$, the asymptotics of the integrated density of states has the form*

$$lim_{E\to0}E^{2/(\alpha-2)}logN(E_0(B) + E) = -C(\alpha, \mu, \rho)$$

where

$$C(\alpha, \mu, \rho) = \frac{1}{2}(\alpha - 2)\mu^{2/(\alpha-2)}(\frac{2\pi\rho}{\alpha}\Gamma(\frac{\alpha - 2}{\alpha}))^{\alpha/(\alpha-2)}.$$

One notes that the asymptotic decay at the lower spectral boundary coincides with the behavior for $B = 0$ if $2 < \alpha < 4$ but it differs in the case $\alpha > 4$ (v., Nakao (1977) and Pastur (1977)).

6.12.2   Lifshitz Tails

Klopp (1999b) considered the Lifshitz tails for the Anderson model

$$H_\omega = H + V_\omega$$

where $H = -\Delta + W$ is a $\Gamma-$periodic Schrödinger operator acting on $L^2(\mathbf{R}^d)$. In this case H has the Floquet decomposition with Floquet eigenvalues $E_n(\theta)$:

$$\sigma(H) = \cup_{n\in\mathbf{N}}E_n(\mathbf{T}^*).$$

In this section, let $n(E)$ denote the IDOS for H and let $N(E)$ denote the IDOS for H_ω. One notes that

$$n(E) = \frac{1}{(2\pi)^d} \sum_{n\in\mathbf{N}} \int_{\{\theta\in\mathbf{T}^*;E(n)\leq E\}} d\theta.$$

It is a continuous, positive, increasing function which is constant in the gaps of the spectrum of H. The IDOS n is said to be non-degenerate at 0 if

$$lim_{E\to0+} \frac{log(n(E) - n(0))}{logE} = \frac{d}{2}.$$

Klopp has shown that at the edge of a gap of H that is not filled in for H_ω, the IDOS of H_ω has a Lifshitz tail behavior if and only if the IDOS of H is non-degenerate.

Let $\sigma = \sigma(H)$ denote the spectrum of H and assume that σ has a gap below energy 0 of length at least δ:

(K1) for some $a > 0$ and $\delta > 0$, $\sigma \cap [0, a) = [0, a)$ and $\sigma \cap [-\delta, 0) = \emptyset$

(K2) V is nonnegative, $V \in l^1(L^p(\mathbf{R}^d))$ where $p = 2$ if $d \leq 3$, $p > 2$ if $d > 4$ and $p > d/2$ if $d \geq 5$; there exist $g_+ \geq 0$ in $L^p(C_0)$ such that $V \not\equiv 0$ and

$$0 \leq V(x + \gamma)(1 + |\gamma|)^{d+2} \leq g_+(x).$$

Here C_0 is a fundamental cell of the integer lattice $\Gamma = \bigoplus_{1 \leq j \leq d} \mathbf{Z} e_j$.

(K3) $(\omega_\gamma)_{\gamma \in \Gamma}$ is a collection of independent, identically distributed bounded, nonnegative random variables, their essential support contains 0 and is not reduced to a single point; and

$$lim_{\epsilon \to 0+} \frac{log|log\mathbf{P}(\{\omega_0 \leq \epsilon\})|}{|log\epsilon|} = 0$$

(K4) for $H_t = H + tV_\omega$ with spectrum Σ_t, $t \in [0, 1]$ assume that there is a $\delta' > 0$ such that $\Sigma_t \cap [-\delta', 0) = \emptyset$.

Theorem 21 *(Klopp) Under the assumptions (K1) - (K4)*

$$lim_{E \to 0+} \frac{log|log(N(E) - N(0))|}{logE} = -\frac{d}{2}$$

if and only if n is nondegenerate.

In dimension one, if H is a periodic Schrödinger operator, all the bands of the spectrum of H are simple and their extrema are nondegenerate. Now assume V satisfies (K3), assume $(\omega_\gamma)_{\gamma \in \Gamma}$ are independent, identically distributed random variables with essential support given by $[-p, p]$ and assume

$$lim_{\epsilon \to 0+} sup \left| \frac{log|log\mathbf{P}(\{\omega_0 \leq -p + \epsilon\})|}{log\epsilon} \right| + \left| \frac{log|log\mathbf{P}(\{\omega_0 \geq p - \epsilon\})|}{log\epsilon} \right| = 0.$$

Theorem 22 *(Klopp) If H_ω satisfies the above conditions, then*

$$lim_{E \in \sigma(H_\omega), E \to E_0} \frac{log|log(|N(E) - N(E_0)|)|}{logE} = -\frac{1}{2}.$$

The bottom of the spectrum for any dimension d for a periodic Schrödinger operator is a simple Floquet eigenvalue and its minimum is nondegenerate. Klopp has shown:

Theorem 23 *(Klopp) If 0 is the infimum of the spectrum of H and (K2) and (K3) are satisfied, then*

$$lim_{E \to 0+} \frac{log|log(N(E))|}{logE} = -\frac{d}{2}.$$

In this case, one says that the Lifshitz exponent is d.

6.12.3   Poisson with Random Coupling

Combes and Hislop (1994) considered the Poisson-type Hamiltonian with random coupling constants:

$$H_{\omega,\omega'} = -\Delta + \sum_{i \in \mathbf{Z}} q_i(\omega) f(x - X_i(\omega'))$$

on $L^2(\mathbf{R}^d)$. Assume that $\{q_n(\omega) | n \in \mathbf{Z}^d\}$ is a family of independent, identically distributed random variables, with single-site distribution function $g(\lambda)d\lambda$, with $supp(g) = [0, g_0]$ and $\|g\|_\infty < \infty$. Set $\delta_0 = \|g\|_\infty^{-1}$, which is a measure of the disorder. If the single site potential f satisfies $f > 0$ and there are constants $c_0, c_1 > 0$ and α, β with $9d < \beta < \alpha < \infty$ such that

$$c_0(1 + \|x\|)^{-\alpha} \leq f(x) \leq c_1(1 + \|x\|)^{-\beta},$$

then Combes and Hislop have shown that there is pure point spectra at low energy and the eigenfunctions are polynomially decaying. The Hamiltonian $H_{\omega,\omega'}$ is ergodic, self-adjoint with deterministic spectra $\Sigma = [0, \infty)$.

Theorem 24 *(Combes and Hislop) For δ_0 large enough, there is a $E_P > 0$ such that $\sigma(H_{\omega,\omega'}) \cap [0, E_P]$ is pure point almost surely and the eigenfunctions are bounded above by $C\|x\|^{-(\beta-d)/2}$ for $\beta > 9d$.*

We note that their result requires the random coupling and does not treat the pure Poisson case.

6.13   Multiscale Analysis

The approach to multiscale analysis due to Frölich and Spencer (1983) and von Dreifus and Klein (1989) in the lattice case with the extension to the continuum case by Holden and Martinelli (1984). The reader should also note Figotin and Klein (1994), Barbaroux, Combes and Hislop (1999a), Combes and Hislop (1994) and Klopp (1995a). For a detailed summary, see Stollmann (1999c). The basic step in multiscale analysis is to verify the exponential decay of the random Green's function for a given finite scale with high probability. This result is then shown to hold for increasing scales. Finally, the exponential decay of the Green's function at all scales is used to show Anderson localization.

6.13.1   Wegner Estimate

Let Λ be a hypercube and let H^Λ denote the restriction of the Hamiltonian H to Λ with Dirichlet boundary conditions. Multiscale analysis requires control of the norm of the Green's function of the operators H and their restriction H^Λ with high probability. The first step is given by the Wegner estimate which says that the probability that the corresponding Green's function norms are bigger than a given number $1/\eta, 0 < \eta < E$ is no more than proportional to $\eta|\Lambda|^2$, viz.

$$\mathbf{P}\{\|G_\Lambda(E)\| \geq 1/\eta\} \leq CE\eta|\Lambda|^2.$$

Usually one takes Λ as a cube of size L, let $\eta = L^{-s}$ for suitably large s, so that $1/\eta = L^s$ is large and $\eta|\Lambda|^2$ is small. In other words, the Wegner estimate states that with high probability the eigenvalues of H^Λ do not want to be too close to any given E. This is a precursor to Anderson localization.

For proofs of the Wegner estimate, the reader is directed to Stollmann (1999a) and Stollmann (1999c).

6.13.2   Regular Cubes

Let $\Lambda_l(x)$ be a cube of side l centered at x. Let $\Lambda_l^{inn}(x) = \Lambda_{l/3}(x)$ and $\Lambda_l^{out}(x) = \Lambda_l \backslash \Lambda_{l-2}$ which represents a thick boundary. Set $\chi_l^{inn} = \chi_{\Lambda_l^{inn}}$ and $\chi_l^{out} = \chi_{\Lambda_l^{out}}$. For $\gamma > 0, E > 0$ with $E \notin \sigma(H^{\Lambda_l})$, the cube $\Lambda_l(x)$ is said to be (γ, E)-regular if

$$\|\chi_l^{out} R_{\Lambda_l}(E) \chi_l^{inn}\| \leq e^{-\gamma l/2}$$

This says that the finite cube Green's function decays exponentially from the center of the cube to the boundary. As it turns out, regularity of the finite cube Green's function is an indication of exponential localization for a random operator. One needs to prove that it holds with high probability at a sufficiently large scale.

6.13.3   Starting Hypothesis

For a given scale L_0, let $P(L_0)$ represent the probability statement

$$\mathbf{P}\{\Lambda_{L_0}(0) \text{ is } (\gamma_0, E_0)\text{-regular}\} \geq 1 - \frac{1}{L_0^p}$$

where $p > d$; i.e., we have localization in a large but finite volume with high probability.

Multiscale analysis states that if we can verify $P(L_0)$ for some sufficiently large scale L_0, then $P(L)$ is also true for all scales $L = L_k$ where $L_{k+1} = L_k^\alpha$ for $\alpha > 1$ and $k = 0, 1, 2, ...$ This step is shown by induction using the Wegner estimate. Thus, multiscale analysis depends on proving $P(L_0)$, which is called the starting hypothesis or initial length scale estimate, $[H1](\gamma_0, E_0)$.

The multiscale analysis theorem has the form:

Theorem 25 *(Multiscale Analysis) If H_ω is an Anderson model for which the starting hypothesis is true, then given a γ with $0 < \gamma < \gamma_0$ there is a B such that if $L > B$ there is a δ such that with probability one H_ω has pure point spectrum in $(E_0 - \delta, E_0 + \delta)$ and the corresponding eigenfunctions decay exponentially fast. More specifically one shows that if $[H1](\gamma_0, E_0)$ is true, then (SW1) and (SW2) hold.*

6.13.4   Eigenfunction Decay Inequality

Let $R_\Lambda(z) = (H_\Lambda - z)^{-1}$. Two elements of the multiscale analysis proof are the eigenfunction decay inequality (EDI) and the geometric resolvent inequality or geometric resolvent equality (GRE). The EDI states that for every every bounded subset $U \subset \mathbf{R}$ and every generalized eigenfunction ψ of H with eigenvalue $E \in U$, there is a constant $C_{U,M}$ such that

$$\|\chi_\Lambda^{inn}\psi\| \leq C_{U,M}\|\chi_\Lambda^{out}R_\Lambda(E)\chi_\Lambda^{inn}\|\|\chi_\Lambda^{out}\|.$$

The geometric resolvent inequality states that if there are cubes $\Lambda \subset \Lambda'$ with centers in $\mathbf{Z}^d$ and sidelengths in $2\mathbf{N}+1$ and if $A \subset \Lambda^{inn}, B \subset \Lambda'\backslash\Lambda$ and $E \in U$ then

$$\|\chi_B R_\Lambda(E)\chi_A\| \leq C_{U,M}\|\chi_B R_\Lambda(E)\chi_\Lambda^{out}\|\|\chi_\Lambda^{out}R_\Lambda(E)\chi_A\|.$$

Let $W(\phi)$ denote the first order differential operator

$$W(\phi) = [\Delta, \phi],$$

where $[\cdot, \cdot]$ denotes the commutator. The geometric resolvent equation is the following relationship between resolvents of Hamiltonians restricted to $\Lambda \subset \Lambda'$: if ϕ_Λ is a C^2–function on Λ, then

$$\phi_\Lambda R_{\Lambda'}(z) = R_\Lambda(z)\phi_\Lambda + R_\Lambda(z)W(\phi_\Lambda)R_{\Lambda'}(z)$$

for z in the resolvent set of H^Λ and $H^{\Lambda'}$.

6.13.5   Scholium: Initial Length Scale Estimate

A more refined statement of the initial scale estimate is the following. Let $\Lambda = \{x \in \mathbf{R}^d \mid |x_i| \leq l/2\}$ and for $\delta > 0$ let $\Lambda_{l,\delta} = \{x \in \Lambda_l \mid dist(x, \partial\Lambda_l) > \delta\}$. Let $\chi_{l,\delta}$ be a C^2-function which has $supp(\chi_{l,\delta}) \subset \Lambda_l, \chi_{l,\delta} \geq 0, |\nabla\chi_{l,\delta}| \subset \Lambda_l \backslash \Lambda_{l,\delta}$ and $\chi_{l,\delta}|\Lambda_{l,\delta} = 1$. Let $W(\chi) = [\chi, H_0]$. The initial scale estimate $[H1](\gamma_0, l_0)$ states that for some $\gamma_0 > 0, l_0 > 1$ there is a $\xi > 2d$ such that

$$\mathbf{P}\{sup_{\epsilon>0}\|W(\chi_{l_0,\delta})R_{\Lambda_{l_0}}(E + i\epsilon)\chi_{l_0/3}\| \leq e^{-\gamma_0 l_0}\} \geq 1 - l_0^{-\xi}$$

for E near the band edges. For more details see Combes and Hislop (1994) and Stollmann (1999c).

6.14   Bethe Lattice

Based on Anderson's model, it was argued early on that the motion of an electron in a one dimensional or two dimensional crystal with impurities should show exponential localization, i.e. pure point spectrum with exponentially decaying eigenfunctions and thus no conductivity. It has been shown that this is correct in one dimension, i.e. there is exponential localization for any disorder. In three or more dimensions both localized and extended states – i.e. absolutely continuous spectrum, are expected for small disorder, where the separation boundary is called the mobility edge. Dimension two is still somewhat of a mystery.

As we have discussed above, in the multidimensional continuum case exponential localization has been shown for large disorder and low energy. The occurrence of absolutely continuous spectrum or extended states in the Anderson model on $\mathbf{Z}^d$

has been an open question. However, for the Bethe lattice or Cayley tree , with connectivity K, the numerical work by Abou-Chacra, Anderson and Thouless (1973) and Abou-Chacra and Thouless (1973) indicated more interesting behavior. Their work showed that for sufficiently low disorder there should be an energy at which localization breaks down, which converges to $\frac{K+1}{2}$ in the zero disorder limit. Work of Miller and Derrida (1994) indicated that there is an energy which in the zero disorder limit converges to $\sqrt{K}$, above which conducting properties of the system vanish. Thus, there is a mobility edge or actually an interval being predicted where near $E = \pm\frac{K+1}{2}$ there is a transition from pure point spectrum to singular continuous or absolutely continuous spectrum, however, without conducting properties. Then near $E = \pm\sqrt{K}$ there is a transition to absolutely continuous spectrum with conducting properties.

The Bethe lattice B is an infinite connected graph with no closed loops and a fixed degree or number, $K + 1$, of nearest neighbors for each vertex or site. The number K is called the connectivity. The distance between any two sites x, y is denoted $d(x, y)$. It is the length of the shortest path connecting x and y. The Anderson Hamiltonian on the Bethe lattice is given by

$$H(g) = \frac{1}{2}\Delta + gV$$

where the Laplacian is

$$(\Delta u)(x) = \sum_{y \mid d(x,y)=1} u(y)$$

acting on $l^2(B) = \{u : B \to \mathbf{C} \mid \sum_{x \in B} |u(x)|^2 < \infty\}$. The spectrum of $H_0 = \frac{1}{2}\Delta$ is

$$\sigma(H_0) = [-\sqrt{K}, \sqrt{K}].$$

As usual, V is a random potential with $V(x), x \in B$ being independent, identically distributed random variables with a common probability distribution h.

Let $G(x, y, z) = \langle x|(H - z)^{-1}|y \rangle$ denote the Green's function for $x, y \in B$ and $Im(z) > 0$. Set $G(z) = \mathbf{E}(G(0, 0, z))$. The integrated density of states (IDOS) is given by

$$N(E) = \mathbf{E} \langle 0|P(-\infty, E)|0 \rangle$$

where P is the spectral projection of H. And the density of states (DOS) is

$$\rho(E) = \frac{dN(E)}{dE}.$$

Simon (1983a) has shown that $G(z)$ is the Borel transform of $N(E)$:

Theorem 26 *(Simon)*

* $G(z) = \int \frac{dN(E)}{E-z}$

* $G(E + i0) = lim_{\eta \to 0} G(E + i\eta)$ *exists for almost every $E \in \mathbf{R}$*

- if $G(E + i0)$ exists for every E in an open interval, then $N(E)$ is absolutely continuous and

$$\rho(E) = \frac{1}{\pi} Im \, G(E + i0).$$

In particular one sees that the analyticity of $\rho(E)$ will follow from the analyticity of $G(E + i0)$.

Let H^{Λ_l} denote the restriction of H to $l^2(\Lambda_l)$ with Dirichlet boundary conditions. Here Λ_l is the set of all sites whose distance from an origin is less than or equal to l. Define the Green's function

$$G_l(x, y, z) = < x|(H^{\Lambda_l} - z)^{-1}|y >$$

for $x, y \in \Lambda_l$ and $Im(z) > 0$. Then one has:

Theorem 27 *(Acosta and Klein) If $Im(z) > 0$ then*

$$lim_{l \to \infty} G_l(x, y, z) = G(x, y, z).$$

6.14.1   Free Hamiltonian on Bethe Lattice

Consider first the free Hamiltonian, where $V = 0$. Then one can check that

$$G^{free}(z) = \frac{-1}{z + [1 + (1/K)]\gamma_+(z)}$$

for $Im(z) > 0$ where

$$\gamma_+(z) = \frac{1}{2}[-z + (z^2 - K)^{1/2}].$$

The integrated density of states for the free Hamiltonian, $N^{free}(E)$, is absolutely continuous in this case and the density of states is given by

$$\rho^{free}(E) = \begin{cases} \frac{2}{\pi} \frac{(K+1)(K-E^2)^{1/2}}{(K+1)^2 - 4E^2}, & \text{if } E^2 < K \\ 0, & \text{otherwise.} \end{cases}$$

In this case one sees that $\sigma(H_0)$ is contained in $[-\sqrt{K}, \sqrt{K}]$.

6.14.2   Cauchy Model on a Bethe Lattice

In the case of a Cauchy distribution

$$d\mu = (\frac{\lambda}{\pi}) \frac{1}{(\lambda^2 + x^2)} dx,$$

Simon (1983b) has shown:

Theorem 28 *(Simon) If μ is the Cauchy distribution, then*

$$G(z) = G^{free}(z + i\lambda).$$

And by the result above the density of states for the Cauchy model is

$$\rho(E) = \frac{1}{\pi} Im \left(\frac{-1}{E + i\lambda + [1 + (1/K)]\{-(E + i\lambda) + [(E + i\lambda)^2 - K]^{1/2}\}} \right).$$

In particular for the Cauchy model $\sigma(H) = \mathbf{R}$ and the density of states is analytic in the strip $|Im(E)| < \lambda$.

Acosta and Klein have shown that for a distribution sufficiently close to the Cauchy distribution, the density of states is analytic in a strip about the real axis. Let S_α denote the Banach space

$$S_\alpha = \{f : [0, \infty) \to \mathbf{C} \text{ absolutely continuous with } \|f\|_{S_\alpha} < \infty\}$$

where $\|f\|_{S_\alpha} = \sum_{k=0}^{1} \|e^{\alpha t} f^{(k)}(t)\|_\infty$.

Theorem 29 *(Acosta and Klein) Let φ_0 be the characteristic function of the Cauchy distribution and let H_ω be an Anderson model with characteristic function φ. Then for any $\delta > \frac{1}{2}(K + 1)$ there is a neighborhood U around 0 in S_α such that if $\varphi - \varphi_0 \in U$ then $\rho(E)$ is analytic in the strip $|ImE| < \epsilon$ for $\epsilon > 0$.*

6.14.3 Anderson Model Spectra – Bethe Lattice

Returning to the general Anderson model, $H(g) = H_0 + gV$, Acosta and Klein (1992) have shown:

Theorem 30 *(Acosta and Klein) The spectrum of $H(g)$ is given by*

$$\sigma(H(g)) = \sigma(H_0) + g \, supp(\mu) = [-\sqrt{K}, \sqrt{K}] + g \, supp(\mu)$$

with probability one. There are sets $\Sigma_{g,pp}, \Sigma_{g,sc}, \Sigma_{g,ac} \subset \mathbf{R}$ such that $\sigma_{pp}(H(g)) = \Sigma_{g,pp}, \sigma_{sc}(H(g)) = \Sigma_{g,sc}, \sigma_{ac}(H(g)) = \Sigma_{g,ac}$ with probability one.

The first step in proving the conjectures of Abou-Chacra and Thouless was provided by Aizenman (1994). He showed:

Theorem 31 *(Aizenman) For energies beyond $\frac{K+1}{2}$ at weak disorder, exponential localization occurs for the Bethe lattice.*

Klein (1995a) provided the second step in showing that the Anderson model on the Bethe lattice has extended states for small disorder inside the spectrum of H_0. First, define a probability measure μ to be p-admissible if its characteristic function

$$\varphi(t) = \int e^{-itv} d\mu(v)$$

is $2p$-differentiable on $(0, \infty)$ with all $2p$ derivatives bounded on $(0, \infty)$. E.g., any probability distribution with

$$\int |v|^{2p} d\mu(v) < \infty$$

is p-admissible. Klein (1995a) in the 1-admissible and then Klein (1995b) in the general cased showed that the conjectures of Miller and Derrida hold:

Theorem 32 *(Klein) Let $H(g)$ denote the Anderson Hamiltonian on the Bethe lattice with μ $p-$admissable, for $p \in \mathbf{N}$. Then for any $E \in (0, \sqrt{K})$ there exists a $g(E) > 0$ such that for any g with $|g| < g(E)$ with probability one the spectrum of $H(g)$ in $[-E, E]$ is purely absolutely continuous, i.e. $\Sigma_{g,ac} \cap [-E, E] = [-E, E]$ and $\Sigma_{g,pp} \cap [-E, E] = \Sigma_{g,sc} \cap [-E, E] = \emptyset$.*

The proof of this theorem follows from an examination of the Green's function of $H(g)$:
$$G_g(x, y, z) = < \delta_x |(H(g) - z)^{-1}| \delta_y >$$
for $x, y \in B$ and $z = E + i\eta$ with $E \in \mathbf{R}$ and $\eta > 0$.

In terms of the integrated density of states (IDOS), for the case $p = 1$, Klein has shown:

Theorem 33 *(Klein) If $H(g)$ is 1-admissable, then for any $E, 0 < E < \sqrt{K}$ there is an $g(E) > 0$ such that for $|g| < g(E)$ the integrated density of states, $N_\lambda(E')$, is continuously differentiable on $(-E, E)$ with*

$$N'_g(E') = lim_{\eta \to 0} \frac{1}{\pi} Im\mathbf{E}(G_g(x, x, E' + i\eta))$$

for any $x \in \mathbf{B}$.

6.14.4   Spreading of a Wave Packet – Bethe Lattice

A measure of the spread of a wave packet at time t is its mean squared displacement

$$r_g^2(t) = \sum_{x \in B} |x|^2 |\psi_t^g(x)|^2 = \sum_{x \in B} |x|^2 |e^{itH(g)}(0, x)|^2$$

where $|x| = d(0, x)$ and $e^{itH(g)}(0, x) = < \delta_0 |e^{itH(g)}| \delta_x >$. Once can show that

$$r_g^2(t) \leq Ct^2$$

(v. Simon (1990)). In case of free motion, i.e. $g = 0$, then $r_g^2(t) \sim t^2$.

Anderson conjectured that for large disorder there are only localized states so that $sup_t r_g^2(t) < \infty$.

Aizenman (1994) showed that for large disorder, both in the multidimensional lattice case and in the Bethe lattice case $\mathbf{E}(sup_t r_g^2(t)) < \infty$.

Set
$$r_{x,g}^2(t) = \sum_{y \in B} d(x, y)^2 |(e^{itH(g)} \delta_x)(y)|^2.$$

Theorem 34 *(Klein)*

$$\mathbf{P}(lim_{t \to \infty} sup \frac{r_{x,g}(t)}{t^2} > 0 \text{ for some } x \in B) = 1$$

for the Anderson model on the Bethe lattice with small disorder; i.e. the mean square distance travelled by a particle in time t is found to grow as t^2 for large time t.

A measure of d.c. electrical conductivity for the Bethe lattice is given by

$$\sigma_g(E) = \lim_{\eta \to 0} \eta^2 \sum_{x \in B} |x|^2 \mathbf{E}(|G_g(0, x, E + i\eta)|^2).$$

One expects that $\sigma_g(E) = 0$ in regions with localization and $\sigma_g(E) > 0$ in regions with extended states. Klein shows:

Theorem 35 *(Klein) For the Bethe lattice there exists a $g_0 > 0$ such that for $|g| < g_0$ there are energies $E_g^{\pm} \in (-\sqrt{K}, \sqrt{K})$ with $\lim_{g \to 0} E_g^{\pm} = \pm\sqrt{K}$ such that $H(g)$ has purely absolutely continuous spectrum in the interval $I_g = (E_g^-, E_g^+)$ and $\sigma_g(E) = \infty$ for all $E \in I_g$.*

6.15   Spectral Statistics – the Discrete Model

Let $V = \{V_x, x \in \mathbf{Z}^d\}$ consist of independent, identically distributed random variables and assume their common distribution has a bounded density

$$P(V_x \in dv) = h(v)dv$$

with $\|h\|_\infty < \infty$. Consider the Anderson tight binding model

$$H = -\Delta + V$$

on $l^2(\mathbf{Z}^d)$ and its restriction H^Λ to finite hypercubes $\Lambda \subset \mathbf{Z}^d$ with Dirichlet boundary conditions. Here Δ is the discrete Laplacian defined by

$$(\Delta u)(x) = \sum_{|y-x|=1} u(y).$$

Let $\{E_j(\Lambda)\}$ denote the eigenvalues of H^Λ. Since $V = \{V_x\}$ is an ergodic random field for almost all realizations of V, the limit

$$N(E) = \lim_{\Lambda \uparrow \mathbf{Z}^d} \frac{1}{|\Lambda|} |\{j | E_j(\Lambda) \le E\}|$$

exists for all $E \in \mathbf{R}$; v. Carmona - Lacroix (1990). This limit is called the integrated density of states. Since $N(E)$ is a non-decreasing function of E, it is differentiable almost everywhere. If the derivative

$$\rho(E) = dN(E)/dE$$

exists at $E \in \mathbf{R}$, then it is called the density of states (DOS) at E. If the random potential V has a bounded density, then $N(E)$ is absolutely continuous and the average spacing of eigenvalues is of order $|\Lambda|^{-1}$ near energy E where $\rho(E)$ exists and is positive. As in random matrix theory, we rescale the eigenvalues and consider the spectrum

$$\xi_j(\Lambda, E) = |\Lambda|(E_j(\Lambda) - E)$$

for $J = 1, ... |\Lambda|$ as Λ gets large.

Let $\Lambda = [1, L]^d, L = 1, 2, ...$ be hypercubes in $\mathbf{Z}^d$ and define the Green's function

$$G^\Lambda(x, y, z) = < \delta_x|(H^\Lambda - z)^{-1}|\delta_y > .$$

Let $M(\mathbf{R})$ denote the space of all Radon measures on $\mathbf{R}$ and let $M_p(\mathbf{R})$ denote the subspace of all integer values Radon measures on $\mathbf{R}$. Each $\xi \in M_p(\mathbf{R})$ can be written as

$$\xi(dx) = \sum_j \delta_{\xi_j}(dx)$$

with sequence $\{\xi_j\}$ having not finite accumulate point. A random variable ξ_ω taking values in $M_p(\mathbf{R})$ is called a point process. We say $\mu(dx)$ is the intensity measure of ξ_ω if

$$\mu(A) = \mathbf{E}[\xi_\omega(A)]$$

for each Borel set A.

The standard point process is the Poisson point process. Recall that ξ_ω is said to be a Poisson point process with intensity measure $\mu(dx)$ if for each bounded Borel set A, $\xi_\omega(A)$ obeys the Poisson distribution

$$P(\xi_\omega(A) = k) = e^{-\mu(A)}\frac{\mu(A)^k}{k!}$$

for $k \geq 0$ and if $A_1, ..., A_n$ are disjoint, then $\xi(A_1), ...\xi(A_n)$ are independent random variables.

Define the point process $\xi(\Lambda, E)$ by

$$\xi(\Lambda, E)(dx) = \sum_j \delta_{|\Lambda|(E_j(\Lambda)-E)}(dx).$$

We define the following condition:

(AM) the fractional moment of the Green's function is said to decay exponentially fast if there is an $s_0 \in (0, 1), C > 0, m > 0, r > 0$ such that

$$\mathbf{E}[|G^\Lambda(x, y, z)|^{s_0}] \leq Ce^{-m|x-y|}$$

for any hypercube $\Lambda \subset \mathbf{Z}^d, x \in \Lambda, y \in \partial\Lambda$ and $z \in \{z|Im(z) > 0, |z - E| < r\}$.

Extending the earlier results of Molchanov (1981) for the one dimensional random Schrödinger operator, Minami (1996) has shown:

Theorem 36 *(Minami) If the density of states $\rho(E)$ exists at E and is positive and condition (AM) holds, then the point process $\xi(\Lambda, E)$ defined above, converges weakly as $L \to \infty$ to the Poisson point process ξ with intensity measure $\rho(E)dx$.*

The result of Minami shows that for this model we have Anderson localization; i.e. with probability one the spectrum of H is pure point in the vicinity of E and the corresponding eigenfunctions decay exponentially fast. And for the energy E near where Anderson localization is expected, there is no correlation between the eigenvalues of H^Λ if Λ is large.

Sufficient conditions for (AM) have been given by Aizenman and Molchanov (1993), Graf (1994) and Minami (1996). E.g., Graf showed that if $\|\rho\|_\infty$ is sufficiently small, then (AM) holds for all E; and Minami showed that if $d = 1$ and $\|\rho\|_\infty$ and $\mathbf{E}[|V_x|^\epsilon]$ are finite for some $\epsilon > 0$, then (AM) holds for all E.

6.16 Mobility Edge – the Discrete Model

Kirsch, Krishna and Obermeit (1999) have studied the problem of mobility edges for a discrete Anderson model, viz. assume

$$H_\omega = H_0 + V_\omega$$

acting on $l^2(\mathbf{Z}^d)$ where $V_\omega(n) = a_n q_n(\omega)$ and where q_n are independent, identically distributed random variables with an absolutely continuous probability distribution μ. We assume $a_n > 0, n \in \mathbf{Z}^d$ and a_n are bounded.

A probability measure μ on $\mathbf{R}$ is said to be τ–regular $(0 < \tau \leq 1)$ if for some $v > 0$ and $C < \infty$

$$\mu([z - \delta, z + \delta]) \leq C|\delta|^\tau \mu([z - v, z + v])$$

for all $0 < \delta < 1$ and $z \in \mathbf{R}$.

Theorem 37 *(Kirsch, Krishna and Obermeit) For $d \geq 1$, if μ is 1-regular, $\int |x| \mu(dx) < \infty$ and $a_n \to 0$ as $|n| \to \infty$, then $\sigma_c(H_\omega) \subseteq [-2d, 2d]$ **P**-almost surely.*

This implies that any spectrum outside $[-2d, 2d]$ is pure point almost surely. Delyon, Simon and Souillard (1985) showed that if $\sum_n |a_n|^p < \infty$ with

$$p = sup\{k \in \mathbf{R}^+ | \int |x|^k \mu(dx) < \infty\}$$

there is no essential spectrum outside $[-2d, 2d]$.

As noted above, Klein demonstrated the existence of absolutely continuous spectrum for sufficiently small coupling constant in the Bethe lattice. Krishna (1990) showed that for the case of decaying randomness with sufficiently rapid decay, there is absolutely continuous spectra in $[-2d, 2d]$ for $d \geq 3$. Viz.,

Theorem 38 *(Krishna) if $d \geq 3, |a_n| \leq |n|^{-\alpha}$ for $\alpha > 1$ as $n \to \infty$ with $\int x d\mu(x) = 0$ and $\int x^2 d\mu(x) = \sigma^2 < \infty$, then*

$$[-2d, 2d] \subseteq \sigma_{ac}(H_\omega).$$

Theorem 39 *(Kirsch, Krishna and Obermeit) If $d \geq 3, |a_n| \leq |n|^{-\alpha}$ for $\alpha > 1$ as $n \to \infty$ and μ is 1-regular asymptotically large (v. Kirsch, Krishna and Obermeit) with respect to $\{a_n\}$ with a probability measure with $\int |x|^2 d\mu(dx) < \infty$, then $\sigma(H_\omega) = \mathbf{R}, \sigma_c(H_\omega) \subseteq [-2d, 2d], \sigma_{ac}(H_\omega) = [-2d, 2d]$ **P**-almost surely, and the mobility edges are precisely $\{-2d, 2d\}$.*

This theorem does not rule out singular continuous spectrum or the presence of pure point spectrum in $[-2d, 2d]$.

The proof uses the Simon-Wolff criterion which asserts that for the absence of continuous spectrum of H_ω on (a, b), the resolvent kernel $(H_\omega - E + i0)^{-1}(n, m)$ must be square summable in m for each $n \in \mathbf{Z}^d$ for almost every (ω, E). Viz., in the discrete case, let $G(n, m, z) = < \delta_n |(H_\omega - z)^{-1}| \delta_m >$ for $z = E + i\epsilon$.

Theorem 40 *(Simon and Wolff) If for all $n \in \mathbf{Z}^d$ and Lebesgue almost every $E \in (a, b)$, the condition*

$$lim_{\epsilon \to 0} \sum_{m \in \mathbf{Z}^d} |G(n, m, E + i\epsilon)|^2 < \infty$$

for almost every realization of $\{V_n\}_{n \in \mathbf{Z}^d}$ holds, then there is no continuous spectrum for H_ω on (a, b).

Kirsch, Krishna and Obermeit use the Aizenman-Molchanov (1993) technique to verify the Simon-Wolff criterion outside $[-2d, 2d]$ to rule out a continuous component of the spectrum there. The reader should also note the more recent work of Krishna and Sinha (1999) and its relation to work of Jaksic and Molchanov (1999) on localization of surface spectra and the models of surface randomness of Jaksic and Last (1999).

6.17   Gaussian Potentials

Fischer, Leschke and Müller (1999) have examined Gaussian models on $\mathbf{R}^d$ where

$$H_\omega = -\Delta + V_\omega(x)$$

with V_ω a Gaussian process. Assuming certain continuity properties for the covariance function of V, the authors show that these Schrödinger operators have almost surely an absolutely continuous integrated density of states and no absolutely continuous spectrum at sufficiently low energies. Viz.,

Theorem 41 *(Fischer, Leschke and Müller) In the case of the multidimensional random Schrödinger operator, H_ω, which satisfies the conditions above, there exists an energy $E_0 < 0$ such that*

$$\sigma_{ac}(H_\omega) \cap (-\infty, E_0] = \emptyset$$

almost surely.

6.18   Correlated Potentials

Models with correlated potentials have been considered by von Dreifus and Klein (1989) and Combes, Hislop and Mourre (1998). In this case the Hamiltonian has the form

$$H_\omega = -\Delta + \sum_{i \in \mathbf{Z}^d} q_i(\omega) f(x - i)$$

where the random potentials $q_i(\omega) f(x - i)$ at each site i are correlated; i.e., the coupling constants q_i are not independently distributed. For this model it has been shown that the bottom of the spectrum is dense pure point.

6.19 D. C. Electrical Conductivity – the Discrete Model

Kunz and Souillard (1980) studied the electrical conductivity for one dimensional random Schrödinger operator in the lattice case, i.e. acting on $l^2(\mathbf{Z})$ following earlier work of Mott and Twose (1961), Halperin (1967), and Thouless (1974). Kunz and Souillard examined the discrete Hamiltonian

$$(H_\omega)\psi(x) = -\psi(x+1) - \psi(x-1) + V_\omega(x)\psi(x)$$

for $x \in \mathbf{Z}$ where $V(x)$ are independent, identically distributed random variables with density h.

Theorem 42 *(Kunz and Souillard) In this case H_ω is deterministic; with probability one (1) the spectrum of H is $\Sigma = \sigma(H_\omega) = [-2,2]+supp(h)$, (2) the diffusion constant is zero, and (3) the d.c. electrical conductivity, σ_H, associated with H is zero for zero temperature. If the support of h is bounded, then σ_H is zero for all temperatures.*

Aizenman and Graf (1999) have also examined the d.c. electrical conductivity of an electron gas with Fermi energy E_F at zero temperature and zero magnetic field for the multidimensional discrete model on $l^2(\mathbf{Z}^d)$. It is given by

$$\sigma_{i,j}(E_F) = lim_{\eta\to 0} \frac{\eta^2}{\pi} \sum_{x\in\mathbf{Z}^d} x_i x_j \mathbf{E}(|G(0,x,E_F+i\eta)|^2).$$

Based on their work on bounds on the fractional moments of the Green's function $G(x,y,E+i\eta)$, Aizenman and Graf have shown:

Theorem 43 *(Aizenman and Graf) If for some $0 < s < 1$ for energies $E \in (a,b)$*

$$\mathbf{E}(|G(x,y,E+i\eta)|^s) \le C^s e^{-s\mu|x-y|}$$

for some constants $C, \mu > 0$, then (1) the spectrum of H in the interval (a,b) almost surely consists only of non-degenerate eigenvalues with exponentially localized eigenfunctions and (2) the d.c. electrical conductivity vanishes:

$$\sigma_{i,j}(E_F) = 0$$

for $E_F \in (a,b)$.

One notes the similarity to the result of Klein (1995c) discussed above for Bethe lattices.

6.20 Diffusion Coefficient

Barbaroux, Fischer and Müller (1999) have considered the diffusion exponents and the diffusion constant for random Schrödinger operators on both $l^2(\mathbf{Z}^d)$ and $L^2(\mathbf{R}^d)$. The diffusion constant $D(\psi)$ is defined by

$$D(\psi) = lim_{\epsilon\to 0}\mathbf{E}(\epsilon^2 \int_{\mathbf{R}} \| |X|(H_\omega - E - i\epsilon)^{-1}\psi\|^2 dE).$$

Their result subsumes the earlier result of dynamical localization of Aizenman and Graf (1998) for the discrete Anderson model. Based on their results Barbaroux, Fischer and Müller are able to show that for the Anderson model of the periodic continuous Schrödinger operator of Barbaroux, Combes and Hislop (1997) the diffusion constant vanishes. Similarly, Barbaroux, Fischer and Müller are able to show that the diffusion constant vanishes for a variation of the long range, single site model of Kirsch, Stollmann and Stolz (1998). They also treat the case of an Anderson model with correlated potentials. We refer the reader to the paper for the details.

6.21 Deformed Quantum Wires

The common feature of a curved, bulged or windowed quantum wire is the occurrence of bound states with eigenvalues below the essential spectrum. What about deformations along the entire quantum wire? Kleespies and Stollmann (1999) have shown that in this case there are infinitely many eigenvalues or dense point spectrum with exponentially localized eigenfunctions at the bottom of the spectrum.

The model of Kleespies and Stollmann is a quantum wire which is dented along one entire side. (The extension of this result to a quantum wire dented on both sides is elementary.) For $d_{max} > 0$ and $0 < d < d_{max}$ let $\Omega = [0, d]^{\mathbf{Z}}$ with ith coordinate $\omega(i)$ of $\omega \in \Omega$ giving the deviation of the width of the random strip from d_{max}

$$d_i(\omega) = d_{max} - \omega(i).$$

Let $\gamma(\omega) : \mathbf{R} \to [d_{min}, d_{max}]$ denote the polygon on $\mathbf{R}^2$ giving $\{(i, d_i(\omega)\}_{i \in \mathbf{Z}}$ and

$$D(\omega) = \{(x_1, x_2) \in \mathbf{R} | 0 < x_2 < \gamma(\omega)x_1\}.$$

Fix a probability measure μ on $[0, d]$ with $0 \in supp(\mu) \neq \{0\}$ and set $\mathbf{P} = \mu^{\mathbf{Z}}$. Let

$$H_\omega = -\Delta_{D(\omega)}$$

denote the Laplacian on $D(\omega)$ with Dirichlet boundary conditions. H_ω is a self-adjoint operator on $L^2(D(\omega))$. Ergodicity of $(H_\omega)_{\omega \in \Omega}$ implies that the spectrum and spectral parts are deterministic.

The deformed quantum wire model assumptions are:

(W1) there exists a $a, \delta > 0$ such that $\mu[0, \epsilon] \geq a\epsilon^\delta$

(W2) μ is Hölder continuous, i.e. there are $b, \alpha > 0$ such that $\mu(I) \leq b|I|^\alpha$ for all intervals $I \subset [0, d]$.

We note that $inf\,\sigma(H_\omega)$ lies below the first eigenvalue of the Dirichlet Laplacian on the box of length L (in the x_1 direction, i.e. along the axis of the quantum wire) and width $d_{max} - \epsilon$. Thus

$$inf\,\sigma(H_\omega) = E_0 = \pi^2/d_{max}^2$$

for $\mathbf{P}$-almost every $\omega \in \Omega$. In case $supp(\mu) = [0, d]$, then

$$\sigma(H_\omega) = [E_0, \infty)$$

for **P**-almost every $\omega \in \Omega$.

The number of electrons per unit volume decreases very rapidly near the bottom of the spectrum - i.e. the Lifshitz asymptotics has the form:

Theorem 44 *(Kleespies and Stollmann) If (W1) holds, then the integrated density of states $N(t)$ for H_ω satisfies:*

$$lim_{t \to 0} \frac{log(-logN(E_0 + t))}{log(t)} = -\frac{1}{2}.$$

That is, the dented quantum wire has Lifshitz exponent of one.

Localization at the bottom of the spectrum holds for the dented quantum wire:

Theorem 45 *(Kleespies and Stollmann) If (W2) holds, then there exists a $\delta > 0$ such that **P**-almost surely the spectrum of H_ω is pure point in $[E_0, E_0+\delta]$ with exponentially decreasing eigenfunctions where exponential decay is in the x_1-direction, along the axis of the quantum wire; in other words, E_0 is the fluctuation boundary.*

The elements of the proof are to show for $I_0 = [E_0, E_0 + \delta]$ that there is a subset $\Omega_0 \subset \Omega$ of full measure such that for every $\omega \in \Omega_0$ for $E \in \sigma(H_\omega) \cap I_0$ that every generalized eigenfunction u of H_ω is in L^2 and that this implies that the spectrum of H_ω is pure point.

Let $\Lambda = \Lambda_l(i)$, $D_\Lambda(\omega) = D(\omega) \cap (\Lambda \times \mathbf{R})$ and $H_\omega^\Lambda = -\Delta$ on $D_\Lambda(\omega)$ with Dirichlet boundary conditions. Set $\Lambda^{inn} = \Lambda_{l/3}(i)$ and $\Lambda^{out} = \Lambda_l(i) \backslash \Lambda_{l-2}(i)$ and let χ_Λ^{inn} denote the characteristic function of $\Lambda^{inn} \times \mathbf{R}$.

The eigenfunction decay inequality (EDI) provides the connection between the decay estimate for resolvents and the decay estimate for eigenfunctions. EDI is given by:

Theorem 46 *There is a $C = C(d_{min}, d_{max})$ such that for all $\omega \in \Omega$ and for all generalized eigenfunctions u of H_ω with eigenvalue $E \in [E_0, E_0 + 1]$,*

$$\|\chi_\Lambda^{inn} u\| \le C \|\chi_\Lambda^{out} R_\Lambda(E) \ \chi_\Lambda^{inn}\| \|\chi_\Lambda^{out} u\|$$

where $R_\Lambda(E) = (H_\omega^\Lambda - E)^{-1}$.

This theorem shows that an estimate for $\|\chi_\Lambda^{out} R_\Lambda \chi_\Lambda^{in}\|$ can be turned into an exponential estimate in the x_1- direction for generalized eigenfunctions.

The next step is standard multiscale analysis:

Theorem 47 *(Kleespies and Stollmann) Assume H_ω is as above. Assume the Wegner estimate is true:*

(1) there is an $\alpha > 0$ and $C > 0$ such that for all intervals $I \subset [E_0, E_0 + 1]$ and $\Lambda = \Lambda_l(i)$

$$\mathbf{P}\{\sigma(H_\omega^\Lambda) \cap I \ne \emptyset\} \le C|\Lambda|^2|I|^\alpha$$

Assume the initial length estimate is true:
(2) there is a $\beta \in (0, 1)$ and $\xi > 0$ such that for all $\Lambda = \Lambda_l(i)$

$$\mathbf{P}\{E_1(H_\omega^\Lambda) \le E_0 + l^{\beta-1}\} \le l^{-\xi}.$$

Then there is a $\delta > 0$ such that $\mathbf{P}$-a.s. the spectrum of H_ω is pure point in $I_0 = [E_0, E_0 + \delta]$ and there is a $\gamma > 0$ such that for $\mathbf{P}$-a.e. ω and for every $E \in I_0$ there is a $C = C(\omega, E)$ such that

$$\|u\chi_x\| \leq C exp(-\gamma|x|)$$

for every generalized eigenfunction u of H_ω with eigenvalue E. Here χ_x is the characteristic function of $(x - 1/2, x + 1/2) \times \mathbf{R}$.

Kleespies and Stollmann prove that the initial length scale estimate holds and that the Wegner estimate is true for this model. The proof of the Wegner estimate in this case is very easy. Viz., if $E_n(H_\omega^\Lambda)$ denotes the nth eigenvalue of H_ω^Λ then

$$\mathbf{P}\{\sigma(H_\omega^\Lambda) \cap I \neq \emptyset\} \leq \sum \mathbf{P}\{E_n(H_\omega^\Lambda) \in I\}.$$

And $H_\omega^\Lambda \geq -\Delta_{\Lambda \times (0, d_{max})}$ which has eigenvalues

$$\{E_{n,m} = \pi^2(\frac{n^2}{d_{max}^2} + \frac{m^2}{|\Lambda|^2})|n \in \mathbf{N}, m \in \mathbf{N}_0\}.$$

Thus $n \leq \sqrt{1 + \frac{d_{max}^2}{\pi^2}}$ and $m \leq \frac{|\Lambda|}{\pi}$ is a necessary condition on n, m for $E_{n,m}$ to fall into $[E_0, E_0 + 1]$. One sees that the number of such eigenvalues is restricted above by a bound proportional to $|\Lambda|$; one is left to show that

$$\mathbf{P}\{E_n(H_\omega^\Lambda) \in I\} \leq C|\Lambda||I|^\alpha$$

for which we refer the reader to Kleespies and Stollmann.

For the literature on mathematical modelling in this area, we direct the readers to the discussion Chapter 4 and to the papers of Takagaki and Ferry (1992a, 1992b, 1992c), Timp and Howard (1991), Kumar and Bagwell (1992) and Makarov and Yurkevich (1989).

6.22   Anderson Localization in Microwave Cavities

Dembowski, Gräf et al. (1999) have recently demostrated Anderson localization in a string of deformed microwave cavities. They have measured the field distributions and eigenvalues for a periodic three dimensional microwave resonator which was composed of 18 to 20 nearly identical cells. External screws which penetrate the cavities provided the perturbations. Signatures of Anderson localization were observed at both the upper and lower edge of the first band. They also observed extended states in the middle of the band. In fact they observed the co-existence of both extended modes and localized modes. For a similar case in an acoustic system, see He and Maynard (1986).

Chapter 7

Landau Models and the Quantum Hall Effect

7.1 Introduction

The work of von Klitzing, Dorda and Pepper (1980) showed that the Hall conductivity is quantized at very low temperatures at integer multiples of e^2/h, where e is the electron charge and h is Planck's constant. Later work by Laughlin (1981) and Thouless, Kohmoto et al. (1982) began to explore the geometric origins of the quantum Hall effect and the work of Prange (1981), Joynt and Prange (1984) and Thouless (1981) related the plateaux of the Hall conductance which appear when the magnetic field or charge carrier density is changed to Anderson localization.

Consider a disordered two dimensional electron gas (2DEG). It is generally thought that all electrons in such a two dimensional system are localized. In Abrahams, Anderson, Licciardello and Ramakrishnan (1979) it was argued (based on scaling theory and renormalization group calculations) that in the absence of a magnetic field, two is a critical dimension for localization. As we have noted, in dimension one all states are exponentially localized for any strength of disorder. And for dimensions greater than two, there exist extended states for low disorder strengths. In two dimensions all states are localized except for states corresponding to isolated critical energies where the localization length diverges. When a two dimensional electron system is subjected to a strong perpendicular magnetic field, the energy spectrum becomes a series of Landau levels, which are broadened by the impurities. The external magnetic field breaks the time reversal symmetry, it suppresses the enhanced back scattering, extended states (i.e., nonlocalized states) appear in the center of each Landau band, and disorder continues to localize electronic states away from the centers of the Landau bands. The disorder is responsible for the Hall conductance to be quantized in the case that the Fermi energy lies in the localized region of the density of states.

Nonzero Hall conductance means that there is a delocalized state below the Fermi energy. And when the strength of disorder is sufficiently strong, the Hall system is expected to become an Anderson insulator, which means the vanishing of the delocalized states below the Fermi energy. The process by which these states disappear recently has been the subject of much research both experimentally and numerically. We will review this problem below.

The integer quantum Hall transition is driven by varying the location of the chemical potential or Fermi energy with respect to a critical value E_c. Let $\delta = |E_F - E_c|$. The critical value E_c depends on B. Thus the transition can be achieved by changing B while keeping the electron density fixed; that is, transitions between different quantum Hall levels occur when the Fermi energy "sweeps" through the the critical energy or energies in the middle of the Landau bands, where the states are delocalized.

The problems which need to be explained in any theory of the quantum Hall effect are:

- why does the quantum Hall effect occur at integer values n, i.e. the integer quantum Hall effect (IQHE)? of course, there is also the fractional quantum Hall effect to describe;

- what causes the plateaux?

- how are the plateaux related to the vanishing of the direct conductivity?

- are localized states required for the IQHE?

- between two different plateaux must there be an energy where the localization length diverges?

- how do edge states contribute to the quantum Hall effect? in particular how does one reconcile localized states and extended states, both of which seem to be required for the IQHE?

Mathematical models of edges states are developed in this chapter. One can view the quantization of the Hall conductance as following from the contribution of N one-dimensional edge channels, i.e. the number of extended states below E_F, each contributing a conductance of e^2/h. In an ideal 2DEG in the quantum Hall regime, the current flows along the edge (v., MacDonald and Streda (1984)). However, in real systems the situation is more complex. The edge states have been "imaged" recently using scanned potential microscopy (McCormick et al. (1999)). They have found that in high magnetic fields in the transition regions between quantum Hall plateaux the current is concentrated near the edges. However, on the quantum Hall plateaux, the current is distributed throughout the bulk in a complex manner. For further information on measurements in this area, the reader is directed to McCormick et al. (1999), Tessmer et al. (1998) and Kang et al. (1999); see also the discussion in Thouless (1993), Khmelnitskii (1984, 1992), Büttiker (1988) and Schulz-Baldes, Kellendonk and Richter (1999). For background reading on quantum Hall effect, see Thouless (1998), Richter and Seiler (1999), Huckestein (1995) and Girvin (1998).

The results in this chapter explore what is theoretically known to date regarding these areas of study. The theory today is incomplete and the details of related work such as Connes' (1986) non-commutative geometry and its relationship to IQHE would require a larger discussion. The focus will be on spectral theory of operators and the properties related to IQHE. In this connection the reader should also note the recent work of Marcolli and Mathai (1999) where they relate the work of Connes to the fractional valued topological invariant, viz. the orbifold Euler characteristic.

The first observation is that for a spinless, free two dimensional system of electrons in a perpendicular magnetic field, no quantization of the Hall conductance is observed:

Theorem 1 *(Bellissard, van Elst and Schulz-Baldes) The classical Hall formula holds for a quantum mechanical system composed of a free fermion gas at all temperatures.*

In terms of localization length and the quantum Hall effect, Bellissard and co-workers have shown:

Theorem 2 *(Bellissard, van Elst and Schulz-Baldes) If the quantum Hall conductivity jumps by an integer (as a function of the filling factor), then the localization length must diverge at some energy between the localized state regions.*

Combes and Hislop (1996) have shown that the integrated density of states is Lipshitz continuous at all energies but the Landau energies. Thus, the localization length may diverge at the Landau energies but to date there is no proof of this.

The chemical potential at zero temperature is called the Fermi energy E_F. The density of electrons is given by

$$n = \int_{-\infty}^{E_F} dE\, n(E) = N(E_F).$$

One sees that without spectral gaps, changing continuously the electron density is equivalent to changing the Fermi level.

Theorem 3 *(Bellissard, van Elst and Schulz-Baldes) If the Fermi level is in a region of localized states, then the direct conductivity must vanish and the Hall conductivity cannot change since localized states do not participate in the current. And if there were no localized states, changing the filling factor will cause the Hall conductance to change (as long as the Fermi level is in the spectral bounds).*

Thus, changing the filling factor will force the Fermi level to change continuously in a region of localized states while the Hall conductance will remain constant (hence a plateaux). And if there is a spectral gap, with the IDOS equal to say n_0 on that gap, changing the filling factor from $n_0 - \epsilon$ to $n_0 + \epsilon$ will cause the Fermi level to jump discontinuously across the gap. And in this case no plateaux would be observed.

For a proof of these last three theorems, see Bellissard, van Elst and Schulz-Baldes (1994).

7.2 Landau Hamiltonian

The Landau Hamiltonian

$$H_A = (p - A)^2$$

acting on $L^2(\mathbf{R})$ where $p = -i\nabla$ and A is the vector potential

$$A = \frac{B}{2}(x_2, -x_1).$$

The spectrum of H_A is $\{E_n(B)\}$, each of infinite multiplicity due to translation invariance, where

$$E_n(B) = (2n + 1)B$$

for $n = 0, 1, 2, \ldots$.

7.3 Landau Hamiltonian with Impurities

Consider now a random Schrödinger operator in an external magnetic field with a random potential,

$$H_\omega = H_A + V_\omega.$$

The random potential V_ω is Anderson like:

$$V_\omega(x) = \sum_{i \in \mathbf{Z}^2} q_i(\omega) f(x - i)$$

with the assumptions:

(V1) $f \geq 0, f \in C^2, supp(f) \subset B(0, 1/\sqrt{2})$, where $B(x, r)$ is the ball of radius r and center x; and there is a constant C_0 and r_0 such that $f|B(0, r_0) > C_0$;

(V2) $\{q_i(\omega)\}$ is independent, identically distributed family of random variables with common distribution $h \in C^2([-M, M])$ for some $0 < M < \infty$ such that $\int \lambda h(\lambda) d\lambda = 0$ and $h(\lambda) > 0$ for Lebesgue almost every $\lambda \neq 0$.

Without any disorder the Landau Hamiltonian has a spectrum of evenly spaced Landau levels. When the random potential is added, these Landau levels are broadened into bands. What can be said about the spectrum at the band edges? And what is the nature of the spectrum on the interior of these bands?

7.4 Landau Bands

Let $M_0 = sup_{x,\omega} \|V_\omega\| < \infty$. Then one can show

$$\sigma(H_\omega) \subset \cup_{n \geq 0} \sigma_n$$

where $\sigma_n = [E_n(B) - M_0, E_n(B) + M_0]$ which is called the nth Landau band. One notes that $\sigma(H_\omega)$ is not necessarily equal a.s. to $\cup_{n \geq 0} \sigma_n$.

7.5 Magnetic Translations

The magnetic translations are given by

$$U_a = e^{-iBx \wedge a} e^{-ip \cdot a}$$

where $x \wedge a = x_2 a_1 - x_1 a_2$. Let

$$T_a : \Omega \to \Omega$$

denote the $\mathbf{Z}^2$-translations. Then one has

$$U_a H_\omega U_a^{-1} = H_{T_a \omega}.$$

For additional discussion of the magnetic translation group, see Zak (1964).

7.6 Deterministic Spectra

The random Landau model has deterministic spectra:

Theorem 4 H_ω *is a* $\mathbf{Z}^2$ − *ergodic, self-adjoint family of operators. Thus, the spectrum of* H_ω *is deterministic.*

7.7 Localization at the Band Edges

Let H_ω satisfy (V1) and (V2) above. Then Combes and Hislop (1996) have shown that for sufficiently large magnetic field there is localization at the band edges:

Theorem 5 *(Combes and Hislop) Given* H_ω *as above, let*

$$I_n(B) = [E_n(B) + \mathcal{O}(B^{-1}), E_{n+1}(B) - \mathcal{O}(B^{-1})].$$

Then there is sufficiently large magnetic field $B_0 \gg 0$ *such that for* $B > B_0$ *and all* $n = 0, 1, 2, \dots$

$$\sigma(H_\omega) \cap I_n(B)$$

is pure point almost surely and the corresponding eigenfunctions decay exponentially. The integrated density of states is Lipshitz continuous away from $\sigma(H_A)$.

Is there spectrum near the band edges? Combes and Hislop have shown that for the Anderson model, if $\lambda_\pm(B) = B \pm M_0$, are the upper and lower first Landau band edges, then for B sufficiently large there is a C_0 such that with probability one

$$\sigma(H_\omega) \cap [E - C_0 B^{-1/2}, E + C_0 B^{-1/2}] \neq \emptyset$$

for all $E \in [\lambda_-(B), \lambda_+(B)]$.

7.8   Landau Model and Unbounded Random Potentials

Barbaroux, Combes and Hislop (1997b) have extended the results of Combes and Hislop (1996) to show almost sure existence of pure point spectrum of two dimensional Landau Hamiltonian with an unbounded Anderson like random potential, provided the magnetic field is sufficiently high. The model in this case is

$$H_\omega = (p - A)^2 + V_\omega$$

on $L^2(\mathbf{R}^2)$. The random potential V_ω has the form

$$V_\omega(x) = \sum_{i \in \mathbf{Z}^2} q_i(\omega) f(x - i)$$

where

(U1) = (V1)

(U2) $\{q_i(\omega)\}$ is an independent, identically distributed family of random variables with common distribution $h \in L^\infty([-M, M])$ for some $0 < M \le \infty$ with $h(-\lambda) = h(\lambda)$, $h(\lambda) > 0$ for Lebesgue almost every $\lambda \in [-M, M]$ and for some $\epsilon > 0$

$$sup_{\lambda \in \mathbf{R}}\{\lambda^{3+\epsilon}h(\lambda)\} < \infty.$$

The decay condition on h implies that the first two moments of h are finite:

$$\int |\lambda|^k h(\lambda) d\lambda < \infty$$

for $k = 0, 1, 2$. E.g, it holds for the Gaussian distributed coupling constants with

$$h(\lambda) = (\alpha\pi)^{-1/2} e^{-\alpha\lambda^2}.$$

The family H_ω of Schrödinger operators is self-adjoint with probability one and is an ergodic measurable family.

Theorem 6 *The almost sure spectrum $\Sigma(B)$ of H_ω is a closed subset of $\mathbf{R}$ and may have spectral gaps of width at most $\mathcal{O}(B^{-1/2})$.*

Barbaroux, Combes and Hislop (1997b) have shown:

Theorem 7 *(Barbaroux, Combes and Hislop) For the Anderson model with (U1) and (U2) and supp$(h) = \mathbf{R}$, then $\Sigma(B) = \mathbf{R}$ for all $B \ne 0$.*

In terms of localization we have seen that there exist localized states near the band edges for Landau Hamiltonians and to within $\mathcal{O}(B^{-1})$ of the Landau energies in the case of fixed disorder and large magnetic field. For these models there are spectral gaps between the bands of length $\mathcal{O}(B)$. Barbaroux, Combes and Hislop have extended these localization results to unbounded random potentials; they prove localization between the Landau energies and to within $\mathcal{O}(B^{-1})$ of the Landau energies for large magnetic fields. For these models there is typically no spectral gaps as we have noted.

Theorem 8 *(Barbaroux, Combes and Hislop) For the Landau Hamiltonian H_ω satisfying (U1) and (U2) let $I_n(B)$ denote the unbounded set of energies:*

$$I_n(B) = (-\infty, B - \mathcal{O}(B^{-1})] \cup \bigcup_{n=1}^{\infty} [E_j(B) + \mathcal{O}(B^{-1}), E_{j+1}(B) - \mathcal{O}(B^{-1})]$$

where $\mathcal{O}(B^{-1})$ depends on n. For each $n > 0$ there is a $B_n \gg 0$ such for $B > B_n$

$$\Sigma \cap I_N(B)$$

is pure point and the corresponding eigenfunctions decay exponentially.

Theorem 9 *(Barbaroux, Combes and Hislop) For H_ω as above, the IDOS is Lipschitz continuous on $\mathbf{R}\backslash\sigma(H_A)$.*

Thus, the IDOS is Lipschitz continuous at all energies but the Landau energies, which indicates that the localization length may diverge there.

The proof involves extending the Wegner estimate to hold for any $E \notin \sigma(H_A)$ for any field strength $B > 0$ where $supp(g) = \mathbf{R}$. Viz., they have shown:

Theorem 10 *(Wegner Estimate) If (V1) and (V2) hold for any $E_0 \notin \sigma(H_A)$ and for any $\eta > 0$ with*

$$\eta < \frac{1}{2}dist(E_0, \sigma(H_A))$$

and for all $B > 0$, we have

$$\mathbf{P}_\Lambda\left(dist(\sigma(H_A), E_0) < \eta\right) \leq \frac{\eta}{\pi B}c_g|\Lambda|C(B, E_0, \eta)$$

where $\mathbf{P}_\Lambda$ is the probability with respect to the random variables in Λ.

Let V_Λ denote the potential depending only on the q_i in a region Λ. We note that for $H^\Lambda = H_A + V_\Lambda$ on $L^2(\mathbf{R}^2)$, the spectrum $\sigma_{ess}(H^\Lambda) = \sigma_{ess}(H_A)$ since V_Λ has compact support and it is a relatively compact perturbation of H_A.

The absence of continuous spectrum is proven using the Simon-Wolff argument. Multiscale analysis of Combes and Hislop is extended to the case of unbounded potentials. One of the steps of the proof is to verify the initial decay hypothesis on the localized resolvent of the finite area Hamiltonian H^Λ. This is shown using percolation theory to prove the existence of effective magnetic barriers with large probability.

7.9 Localization for Any Magnetic Field with a Band Gap

The restriction of high magnetic field in the result of Combes and Hislop (1996) has been removed in the work of Barbaroux, Combes and Hislop (1997a). They prove exponential localization for a wide class of Landau Hamiltonians at the edges of unperturbed spectral gaps. Consider the random Schrödinger operator of the form

$$H_\omega = H_0 + V_\omega$$

on $L^2(\mathbf{R}^d)$ for $d \geq 1$ where

$$H_0 = (-i\nabla - A)^2 + V_0.$$

Here A is a vector potential and V_0 is a background electrostatic potential. The assumptions for this model are:

(H1) H_0 is essentially self-adjoint on $C_0^\infty(\mathbf{R}^d)$.

(H2) the spectrum of $H_0, \sigma(H_0)$, is semibounded and contains an open gap – i.e. there is a $C_0 \geq 0$ with $-C_0 \leq B_- \leq B_+ \leq \infty$ such that

$$\sigma(H_0) \subset (-C_0, B_-] \cup [B_+, \infty).$$

(H3) H_0 is strongly locally compact

(H4) The operator $H_0(\alpha) = e^{i\alpha\rho} H_0 e^{-i\alpha\rho}$ admits an analytic continuation as a type-A analytic family to the strip

$$S(\alpha_0) = \{x + iy \in \mathbf{C} | |y| < \alpha_0\},$$

where $\rho(x) = (1 + \|x\|^2)^{1/2}$, $\alpha \in \mathbf{R}$ and $\alpha_0 > 0$.

The random perturbation V_ω is assumed to be of Anderson type of the form

$$V_\omega(x) = \sum_{i\in\mathbf{Z}^d} \lambda_i(\omega)u(x - i)$$

where

(H5) the coupling constants $\{\lambda_i(\omega) | i \in \mathbf{Z}^d\}$ form a family of independent, identically distributed random variables with a common distribution having a density h such that $0 \leq h \in L^\infty(\mathbf{R}) \cap C(\mathbf{R})$ and not necessarily finite support, $supp(h) \subset [-m, M], m, M > 0$ and $h(0) > 0$

(H6) h decays sufficiently rapidly near $-m$ and M (v., Barbaroux, Combes and Hislop)

The probability space $(\Omega, \mathbf{P})$ for this model is given by $\Omega = [supp(h)]^{\mathbf{Z}^d}$ and $\mathbf{P}$ is given by the finite product measure.

The single site potential f is assumed to satisfy:

(H7) f has compact support and $0 \leq f \in L^\infty(\mathbf{R}^d)$.

The spectral properties of H_ω are assumed to be:

(H8) H_ω has deterministic spectra Σ, which holds automatically if H_ω is ergodic.

(H9) if $supp(h)$ is bounded there are constants $B'_\pm$ such that $B_- < B'_- < B'_+ < B_+$ such that

$$\Sigma \cap \{(B_-, B'_-) \cup (B'_+, B_+)\} \neq \emptyset.$$

Note that in the unbounded case (H5) and (H7) imply that the deterministic spectrum Σ fills the gap (B_-, B_+) entirely.

Define the perturbed band edges by $\tilde{B}_\pm$ where $B'_- < \tilde{B}_- \le B_-$ and $B'_+ < \tilde{B}_+ \le B_+$ by

$$\tilde{B}_- = sup\{E \in \Sigma | E \le B'_-\}$$
$$\tilde{B}_+ = inf\{E \in \Sigma | E \ge B'_+\}.$$

With these assumptions, we have:

Theorem 11 *(Barbaroux, Combes and Hislop) Assume (H1)-(H9) and suppose supp(h) is bounded. Then there exist constants $E_\pm$ such that $B_- \le E_- < \tilde{B}$ and $\tilde{B}_+ \le E_+ < B_+$ such that $\Sigma \cap (E_-, E_+)$ is pure point with exponentially decaying eigenfunctions.*

If $supp(h)$ is unbounded, one needs to add a coupling constant g and work in the weak disorder regime:

Theorem 12 *(Barbaroux, Combes and Hislop) Let $H_\omega(g) = H_0 + gV_\omega$. Suppose supp(h) is unbounded and assume (H1) - H(8). For any energies $E_\pm$ such that $B_- < E_- < E_+ < B_+$ there is a $g_0 = g_0(E_\pm)$ such that for all $0 < g < g_0$ the region $\Sigma \cap (E_-, E_+)$ is pure point with exponentially decaying eigenfunctions.*

The Wegner estimate in this case is has been shown to have linear dependence in $|\Lambda|$:

Theorem 13 *(Wegner Estimate) Assume (H1)-(H3), (H5) and (H7)-(H8). For any $E_0 \in (B_-, B_+)$ and any $\eta < \frac{1}{2}dist(E_0, \sigma(H_0))$ there is a finite constant C_{E_0} depending on $[dist(E_0, \sigma(H_0))]^{-1}$ such that*

$$\mathbf{P}_\Lambda\{dist(\sigma(H_\omega^\Lambda), E_0 < \eta\} \le C_{E_0}\eta|\Lambda|.$$

In the unperturbed spectral gap the Wegner estimate in this case implies:

Theorem 14 *(Barbaroux, Combes and Hislop) Assume (H1) - (H9) and the support supp(h) is bounded or (H1) - (H8) and supp(h) is unbounded. Then the integrated density of states (IDOS) $N(E)$ is Lipschitz continuous on (B_-, B_+).*

The earlier result of Combes and Hislop (1996) was the case $d = 2$, $V_0 = 0$ and $A = \frac{B}{2}(-x_2, x_1)$ and they treated the case that $supp(h)$ is compact. They showed the existence of localized states away form a region of size $\mathcal{O}(B^{-1})$ centered at the Landau energies $E_n(B)$ for large B.

The main step in the proof of localization in this case is to prove the inital length scale estimate $[H1](\gamma_0, l_0)$ for the local Hamiltonian H_ω^Λ. This is done in two stages. One needs to control the distance of the eigenvalues of H_ω^Λ to the band edges of Σ with good probability. One shows that for $\delta > 0$ and small, then $dist(\sigma(H_\omega^\Lambda), \tilde{B}_\pm) > \delta$ with good probability. Then one uses the Combes-Thomas estimates to deduce exponential decay at energies $E \in (\tilde{B}_- - \delta/2, \tilde{B}_-) \cup (\tilde{B}_+, \tilde{B}_+ + \delta/2)$ with good probability. Using these results one then shows $[H1](\gamma_0, l_0)$ for an appropriate γ_0 and l_0.

The approach of Barbaroux, Combes and Hislop allows them to prove localization at energies between the Landau levels for a fixed nonzero $B > 0$ for the family $H_\omega(g) = H_A + gV_\omega$ for small coupling constant $g > 0$. However, they are not able to address the spectrum to within $\mathcal{O}(B^{-1})$ of the Landau levels in this case.

If $supp(h) = \mathbf{R}$ for this model, then $\Sigma(g)$ fills in the spectral gaps of H_0. In particular

Theorem 15 *(Barbaroux, Combes and Hislop) If H_0 satisfies (H1)-(H3), (H5) and (H7)-(H8) hold, if $\Sigma(g)$ is the a.s spectrum of $H_\omega(g)$, and if $supp(h) = \mathbf{R}$, then $\mathbf{R}\backslash\sigma(H_0) \subset \Sigma(g)$ for $g \neq 0$.*

7.10   Zero Range Model

We continue the review of the nature of the spectrum on the interior of the Landau bands. Dorlas, Macris and Pulé (1999) have shown that for the zero range model the spectrum in the lowest N Landau bands of a random Landau Hamiltonian is entirely pure point when the magnetic field is sufficiently strong and the energies corresponding to the Landau levels are infinitely degenerate. This extends their earlier work Dorlas, Macris and Pulé (1997). Moreover, the eigenfunctions corresponding to the remainder of the spectrum are localized with a uniform bounded localization length as a function of energy, i.e. the localization length does not diverge at the centers of the bands when the magnetic field is strong enough (at least for the lower bands).

The zero range model is given by

$$H = H_0 + \sum_n v_n \delta(r - n)$$

where $H_0 = \frac{1}{2}(-i\nabla - A(r))^2$ and in the symmetric gauge $A(r) = \frac{1}{2}(r \times B)$.

The random potential consists of delta functions of random strengths located on the sites of a regular two dimensional lattice. The strengths of the impurities are independent, identically distributed random variables. Let $\omega_n, n \in \mathbf{Z}[i] = \{n_1 + in_2\}$ denote the Gaussian integers where ω_n are independent, identically distributed random variables. Assume their distribution is given by an absolutely continuous probability measure μ_0 with support $supp(\mu_0) = [-a, a]$ for $0 < a < \infty$. Assume μ_0 is symmetric about the origin and its density ρ_0 is differentiable and satisfies:

$$sup_{x \in (0,a)} \frac{\rho_0'(x)}{\rho_0(x)} < \infty.$$

Define the probability space $(\Omega, \mathbf{P})$ by $\Omega = supp(\mu_0)^{\mathbf{Z}[i]}$ and $\mathbf{P} = \prod_{n \in \mathbf{Z}[i]} \mu_0$.

Theorem 16 *For $m \in \mathbf{Z}[i]$, let τ_m be the measure preserving automorphism of Ω given by $(\tau_m \omega)_n = \omega_{n-m}$. Then the group $\{\tau_m | m \in \mathbf{Z}[i]\}$ is ergodic for the probability measure $\mathbf{P}$.*

We redefine the Hamiltonian to be

$$H_0 = \frac{1}{8\kappa}(-i\nabla - A(z))^2 - \frac{1}{2}$$

where $A(z) = (-2\kappa Im(z), 2\kappa Re(z))$ and $\kappa = B/4$. With this shift the Landau levels are now $\mathbf{N_0}$, the nonnegative integers. Let $\{U_z, z \in \mathbf{C}\}$ to be the family of unitary operators on $\mathcal{H} = L^2(\mathbf{C})$ corresponding to the magnetic translations:

$$(U_z f)(z') = e^{2i\kappa z \wedge z'} f(z + z').$$

Then

$$U_{z_1} U_{z_2} = e^{2i\kappa z_2 \wedge z_1} U_{z_1 + z_2}.$$

And for $n \in \mathbf{Z}[i]$ one has

$$U_n R_\omega(\lambda) U_n^{-1} = R_{\tau_n \omega}(\lambda)$$

where $R_\omega(\lambda)$ is the resolvent

$$R_\omega(\lambda) = (H - \lambda)^{-1}$$

and we set

$$R_0(\lambda) = (H_0 - \lambda)^{-1}$$

for $\lambda \in \mathbf{C} \backslash \mathbf{N_0}$.

Theorem 17 *The ergodicity of τ implies that the operator H is deterministic.*

Theorem 18 *(Dorlas, Macris and Pulé) The spectrum of the zero range model H contains bands about the Landau levels $\mathbf{N_0}$ and an interval from $-\infty$ to a finite negative point. For each $N \in \mathbf{N}$ there is a magnetic field strength B_0 such that for $B > B_0$ with probability one:*

- $\sigma_{cont}(H) \cap (-\infty, N) = \emptyset$.

- *If $m \in \mathbf{N_0} \cap (-\infty, N)$, then m is an eigenvalue of H with infinite multiplicity, i.e. the Landau levels are still infinitely degenerate eigenvalues.*

- *If $\lambda \in \sigma(H) \cap (-\infty, N) \backslash \mathbf{N_0}$ is an eigenvalue of H with eigenfunction ϕ_λ, then for any compact subset B of $\mathbf{C}$*

$$\int_B |\phi_\lambda(z - z')|^2 dz'$$

decays exponentially in z with exponential length less than $2/\kappa$.

The proof of localization involves reducing the problem to a lattice problem and applying the method of Aizenman and Molchanov. Viz., let $\mathcal{M} = l^2(\mathbf{Z}[i])$ and for $\lambda \in \mathbf{C} \backslash \mathbf{N_0}$ define

$$U_\lambda : \mathcal{H} \to \mathcal{M}$$

by $< n|U_\lambda \phi >= (R_0(\lambda)\phi)(n)$. Then U_λ is a bounded operator. Define the function

$$c_n^\lambda = \frac{2\kappa}{\pi} (\psi(-\lambda) - \frac{2\pi}{\omega_n})$$

where $\psi(a) = \Gamma'(a)/\Gamma(a)$ is the digamma function; and define the operators D^λ, A^λ and M^λ on $\mathcal{M}$ as follows: D^λ is diagonal with

$$< n|D^\lambda|n > = c_n^\lambda;$$

$$< n|A^\lambda|n' > = \begin{cases} 0, & \text{if } n = n' \\ R_0(\lambda)(n, n') & \text{if } n \neq n'; \end{cases}$$

and

$$M^\lambda = D^\lambda - A^\lambda.$$

D^λ is closed on the domain

$$\mathcal{D}(D^\lambda) = \{\xi \in \mathcal{M}| \sum_{n \in \mathbf{Z}[i]} |c_n^\lambda|^2| < n|\xi > |^2 < \infty\}.$$

A^λ is bounded so M^λ is closed on $\mathcal{D}(D^\lambda)$. Once checks that for $\lambda \in \mathbf{R}$ the operator M^λ is self-adjoint. We set $\Gamma^\lambda = (M^\lambda)^{-1}$ for $\lambda \in \mathbf{C}\backslash\mathbf{N}_0$ and $0 \notin \sigma(M^\lambda)$.

Then the generalized eigenfunction ϕ of H with eigenvalue λ which is not a Landau level is related to an eigenvector $v = \Gamma^{\lambda_\kappa}U_{\lambda_\kappa}\phi$ of M^λ with eigenvalue zero; and if v decays exponentially then so does the eigenfunction. Here $\phi = (\lambda - \lambda_\kappa)U_\lambda^* v$.

The proof uses the Simon-Wolff technique, viz. if for all $E \in (-1, 1)$ and almost every ω

$$lim_{\epsilon \to 0} \sum_{n \in \mathbf{Z}[i]} | < n|\Gamma^\lambda(E + i\epsilon)|0 > |^2 < \infty$$

then $\sigma_{cont}(M^\lambda) \cap (-1, 1) = \emptyset$ for almost every ω; and if furthermore for almost every pair (ω, E), $\omega \in \Omega$ and $E \in (-1, 1)$

$$lim_{\epsilon \to 0}| < n|\Gamma^\lambda(E + i\epsilon)|0 > | < C_{\omega, E}e^{-m(E)|n|},$$

then with probability one the eigenvectors v_E^λ of M^λ with eigenvalue $E \in (-1, 1)$ are such that

$$| < v_E^\lambda|n > | < D_{\omega, E}e^{-m(E)|n|}.$$

In particular, they show that if λ is a generalized eigenvalue of H with eigenfunction ϕ_λ and v_λ is the corresponding eigenvector of M^λ with eigenvalue 0, then

$$| < v_\lambda|n > | < D_\omega e^{-2\gamma\kappa|n|/s}.$$

It follows that $\sigma_{ac}(H) \cap (-\infty, N) = \emptyset$. To extend this to $\sigma_{cont}(H)$, they use the approach of Delyon, Lévy and Souillard (1985) and Kotani(1986). For the details, see Dorlas, Macris and Pulé.

Dorlas, Macris and Pulé have shown that if the magnetic length is smaller than the average spacing between impurities, then there is no localization length divergence at least for the first few Landau bands. Does this mean there is no quantum Hall effect in this regime? Dorlas, Macris and Pulé argue that due to the fact that the energies at the band centers are infinitely degenerate eigenvalues, the Chern number is still equal to unity and the arguments of Kunz (1987) and Bellessard et al. (1994) continue to imply that the Hall conductance takes on non-zero quantized value equal to the number of Landau levels below the Fermi energy. One notes that there is another regime where the magnetic length is greater than the spacings between impurities and the localization length diverges.

7.11 Randomly Located Impurities

Scrowston (1999) has generalized the model of Dorlas, Macris and Pulé to consider random impurities with both random positions and random strengths. The model is similar to that of Dorlas, Macris and Pulé with the assumption that the density ρ is bounded by a constant ρ_0. The Hamiltonian is

$$H_\omega(\zeta) = H_0 + V$$

where

$$V(z, (\omega, \zeta)) = \sum_{n \in \mathbf{Z}[i]} \omega_n \delta(z - \zeta_n).$$

Let $H_n = B(n + 1/2)P_n + P_n V P_n$ denote the Hamiltonian restricted to the nth level, where P_n is the projection onto this level. Scrowston treats the case $n = 0$. Specifically, set

$$H_\omega(\zeta) = \frac{\pi}{2\kappa} P_0 V(\cdot, (\omega, \zeta)) P_0$$

where P_0 is the orthogonal projection onto $\mathcal{H}_0$, the eigenspace corresponding to thte first Landau level of the kinetic part of the Hamiltonian. P_0 is an integral operator with the kernel:

$$P_0(z, z') = \frac{2\kappa}{\pi} exp(-\kappa|z - z'|^2 - 2i\kappa z \wedge z').$$

One can show that if $\psi \in \mathcal{H}$ then $\psi \in \mathcal{H}_0$ if and only if $\psi(z) = f(z)e^{-\kappa|z|^2}$ where $f(z)$ is entire. In fact the Hamiltonian can be written as

$$H = \sum \omega_n f_{\zeta_n} \otimes \bar{f}_{\zeta_n}$$

where $f_\zeta(z) = \sqrt{\frac{\pi}{2\kappa}} P_0(z, \zeta)$.

Let $B_n = \{z \in \mathbf{R}^2 | n_i - 1/2 \leq z_i < n_i + 1/2, n \in \mathbf{Z}[i], i = 1, 2\}$ denote the unit square. Let $\zeta_n = n + \tilde{\zeta}_n$ for $n \in \mathbf{Z}[i]$, represent the positions of the impurities in the complex plane where $\tilde{\zeta}_n$ are independent, identically distributed random variables. Assume the probability measure μ has support in B_0 with a density r bounded by a constant r_0. Let $\Omega_2 = \prod_{n \in \mathbf{Z}[i]} B_0$ and $\mathbf{P}_2 = \prod_{n \in \mathbf{Z}[i]} \mu$. By combining the space from Dorlas, Macris and Pulé we have the probability space for this model $(\Omega \times \Omega_2)$ and $(\mathbf{P} \times \mathbf{P}_2)$.

The operator τ_m is defined by $(\tau_m(\omega, \zeta))_n = (\omega_{n-m}, \zeta_{n-m})$. Again this operator is ergodic for the probability measure and

$$U_m H_\omega(\zeta) U_m^{-1} = H(\tau_m(\omega, \zeta)).$$

Theorem 19 *The spectrum of $H_\omega(\zeta)$ and its components are deterministic. With probability one $[4a, 4b] \subset \sigma(H_\omega(\zeta))$.*

Theorem 20 *(Scrowston) When the magnetic field is sufficiently high, then with probability one*

- 0 *is an eigenvalue of* H *with infinite multiplicity*

- $\sigma_{cont}(H) = \emptyset$

- *if* $\lambda \in \sigma(H)\backslash\{0\}$ *is an eigenvalue of* H *with eigenfunction* ϕ_λ, *then* ϕ_λ *decays exponentially with rate greater than or equal to* $(\kappa)^{1/4}$.

Scrowton's proof is a modification of the multiscale analysis approach. He uses Kotani's trick to show exponential decay and pure point spectrum with probability one. Viz., it follows that if the eigenfuctions decay exponentially, then $\sigma_{ac} = \emptyset$. To rule out the existence of singular continuous spectrum one needs to apply the methods of Delyon, Kunz and Souillard (1985) and Kotani (1986).

As with Dorlas, Macris and Pulé we define M_0 and V_ω^λ on $\mathcal{M} = l^2(\mathbf{Z}[i])$ by

$$< m|M_0|n > = \frac{\pi}{2\kappa} P_0(\zeta_m, \zeta_n)(1 - \delta_{mn})$$

and

$$< m|V_\omega^\lambda|n > = (1 - \frac{\lambda}{\omega_n})\delta_{mn}$$

where $\omega_n \psi(\zeta_n) = \xi_n$. Set

$$M^\lambda = M_0 + V_\omega^\lambda.$$

As above, the eigenvectors ξ_n of M^λ are related to eigenfunctions of H. Viz., for $\lambda \neq 0$

$$\psi(z) = \frac{\pi}{2\kappa\lambda} \sum_{n \in \mathbf{Z}[i]} P_0(z, \zeta_n)\xi_n.$$

And if ξ_n decays exponentially, $|\xi_n| \leq Ce^{-m|\zeta_n|}$, then $|\psi(z)| \leq Ce^{-m|z|}$.

For $\Lambda \subset \mathbf{Z}[i]^2$, let M_Λ^λ denote the restriction of M^λ to $l^2(\Lambda)$. If $E \notin \sigma(M_\Lambda^\lambda)$ set

$$\Gamma_\Lambda^\lambda(E) = (M_\Lambda^\lambda - E)^{-1}.$$

The multiscale approach in this case is as follows. Let

$$\Lambda_L(n) = \{n' \in \mathbf{Z}[i] \,|\, |n' - n|_\infty < L/2\}$$

for $n \in \mathbf{Z}[i]$ and $L > 0$. For $\beta \in (0, 1)$ and $s \in (1/2, 1)$ and define the conditions:

(RA) $dist(E, \sigma(M_{\Lambda_L(n)}^\lambda(\omega, \zeta))) > \frac{1}{2}e^{-L^\beta}$

(RB) $| < n|\Gamma_{\Lambda_L(n)}^\lambda(E)|n' > | \leq e^{-mL}$ for all $n' \in \tilde{\Lambda}_L(n) = \Lambda_L(n)\backslash\Lambda_{\tilde{L}}(n)$ where $\tilde{L} = L - L^s$.

A region $\Lambda_L(n)$ is said to be (m, E)-regular for some $m > 0$ and $E \in \mathbf{R}$ if (RA) and (RB) hold.

For $Q_0 > 0$ define the two conditions:

(MS1) there is a $L_0 > Q_0$ and m_0 such that

$$\mathbf{P}\{\Lambda_{L_0}(0) \text{ is } (m_0, 0)\text{-regular}\} \geq 1 - L_0^{-p}$$

(MS2) there is an $\eta > 0$ such that for all $E \in (-\eta, \eta)$ and for all $L > L_0$

$$\mathbf{P}\{dist(E, \sigma(M_{\Lambda_L(0)}^\lambda)) < e^{-L^\beta}\} < L^{-q}.$$

The multiscale analysis theorem is then:

Theorem 21 *(Multiscale Analysis) For β and s as above and $\gamma \in (0,1), p > 2, q > 4p + 12$, there is a $Q_0 > 0$ which is independent of λ and $\kappa > 1$ such that if for λ and κ the conditions (MS1) and (MS2) hold, then for all $m \in (0, m_0)$ there is a $\delta > 0$ such that for all $E \in (-\delta, \delta)$ the eigenvectors of M^λ with eigenvalue E decay exponentially with rate greater than or equal to m.*

Thus, for his theorem Scrowston needs only to prove that (MS1) and (MS2) hold for this model.

Klopp (1993) has also studied the random displacement model for the semi-classical Hamiltonian

$$H(h) = -h^2 \Delta + V_\omega$$

where

$$V_\omega(x) = \sum_{i \in \mathbf{Z}^d} u(x - i - \xi_i(\omega));$$

here the random variables $\{\xi_i(\omega)\}$ are distributed in a ball of radius $R < 1/2$. Klopp has shown that in the case $u \leq 0$, i.e. a potential well, this model exhibits localization at negative energies provided that h is sufficiently small.

7.12 Edge States

In the earlier sections we have seen that for an infinite system with impurities the Landau Hamiltonian will have entirely dense pure point spectrum. And the results of Kunz and Bellissard and co-workers show that localized states are necessary for the existence of the quantum Hall effect. Then what carries the Hall current? Halperin (1982) proposed the idea that the Hall current is supported by edge states for infinite systems with boundary. Viz., Halperin has argued that for a two dimensional system with boundaries in a perpendicular magnetic field, there should be quasi-one dimensional edge states extended along the boundaries. Thus, in this model we need to exclude the point spectrum over certain regions. Macris, Martin and Pulé (1998) have shown that for a semi-infinite system where the electron is confined to a half space by a planar wall given by a smooth increasing potential, the Hamiltonian has a continuous spectrum in some intervals in between the Landau levels provided that both the amplitude and spatial variation of the random potential are sufficiently weak.

Consider the wall potential

$$U(x) = \begin{cases} 0 & \text{for } x \leq 0 \\ \mu x^\gamma & \text{for } x \geq 0 \end{cases}$$

where $\mu > 0$ and $\gamma \geq 1$. Assume there is a bounded differentiable external potential $w(r)$ such that

(W1) $\sup_r |w(r)| = w_0 < \infty$

(W2) $\sup_r |\partial_x w(r)| = w_0' < \infty$

The Hamiltonian for this model is

$$H_\omega = H_0 + w(r)$$

where $H_0 = H_A + U(x)$ and H_A is the Landau Hamiltonian in the Landau gauge. For $w(r)$ take a typical realization of a random potential

$$V_\omega(r) = \sum_{n \in \mathbf{Z}^2} \omega_n v(x - n)$$

where $v(r) = 0$ for $|r| > 1/2$ and ω_n are independent, identically distributed random variables with continuous density supported on $[-1, 1]$.

These Hamiltonians are essentially self-adjoint on $C_0^\infty(\mathbf{R}^2)$.

Theorem 22 *(Macris, Martin and Pulé) The spectrum of the wall Hamiltonian H_0 is absolutely continuous and consists of the set $\sigma(H_0) = [\frac{B}{2}, \infty)$.*

Macris, Martin and Pulé show that if w_0 and w_0' are small enough (depending on B and the steepness of the wall), then H cannot have point spectrum in the intervals

$$\Delta_n(B, \delta) =](n + 1)B - \delta, (n + 1)B + \delta[$$

of size $2\delta > B - w_0$ in between the Landau levels and the intervals $\Delta_n(B, \delta)$ are in the spectrum of H.

Theorem 23 *(Macris, Martin and Pulé) The spectrum of H_ω contains the interval $[B/2, \infty)$ with probability one.*

Theorem 24 *(Macris, Martin and Pulé) If $\frac{B}{2} - w_0 > \delta$ for some $\delta > 0$, and w_0' is sufficiently small, then H has no eigenvalues (i.e. no point spectrum) in the intervals $\Delta_n(B, \delta)$ of size $2\delta > B - w_0$ in between the Landau levels and the whole interval $\Delta_n(B, \delta)$ is included in the spectrum of H. Thus, the spectrum of H on $\Delta_n(B, \delta)$ is purely continuous.*

Macris, Martin and Pulé show that their bounds in fact guarantee the existence of continuous spectrum only in a finite number of intervals, although they expect to find continuous spectrum in between all Landau levels. In any case for a large class of wall potentials Macris, Martin and Pulé have shown that there are intervals of order $B - 2w_0$ centered in between Landau levels where the spectrum is continuous.

7.13   Extended Edge States

Fröhlich, Graf and Walcher (1998) have studied edge states for finite and infinite samples. In their work the extended nature of the edge states is studied using the "guiding center" coordinate of the cyclotron motion. Fröhlich, Graf and Walcher (1998) have proven similar theorems to Macris, Martin and Pulé. Again they consider the half-plane model. However, they use the Mourre theory of positive commutators to prove absolute continuity of the energy spectrum in between the Landau levels.

7.13.1 Mourre Theorem

Let $E_\Delta(H)$ denote the spectral projector of the Hamiltonian H on some energy interval Δ. Assume there is a conjugate operator A such that the commutator of A with H is positive on Δ, i.e.

$$E_\Delta(H)[H, iA]E_\Delta(H) \geq \alpha E_\Delta(H)$$

for some $\alpha > 0$. Clearly, if ψ is an eigenvalue of H, then $(\psi, [H, iA]\psi) = 0$ and the Mourre condition implies that H cannot have an eigenvalue in Δ. Mourre showed that under regularity conditions on H and $[H, iA]$ and $[[H, iA], iA]$, then the spectrum of H is actually purely absolutely continuous on the interval Δ; v., Mourre (1981), Cycon et al. (1987). More precisely,

Theorem 25 *(Mourre) Let H denote a self-adjoint operator acting on the Hilbert space $\mathcal{H}$. Let A be a self-adjoint operator with $\mathcal{D}(A) \cap \mathcal{H}$. Let*

$$ad_A^n(H) = [...[H, A], A], ...A]$$

denote the n-fold commutator. Let $g_\Delta(\lambda)$ denote a smoothed characteristic function of Δ. Assume the operators $g_\Delta(H)ad_A^n(H)g_\Delta(H)$ for $1 \leq n \leq N$ for $N = 2$ can all be extended to a bounded operator on $\mathcal{H}$; and assume $g_\Delta(H)i[H, A]g_\Delta(H) \geq \theta g_\Delta(H)^2 + K$ for some $\theta > 0$ and compact operator K. Then in the interval Δ, H can only have absolutely continuous spectrum with finitely many eigenvalues of finite multiplicity. If $K = 0$, then there are no eigenvalues in the interval Δ.

7.13.2 Fröhlich, Graf and Walcher Model

We again consider the half-plane model infinite in the x-direction with a wall confining the electron gas to a region $y < 0$. The magnetic field B is perpendicular to the plane. The Hamiltonian is given by

$$H = H_0 + V_0 + V_d$$

with $A = (-By, 0)$. Here V_0 is the smooth edge potential which vanishes for $y < 0$ and V_d is the disorder potential. Let $V = V_0 + V_d$.

For the conjugate operator Fröhlich, Graf and Walcher use the guiding center coordinate of the cyclotron:

$$\Pi = BZ_x = p_y + Bx.$$

One checks that in this case

$$[H, i\Pi] = -\partial_y V.$$

The assumptions on V_0 are for a smooth but steep edge potential:
    (F1) V_0 vanishes for $y < 0$;
    (F2) $V_0'(y) \geq 0$ for all y;
    (F3) $inf\{V_0'(y), y \geq b\} > 0$ for all $b > 0$.

Theorem 26 *(Fröhlich, Graf and Walcher) If H is as above and satisfies the conditions (F1) - (F3) and if $E \notin \sigma(H_0) = \{(2n+1)B\}$, then there is a δ such that if the disorder is sufficiently small $|V_d| < \delta$, there is an open interval Δ containing E and a positive constant α such that*

$$-E_\Delta(H)[H, i\Pi]E_\Delta(H) \geq \alpha E_\Delta(H).$$

And based on Mourre's theorem the spectrum of H is absolutely continuous on Δ.

Fröhlich, Graf and Walcher also consider the case of an infinitely steep edge by using Dirichlet boundary conditions. Again for this case they show for weak disorder the edge states are extended states, i.e. they correspond to absolutely continuous spectrum.

7.14   Propagation Along the Edge States

De Bièvre and Pulé (1998) have also examined the half-plane model ($x > 0$) in a perpendicular magnetic field with a disorder potential. They show that there exists a spectral interval of size B between the Landau levels in which there are no bound states and in which the extended states propagate with speed of order $\sqrt{B}$ in the y-direction. As with Fröhlich, Graf and Walcher, de Bièvre and Pulé use Mourre's estimate but for the conjugate operator they use the y−coordinate.

The model Hamiltonian in this case is

$$H = H + W_B$$

where

$$H_0 = \frac{1}{2}p_x^2 + \frac{1}{2}(p_y - Bx)^2$$

with Dirichlet boundary conditions at $x = 0$ and $W_B \in L^\infty(\mathbf{R}_+ \times \mathbf{R}, dxdy)$ and $\|W\|_\infty \leq AB$ for some constant A.

One notes that by translation invariance in the y−direction H_0 has the form

$$H_0 = \int_{\mathbf{R}} dk H(k)$$

where

$$H(k) = -\frac{1}{2}\frac{d^2}{dx^2} + \frac{1}{2}(k - Bx)^2$$

for $x > 0$ acting on $L^2(\mathbf{R}_+ \times \mathbf{R}, dxdk)$.

The spectrum of $H(k)$ consists of nondegenerate eigenvalues $E(k), n \in \mathbf{N}_0$ with normalized eigenfunctions $\phi_n(x, k)$. The nth band space $\mathcal{H}_n$ is defined to be the space consisting of functions of the form $f(k)\phi_n(x, k)$ where $f \in L^2(\mathbf{R}, dk)$. It is an H_0 invariant subspace of $L^2(\mathbf{R}_+ \times \mathbf{R}, dxdk)$.

One notes that the velocity in the y−direction is given by

$$V_y = i[H_0, Y]$$

where Y is the operator given by multiplication by y.

7.14.1 Edge Spaces and Bulk Spaces

The spectrum of H_0 is absolutely continuous and fills the entire half line $[1/2, \infty)$. Define within $\mathcal{H}_n$ the edge spaces and bulk spaces as

$$\mathcal{H}_{n,e}(\sigma, \gamma) = L^2((-\infty, \sigma B^\gamma], dk) \subset \mathcal{H}_n$$

and

$$\mathcal{H}_{n,b}(\sigma, \gamma) = L^2([\sigma B^\gamma, \infty), dk) \subset \mathcal{H}_n$$

so $\mathcal{H}_n = \mathcal{H}_{n,e} \oplus \mathcal{H}_{n,b}$. $\mathcal{H}_{n,e}$ is called an edge space for all $\gamma \leq 1/2$ and $\mathcal{H}_{n,b}$ is called a bulk space for all $\gamma > 1/2$. These spaces are H_0 invariant.

Theorem 27 *(de Bièvre and Pulé) The spectrum of H_0 restricted to $\mathcal{H}_{n,e}$ has the form*

$$[B(n + 1/2 + c_\sigma), \infty).$$

Theorem 28 *(de Bièvre and Pulé) If $k_0 \in (-\infty, k_B)$ where k_B is of order $\sqrt{B}$, then the wave packet $f(k)\phi_n(x, k)$ belongs to the edge space $\mathcal{H}_{n,e}(\sigma, 1/2)$ and the wave packet speeds along the edge in the y-direction with velocity of order $\sqrt{B}$, where the group speed in the y-direction is given by $-\partial_k E_n(k_0)$ by the Feynman-Hellman theorem. The wave packet is exponentially small for x greater than $1/\sqrt{B}$. If $k_0 \in [k_B, \infty)$ with k_B of order B^γ with $\gamma > 1/2$, then the group velocity is exponentially small in B and the wave packet is exponentially small within the region $0 \leq x \leq 1/\sqrt{B}$ (i.e. close to the edge).*

7.14.2 Disorder Potentials

When the disorder potential is added, de Bièvre and Pulé are able to show that the spectrum of the Hamiltonian is purely absolutely continuous in a spectral interval of size γB between the Landau levels.

Theorem 29 *(de Bièvre and Pulé) Let $n \in \mathbf{N}$ be fixed and let λ, λ' be such that $\lambda + \lambda' < 1$. Then there exists a $\delta > 0$ such that if W_B is sufficiently weak, viz. $\|W\|_B < \delta(n, \lambda, \lambda')B$ and $\delta(n, \lambda, \lambda') > \epsilon$, then*

$$\sigma_{sing}(H) \cap (B(n + \frac{1}{2} + \lambda), B(n + \frac{3}{2} - \lambda')) = \emptyset.$$

The proof follows from an application of Mourre's theorem examining

$$E_\Delta(H)i[Y, H]E_\Delta(H).$$

It also follows from de Bièvre and Pulé's work that for δ sufficiently small there exists an interval of absolutely continuous spectrum between the Landau levels for this model.

7.15   Edge Currents without Edges

Halperin's (1982) discussion of the impact of boundaries for systems in the presence of a magnetic field has been mentioned earlier as well as the related problem of edge states for the half plane. What about edge currents without edges? Exner, Joye and Kovarik (1999) have considered the model of a periodic array of point scatterers in a perpendicular magnetic field. For this model they show that the Landau levels remain infinitely degenerate and between them the system has bands of absolutely continuous spectrum. The model Hamiltonian is given by

$$H_{\alpha,l} = (-i\partial_x + By)^2 - \partial_y^2 + \sum_j \tilde{\alpha}\delta(x - x_0 - jl)$$

where $l > 0$ is the array spacing. The boundary conditions are

$$L_1(\psi, \mathbf{a}_j) + 2\pi\alpha L_0(\psi, \mathbf{a}_j) = 0$$

for $j = 0, \pm 1, \pm 2, \ldots$ and $\mathbf{a}_j = (x_0 + jl, 0)$. The Bloch decomposition in the x-direction states that

$$H_{\alpha,l} = \frac{l}{2\pi} \int_{|\theta l| < \pi}^{\oplus} H_{\alpha,l}(\theta)d\theta.$$

$H_{\alpha,l}(\theta)$ has the same form as above on the strip $0 \le x \le l$, with the boundary condition $\partial_x^i \psi(l-, y) = e^{i\theta l}\partial_x^i \psi(0+, y)$ for $i = 0, 1$. The Green's function is given by Krein's formula in terms of the free Green's function G_0 and $\xi(\mathbf{a}, \theta, z)$.

The Bloch condition determines the eigenvalues and eigenfunctions of the transverse part of the free operator

$$\mu_m(\theta) = (\frac{2\pi m}{l} + \theta)^2$$

and

$$\eta_m^\theta(x) = \frac{1}{\sqrt{l}}e^{i(2\pi m + \theta l)x/l}.$$

Let $z_n = B(2n + 1), n = 0, 1, 2, \ldots$ denote the Landau levels. $H_{\alpha,l}(\theta)$ has eigenvalues away from z_n which are denoted by $\epsilon_n(\theta) = \epsilon_n^{(\alpha,l)}(\theta)$. These are given by the implicit equation

$$\alpha = \xi(\mathbf{a}_0, \theta, \epsilon) \qquad (*)$$

with eigenfunctions

$$\psi_n^{(\alpha,l)}(\mathbf{x}, \theta) = G_0(\mathbf{x}, \mathbf{a}_0, \theta, \epsilon_n(\theta)).$$

As noted by Dorlas, Macris and Pulé (1999), each z_n remains infinitely degenerate eigenvalues of $H_{\alpha,l}(\theta)$. Spectral bands are given by the ranges of $\epsilon_n(.)$ The solutions of $(*)$ do not cross the Landau levels because $\xi(\mathbf{a}_0, \theta, .)$ is increasing in the intervals $(-\infty, B)$ and $(B(2n-1), B(2n+1))$ and diverges at the endpoints. The spectrum is continuous away from z_n if $\xi(\mathbf{x}, \theta, z)$ is nowhere constant as a function of θ, which Exner and co-workers demonstrate. In summary we have:

Theorem 30 *(Exner, Joye, Kovarik) For any real α the spectrum of $H_{\alpha,l}$ consists of Landau levels $B(2n+1)$ and absolutely continuous spectral bands situated between adjacent Landau levels and below B.*

Exner and coworkers show that probability current, $j_n(x, \theta)$, is predominantly laminar although vortices may form in the low spectral bands.

As noted in Chapter 3, Albeverio, Geyler and Kostrov (1999) have considered the case of a chain of point scatterers in a three dimensional space with a homogeneous magnetic field. They show that the spectrum is purely absolutely continuous and has at most finitely many gaps. See also Brüning and Geyler (1998).

7.16 Quantum Hall Effect – Kunz's Theorem

The integer quantum Hall effect (IQHE) states that the Hall conductivity $\sigma_{xy} = e^2 n/h$ for $n = 0, 1, 2, \ldots$ and the parallel conductivity $\sigma_{yy} = 0$ when the filling factor stays close to an integer; i.e. there is quantization of the Hall conductivity and there are plateaus. As noted above, the IQHE requires: (1) a magnetic field and the resulting Landau band structure; (2) a random potential which results in edge localization; (3) two dimensional motion. With these ingredients Kunz (1987) was able to show that the two major properties of the IQHE cited above hold. We review his results in this section.

Consider a system composed of an infinite number of independent electrons moving in the (x, y)-plane under the influence of a perpendicular magnetic field. Assume that there is a random potential V_ω due to impurities. In canonical units the one electron Hamiltonian is

$$H = \frac{1}{2}v_x^2 + \frac{1}{2}v_y^2 + V_\omega(x, y)$$

where $v_x = \frac{1}{i}\partial_x + y$ and $v_y = \frac{1}{i}\partial_y$. The probability distribution associated with V_ω is assumed to be translation invariant and V_ω is assumed to be bounded

$$\|V_\omega\| = sup_{x,y}|V(x,y)| < \infty.$$

7.16.1 Kubo's Formula

The Hall conductivities are given by the Kubo formulae:

$$\sigma_{x,y} = lim_{\epsilon \to 0}\frac{1}{i\epsilon}\int_{-\infty}^{\infty} e^{-\epsilon t}M(f[v_x, v_y(-t)]f)dt$$

and

$$\sigma_{y,y} = lim_{\epsilon \to 0}\frac{1}{i\epsilon}\int_{-\infty}^{\infty} e^{-\epsilon t}M(f[v_y, v_y(-t)]f)dt + \frac{1}{\epsilon}M(f)$$

in units of $e^2/\hbar$. Here f is the Fermi projector

$$f(H) = \begin{cases} 1 & \text{if } H < \mu \\ 0 & \text{if } H \geq \mu \end{cases}$$

where μ is the chemical potential. That is, $f = E_H(-\infty, \mu)$. Here M denotes the operator which averages over disorder.

7.16.2 Kunz' Theorem

In the natural units, the eigenvalues of H_0 are $\{n + 1/2, n = 0, 1, ...\}$ and as shown by Thouless (1981) if the disorder is small enough $M_0 = \|V_\omega\| < 1/2$, then the spectrum of H_ω is contained in the intervals $A_n = [n + 1/2 - M_0, n + 1/2 + M_0]$, as we have described.

Define
$$A_n^+ = [n + 1/2 - M_0 - \delta_n, n + 1/2 + M_0]$$
and
$$A_n^- = [n + 1/2 - M_0, n + 1/2 - M_0 + \delta_n]$$

where $0 < \delta_n < M_0$. Kunz makes the following assumptions:

(K1) the spectrum of H_ω in $A^\pm$ is pure point for all $n \geq 0$ and is simple;

(K2) the localization length is finite.

Theorem 31 *(Kunz) Assume (K1) and (K2). Then for each integer $n \geq 0$ there exist densities*
$$\rho_{-,n} = \rho(n + 1/2 + M_0 - \delta_n)$$
and
$$\rho_{+,n} = \rho(n + 3/2 - M_0 - \delta_n)$$
such that if the electron density belongs to the interval
$$\rho \in [\rho_{-,n}, \rho_{+,n}]$$
then
$$\sigma_{x,y} = \frac{1}{2\pi}(n + 1)$$
and
$$\sigma_{y,y} = 0$$
where n is the Landau band index.

In addition Kunz showed that the Hall conductivity is a topological invariant and there is at least one energy in each Landau band with infinite localization length. This is consistent with Halperin's argument that the localization length must diverge in each Landau band.

Kunz' method of proof is to define a periodic model which approximates the quantum Hall problem. Viz., let $\Lambda = \Lambda_l(0)$ be a square of side length l centered at the origin. Let V^l be a periodic potential equal to V in Λ with $V^l(x + l, y) = V^l(x, y + l) = V(x, y)$. Set $H^l = H_0 + V^l$. The Hall conductivity is given by

$$\sigma_{x,y}^l = \frac{1}{i}M(f^l[f_y^l, f_x^l])$$

where
$$f^l = \oint_\Gamma \frac{dz}{2\pi i}G_z^l.$$

Here $G_z = (z - H)^{-1}$ and $G_z^l = (z - H^l)^{-1}$ are the Green's function. Γ is a contour in the complex plane crossing the real axis at $\pm\mu$ and encircling the points $\pm i\epsilon$.

Theorem 32 *(Kunz) The Hall conductivities can be expressed as*

$$\sigma_{x,y} = -2Im\, M(fv_y \oint_\Gamma \frac{dz}{2\pi i} G_z v_y G_z^2 f)$$

and

$$\sigma_{y,y} = -2Im\, M(fv_x \oint_\Gamma \frac{dz}{2\pi i} G_z v_y G_z^2 f))$$

Based on the cyclic properties of M, this can be more simply expressed as:

Theorem 33 *(Kunz)*

$$\sigma_{x,y} = \frac{1}{i} M(f[f_y, f_x])$$

and

$$\sigma_{y,y} = 0.$$

In terms of Λ, we have:

Theorem 34 *(Kunz)*

$$\sigma_{x,y}^l = \frac{1}{i} M(f^l[f_y^l, f_x^l])$$

where

$$f^l = \oint \frac{dz}{2\pi i} G_z^l$$

$$f_x^l = \oint \frac{dz}{2\pi i} G_z^l (v_x + k_x) G_z^l = \partial_{k_x} H^l(k)$$

and

$$f_y^l = \oint \frac{dz}{2\pi i} G_z^l (v_y + k_y) G_z^l = \partial_{k_y} H^l(k).$$

Theorem 35 *(Kunz)*

$$\sigma_{x,y} = lim_{l\to\infty} \sigma_{x,y}^l.$$

Define the operator $\hat{H}^l(k) = \frac{1}{2}(\frac{1}{i}\partial_x + y)^2 + \frac{1}{2}(\frac{1}{i}\partial_y)^2 + V^l(x,y)$ on $L^2(\Lambda)$ with boundary conditions

$$\phi(l,y) = e^{ik_x l}\phi(0,y), \phi(x,l) = e^{ik_y l - ilx}\phi(x,0)$$

$$\phi_x(x,l) + il\phi(x,l) = e^{ik_y l - ilx}\phi_x(x,0), \phi_y(x,l) = e^{ik_y l - ilx}\phi_y(x,0)$$

and

$$\phi_x(l,y) = e^{ik_z l}\phi_x(0,y), \phi_y(l,y) = e^{ik_y l}\phi_y(0,y).$$

Define the projector

$$P^l(k) = \oint_\Gamma \frac{dz}{2\pi i}(z - \hat{H}^l(k))^{-1}.$$

Here $P^l(k)$ is periodic in k:

$$P^l(k_x + k, k_y) = P^l(k_x, k_y + k) = P^l(k_x, k_y).$$

Theorem 36 *(Kunz)*

$$\sigma^l_{x,y} = \frac{1}{(2\pi)^2 i} \int_0^k dk_x \int_0^k dk_y Tr P^l(k)[\partial_{k_y} P^l, \partial_{k_x} P^l]$$

and

$$\sigma^l_{x,y} = \frac{1}{2\pi} n$$

where the integer is a topological invariant.

There are two approaches to the proof here. One is to relate $\sigma^l_{x,y}$ to the first Chern number of a complex vector bundle induced on $P^l(k)$ on the 2-torus, or the direct calculation of Kunz. The direct proof goes as follows. Replace the potential V by λV for $\lambda \in [0, 1]$. When there is no potential the equations of motion are

$$v_x(t) = v_x cos(t) + v_y sin(t)$$

and

$$v_y(t) = -v_x sin(t) + v_y cos(t).$$

Putting this into the Kubo formula for $\sigma_{x,y}$ and using the fact that $[v_x, v_y] = i$, it follows that

$$\sigma_{x,y} = M(f_0) = \frac{1}{2\pi} \sum_{n | n+\frac{1}{2} < \mu} .$$

The electronic density ρ in a gap is given by

$$\rho = M(f) = \frac{1}{2\pi i} \oint_\Gamma dz M(G_z).$$

Replacing the potential V by λV in this last equation and taking the derivative gives:

$$\frac{d\rho}{d\lambda} = \frac{1}{2\pi i} \oint_\Gamma dz M(G_z V G_z) = \frac{1}{2\pi i} M(\oint_\Gamma dz G_z^2 V) = 0$$

using the cyclic property of M. One has

$$\rho = M(f_0) = \frac{1}{2\pi} \sum_{n | n+\frac{1}{2} < \mu} .$$

Thus, we can conclude that

$$\sigma_{x,y} = \frac{1}{2\pi} N$$

when $\rho = \frac{1}{2\pi} N$, where N is the Landau band index.

7.17 Quantum Hall Effect – the Discrete Model

Aizenman and Graf (1999) have demonstrated the analogue of the result of Kunz, in particular the Bellissard et al. theorem for the IQHE for the discrete model. Here the Hamiltonian is given by

$$H = H_0 + \lambda V_x$$

acting on $l^2(\mathbf{Z}^d)$ where H_0 incorporates a uniform magnetic field and λV is the random potential whose probability distribution is invariant and ergodic under translations. Like Avron, Seiler and Simon (1983), Aizenman and Graf define the Hall conductance by

$$\sigma_H = \frac{e^2}{h} \mathbf{E}(Index(P_{E_F}, U_a P_{E_F} U_a^*)).$$

Here P_E is the spectral projection on the energy range $(-\infty, E]$ and U_a is a gauge transformation of the form

$$U_a \psi(x) = e^{-i\theta_a(x)} \psi(x).$$

Recall the definition of $Index(P, Q)$, for a pair of projections P, Q, on Hilbert space $\mathcal{H}$, is

$$Index(P, Q) = dim\{\psi \in \mathcal{H} | P\psi = \psi, Q\psi = 0\} - dim\{\psi \in \mathcal{H} | P\psi = 0, Q\psi = \psi\}.$$

The properties of $Index$ include:

Theorem 37 *If P, Q, R are projections which differ by compact operators, then one has additivity:*

$$Index(P, Q) + Index(Q, R) = Index(P, R)$$

and for unitaries U with compact difference $U - I$, one has

$$Index(P, Q) = Index(P, UQU^*).$$

Finally, if P, Q are projections with compact difference $P - Q$ and if there is some $n \geq 0$ such that $(P - Q)^{2n+1}$ is trace class, then

$$tr(P - Q)^{2n+1} = Index(P, Q).$$

For proofs and more details, see Avron, Seiler and Simon (1994) and Aizenman and Graf (1999).

Theorem 38 *(Aizenman and Graf) For a random Schrödinger operator H as above, the Hall conductance $\sigma_H(E)$ is a constant multiple of e^2/h throughout each interval of energies E, over which for some $q > 2$ the quantity*

$$\xi_q = \sum_{x \in \mathbf{Z}^2} \mathbf{E}(| < 0 | P_E | x > |^q)^{1/q} |x|$$

is uniformly bounded.

For additional discussion, see Connes (1986), Bellissard et al. (1994), and Aizenman and Graf (1999).

7.18 Tight Binding Model

The tight binding model (TBM) has the Hamiltonian on a two dimensional square lattice given by

$$H = - \sum_{<i,j>} e^{i\theta_{ij}} c_i^+ c_j + h.c. + \sum_i w_i c_i^+ c_i$$

where the hopping interval is taken as unity; c_i^+ is a fermionic creation operator, i.e $c_i^+, (c_i)$ creates (resp. annihilates) an electron at site i and $< i, j >$ denotes the two nearest neighbor sites. A uniform magnetic flux per plaquette is given as $\phi = \sum \theta_{ij} = 2\pi/M$, where the sum is over four links around a plaquette. Here w_i is a random potential with strength $|w_i| \leq W/2$, e.g. $w_i = W f_i$ where f_i are uniform random numbers between $[-1/2, 1/2]$. Although the system is infinite, two dimensional periodicity of $L \times L$ is imposed on a_{ij} and w_i. The thermodynamic limit corresponds to $L \to \infty$.

The Hall conductance is calculated based on the Kubo formula and the Chern numbers. When the Fermi energy lies in the lowest jth energy gap, then the Hall conductance σ_{xy} is obtained by summing the Chern numbers $C(n)$ below the Fermi energy:

$$\sigma_{xy} = \sum_{n \leq j} C(n).$$

Using this approach, say with $M = 8$ and $W = 1$, Sheng and Weng (1999) find three well defined IQHE plateaux at $\sigma_{xy} = ne^2/h$ for $n = 1, 2, 3$, corresponding to the four Landau levels centered at the jumps of the Hall conductance at $E \leq 0$. And when W was increased from 1 to 6, they found a destruction of these IQHE plateaux.

For a recent discussion of the relationship of Chern indices and properties of classical trajectories, see Faure (1999).

7.18.1 Tight Binding Models on Hypercubes

In the area of research on the mathematical physics of the tight binding model, Vidal, Mosseri and Bellissard (1998) have examined an exactly solvable anisotropic TBM on an infinite dimensional hypercube. For these models they derive the spectral and diffusion exponents. If $n(E)$ is the local density of states, the spectral exponent, $\alpha(E)$ is given by

$$\int_{E-\delta E}^{E+\delta E} dn(E') \sim \delta E^{\alpha(E)}$$

as $\delta E \to 0$. For an absolutely continuous spectrum in some interval I, $\alpha(E) = 1$ for $E \in I$ in the spectrum. If the spectrum is pure point, then $\alpha(E) = 0$; and if $0 < \alpha(E) < 1$ for E in some part of the spectrum, then the spectrum is singular continuous there.

Let Δ_d denote the set of vertices of a cube of size one in a d-dimensional space. The infinite dimensional hypercube is $\Delta = \cup_{d>1} \Delta_d$. It is a set of sequences

$\varepsilon = (\varepsilon_k)_{k=0}^{\infty}$ where $\varepsilon \in \{0,1\}$ and $\varepsilon_k = 0$ for all but a finite number of k's. Under addition modulo 2, Δ is a discrete countable group with dual group B. So B is set set of all sequences $\sigma = (\sigma_k)_{k=0}^{\infty}$ with $\sigma = \pm 1$. B is a compact abelian group under pointwise multiplication. The duality between Δ and B is given by the characters:

$$\chi_\sigma(\varepsilon) = \prod_{k=0}^{\infty} \sigma_k^{\varepsilon_k}$$

for $\sigma \in B$ and $\varepsilon \in \Delta$.

The Hilbert space of states is $\mathcal{H} = l^2(\Delta)$, i.e. the set of sequences $\psi(\varepsilon)$ indexed by Δ such that

$$\|\psi\|^2 = \sum_{\varepsilon \in \Delta} |\psi(\varepsilon)|^2 < \infty.$$

The Fourier transform of a wave function $\psi \in \mathcal{H}$ is given by

$$\mathcal{F}\psi(\sigma) = \sum_{\varepsilon \in \Delta} \chi_\sigma(\varepsilon)\psi(\varepsilon).$$

The Fourier transform is in $L^2(B)$ and the Parseval identity holds:

$$\|\mathcal{F}\psi\|^2 = \int_B d\sigma |\mathcal{F}\psi(\sigma)|^2.$$

The translation operator, $T(a)$ for $a \in \Delta$, is defined by

$$T(a)\psi(\varepsilon) = \psi(\varepsilon - a).$$

Since $a = -a$ in Δ, it follows that $T(a)^2 = 1$ for all $a \in \Delta$. Also one can check that T is self adjoint, $T(a) = T(a)^\dagger$. Under Fourier transform $T(a)$ becomes multiplication by $\chi_\sigma(a)$.

Consider the tight binding Hamiltonians on $\mathcal{H}$ defined by

$$H = \sum_{k=0}^{\infty} t_k T_k.$$

For H to be self adjoint, $t_k \in \mathbb{R}$. t_k denotes the hopping term in the kth direction. H is bounded if and only if $\sum t_k < \infty$ and it is self adjoint if $\sum t_k^2 < \infty$. We assume that H is self adjoint.

Under Fourier transform H becomes the multiplication operator by $E(\sigma)$ where E is called the band function:

$$E(\sigma) = \sum_{k=0}^{\infty} \sigma_k t_k.$$

E is real and square integrable with

$$\int_B d\sigma E(\sigma)^2 = \sum_{k=0}^{\infty} t_k^2.$$

The autocorrelation function P is given by

$$P(s) = |<0|e^{isH}|0>| = (\int_{\mathbf{R}} d\mu(E)e^{isE})^2$$

where $|0>$ is the site where the wavepacket is initially localized and μ is the spectral measure. One can show that

$$P(s) = \prod_{k=0}^{\infty} \cos^2(st_k).$$

Define the position operator X_k as the multiplication operator by ε_k in $\mathcal{H}$. It commutes with T_l for $l \neq k$ and $T_k X_k T_k^{-1} = 1 - X_k$. The mean square displacement operator is defined by

$$L_E^2(s) = \sum_{k=0}^{\infty} < \phi|(X_k(s) - X_k(0))^2|\phi >$$

where $|\phi>$ is an initial state with energy close to E and

$$X_k(s) = exp(isH)X_k exp(-isH).$$

One can show that for any E

$$L_E^2(s) = L^2(s) = \sum_{k=0}^{\infty} sin^2(st_k).$$

The diffusion exponent β is given by $L(s) \sim s^\beta$ as $s \to \infty$. Strong localization is characterized by $\beta = 0$, weak localization is characterized by $\beta = 1/2$ and other values of β are called anomalous diffusion. E.g., for quasicrystals, numerical modelling shows that β may vary from 0 to 1. The question arises, is there any relationship between the spectral and diffusion exponents, similar to the Guarneri (1993) inequality $\beta(E) \geq \alpha(E)/d$ for hypercubes where $d = \infty$?

Theorem 39 *(Vidal, Mosseri, and Bellissard) If one has algebraic scaling of the hopping parameters, $t_k \sim k^{-\gamma}$, then*

$$P(s) \leq c_1 e^{-c_2 s^{1/\gamma}},$$

the spectral measure is always absolutely continuous and the correlation function decays as $1/t$:

$$L^2(s) \sim s^{1/\gamma}.$$

The spectrum is bounded if $\gamma > 1$ and unbounded if $\gamma \leq 1$. The diffusion exponent $\beta = 1/\gamma$ can take on any value in $]0, 1[$, even though the spectrum is absolutely continuous.

The authors also consider the case of geometrical scaling $t_k = (q-1)/q^{(k+1)}$ for $q > 1$, where $\|H\| = 1$. We direct the reader to this paper for the details.

7.19 Kubo Formula Approach

The tight-binding lattice model of noninteracting electrons under a random magnetic field has been utilized by several groups. Here the finite system is diagonalized under the generalized boundary conditions $\psi(k + L\hat{x}) = e^{i\phi_1}\psi(k)$ and $\psi(k + L\hat{y}) = e^{i\phi_2}\psi(k)$. The lattice size is $L \times L$. The Hall conductance of an individual eigenstate $|m>$ can be calculated based on the Kubo formula:

$$\sigma_{xy}^m = \frac{ie^2h}{2\pi A} \sum_{n \neq m} \frac{<m|v_y|n><n|v_x|m> - <m|v_x|n><n|v_y|m>}{(E_n - E_m)^2}$$

where E_n is the energy of the nth state, v_x, v_y are the velocity operator in the x, y direction and A is the area of the system. The total Hall conductance for the system is $\sigma_H = \sum_{E_m < E_F} \sigma_{xy}^m$.

For a finite system with periodic boundary conditions, σ_{xy}^m depends on the boundary condition phases ϕ_1, ϕ_2. However, Niu, Thouless and Wu (1985) showed that the boundary condition averaged Hall conductance takes the form

$$< \sigma_{xy}^m > = \frac{1}{4\pi^2} \int d\phi_1 d\phi_2 \sigma_{xy}^m(\phi_1, \phi_2) = C(m)e^2/h$$

where $C(m)$ is an integer, viz. the Chern number of the state $|m>$.

States with nonzero Chern number carry Hall current and are extended states. Whereas a zero Hall conductance or zero Chern number state will always be localized in two dimensional systems in the presence of weak impurities. The transition from an extended state to a localized state can be viewed as being caused by a cancellation of Chern numbers, i.e. converting a nonzero Chern number to a zero by cancellation of Chern numbers.

Thus, the approach is to numerically diagonalize the Hamiltonian on a grid of ϕ_1 and ϕ_2 and calculate the Chern numbers of the states of finite sized systems by converting the above integral to a sum over grid points.

7.20 Random Magnetic Fields

The localization problem in two dimensions in the presence of the a random magnetic field with zero mean, in particular whether extended states exist in the thermodynamic limit for this system, has been the subject of a number of papers. We cite only Sheng and Weng (1995), Liu, Xie, Das Sarma and Zhang (1999) and Yang and Bhatt (1996). In particular Yang and Bhatt use their Kubo formula - Chern number approach to show that for a two dimension square lattice in the presence of a random magnetic field with zero mean there appears to be a localization-delocalization transition implying the existence of extended states at weak disorder.

For other work on electron transport in random magnetic fields, see Mancoff et al. (1995).

7.21   Critical Conductance and its Fluctuations

Wang, Jovanovic and Lee (1996) have calculated the ensemble averaged two terminal conductance $< G >$ and the central moments $< (\delta G)^{2n} >$ for $n \leq 4$. In
this model we have two semi-infinite conducting leads connected to a $W \times L$ disordered network. Baranger and Stone (1989) have shown that linear response theory
applied to this lead-network-lead system gives for the multi-channel two-terminal
conductance

$$G = \frac{e^2}{h} Tr(t^\dagger t) = \frac{e^2}{h} \sum_{i=1}^{W} \frac{1}{cosh^2(\gamma_i L)}$$

where t is the transmission matrix through the disordered region and γ_i is the
Lyapunov exponent of $T^\dagger T$, where T is the total transfer matrix for the entire
$W \times L$ system, $T = \prod_{i=1}^{L} T_i$. Here T_i is the $W \times W$ transfer matrix.

In this work they show that at the critical point for integer quantum Hall
plateau transitions $< G >_c = (0.58 \pm 0.03)e^2/h$ and the central moments are
$< (\delta G)^2 >_c = (0.081 \pm 0.005)(e^2/h)^2$. They conjecture that this behavior of the
critical conductance is universal and that the critical conductance obeys a log-
normal distribution. Although the precise relationship of the two-terminal conductance G and σ_{xy} based on a Kubo formula in a closed system without contacts
is unclear, if one assumes that $< G >_c$ is the four probe σ_{xx}^c, then their result
agrees with the conjecture of Kivelson, Lee and Zhang.

7.22   Zero Magnetic Field Limit

Since all states in a 2DEG are localized in the absence of a magnetic field, a 2DEG
must undergo a phase transition from a quantum Hall state to an insulating state as
the strength of the magnetic field goes to zero. Khmelnitskii (1984) and Laughlin
(1984) argue based on semiclassical methods that the delocalized states at the
centers of Landau bands "float" upward as B decreases and disappear at infinite
energy in the $B \to 0$ limit. More precisely, they propose that floating up happens
when the disorder is of the order of the Landau level spacing. They assert that
the extended state for the nth Landau level behaves like

$$E_n^c = (n + 1/2)\hbar\omega_c \frac{1 + (\omega_c \tau)^2}{(\omega_c \tau)^2}$$

where $\omega_c = eB/mc$ is the cyclotron frequency and τ is the impurity scattering
time.

That is, in the limit of weak magnetic field or strong disorder, extended states
"float" up, tending to infinite energy; for a given electron density and hence finite
E_F, for sufficiently low B all extended states are above E_F and the system becomes
insulating. This is consistent with the global phase diagram of Kivelson, Lee and
Zhang (1992).

Numerical work of Xie and Liu (1995) suggested that for the lattice model that
there are direct transitions from all Landau level quantum Hall states to insulator
states. In particular, Xie and Liu claim:

- the extended state energy E_c for each Landau level is always linear in magnetic field;

- for a given Landau level and disorder there exists a critical magnetic field B_c below which the extended state disappears;

- the lower Landau levels are more robust to the metal-insulator transition with smaller B_c.

The tight binding model work of Yang and Bhatt (1999) expresses the disappearance of the current carrying states in terms of motion and merger of various Chern number states. There are subtleties in extending this lattice based model to the continuous models. Experimental data has been examined by Hilke, Shahar, Song, Xie (1999) to produce a phase diagram of the IQHE as a function of density and magnetic field. They observe a floating up of the lowest energy level but no floating of any higher levels. There is a merging of these levels into the insulating state. In terms of the critical resistivities, they observed a peak in the critical resistivity, ρ_{xx}^c, near a filling factor of one. We direct the reader to these papers for the details.

7.23 Features of the Quantum Hall Effect

We now summarize certain facts regarding the quantum Hall effect:

- in an integer plateau transition, the electron conductivity tensor elements $(\sigma_{xx}, \sigma_{xy})$ change from $(0, n)\frac{e^2}{h}$ to the values $(0, n \pm 1)\frac{e^2}{h}$;

- in the limit of strong magnetic field or weak disorder there is a single critical energy, E_c, within each Landau band where the localization length of electronic states diverges; and E_c is near the center of the Landau levels;

- for zero temperature and infinite sample size, the transitions are continuous phase transitions with a single divergent length scale, ξ, the quasi-particle localization length;

- as the Fermi energy, E_F, moves through a critical value E_c,

$$\xi \sim |E_F - E_c|^{-\nu}$$

  where $\nu \simeq 2.33 \pm 0.03$

- the characteristic energy scale behaves like $1/\xi$, which implies the dynamical exponent is $z = 1$;

- the impurity averaged conductivities are expected to be universal at the quantum critical points;

- based on Chern-Simons approach, Kivelson, Lee and Zhang (1992) claimed that $(\sigma_{xx}, \sigma_{xy})$ at the $(0, n)$ to $(0, n+1)$ transition, has the critical values:

$$(\sigma_{xx}^c, \sigma_{xy}^c) = (.5, n - .5)\frac{e^2}{h};$$

- Huo, Hetzel and Bhatt (1993) using the Kubo formula calculated σ_{xx}^c at the $(0, 0) \to (0, 1)$ transition and found $\sigma_{xx}^c = (0.55 \pm 0.05)\frac{e^2}{h}$.

- the four terminal resistivity at critical energies has been conjectured to be universal with $\rho_{xx}^c \simeq h/e^2$ by Shahar et al. (1995); recall that the relation between the resistivity and conductivity tensors is:

$$\rho_{xx} = \frac{\sigma_{xx}}{\sigma_{xx}^2 + \sigma_{xy}^2}$$

and

$$\rho_{xy} = \frac{\sigma_{xy}}{\sigma_{xx}^2 + \sigma_{xy}^2};$$

- electron-electron interaction does not play a significant role in IQHE;

- σ_{xx} vanishes for all values of the Fermi energy at finite temperature;

- there are direct transitions from higher quantum Hall plateau states to the insulating phase;

- for $\delta \neq 0$ the Hall conductivity is quantized and the dissipative conductivity σ_{xx} vanishes;

- at the transition, $\delta = 0$, the Hall conductivity is unquantized and σ_{xx} remains finite so the critical state is conducting.

7.24 Open Problems

In this section we will characterize certain open problems in the area of the spectral theory of quantum Hall effect. All the problems in this section have been examined numerically within the tight binding model or experimentally. However, a detailed treatment within the continuous model has not been performed to date. The problems all really address the work on the global phase diagram (GPD) of the quantum Hall liquid - insulator, which began with the work of Kivelson, Lee and Zhang (1992) and most recently has been reviewed by Sheng and Weng (1999).

Problem 1 *In the continuous model, characterize the quantum Hall effect in the* $B \to 0$ *limit.*

The next two problems want to describe the two insulator regimes characterized by Sheng and Weng (1999), viz. the strong disorder and low magnetic field region (insulator I), where the Anderson insulator is found to be stable against forming quantum Hall states, and the weak disorder and high magnetic field (insulator II), where ρ_{xy} remains quantized near the region of transition. This latter region has been called the "quantized Hall insulator" by Hilke, Shahar, Song, Tsui, Xie and Monroe (1998)

Problem 2 *Describe in the continuous model the quantum Hall liquid to insulator transition; in particular, characterize the different insulator regimes discussed by Sheng and Weng.*

Problem 3 *In the continuous model explain why in the weak disorder and high magnetic field ρ_{xy} remains quantized near h/e^2, while ρ_{xx} increases by an order of magnitude (from h/e^2 to $8h/e^2$)?*

The next problem focusses on behavior at the critical level:

Problem 4 *In the continuous model describe the critical resistivities, ρ_{xy}^c, ρ_{xx}^c: i.e. the resistivities at the quantum Hall liquid - insulator transition.*

7.25 Modular Group Symmetries

Discrete symmetry properties in quantum Hall transitions have been discussed by Kivelson, Lee and Zhang (1992) in terms of the global phase diagram and integer and fractional quantum Hall transitions. If one uses a complexified conductance $\tau = \sigma_{xy} + i\sigma_{xx}$, the transformation laws are $T : \tau \to \tau + 1$, the Landau level addition, and $ST^2S : \frac{1}{\tau} \to \frac{1}{\tau} - 2$, the flux attachment, where $S : \tau \to -1/\tau$. The subgroup of $SL(2, \mathbf{Z})$ generated by these operations is $\Gamma_0(2)$. The critical point of the IQHE of the lowest Landau level is $(\sigma_{xy}^c, \sigma_{xx}^c) = (.5, .5)$ or $\tau^c = \frac{1+i}{2}$; one conjectures that a similar behavior occurs for all equivalent points $\gamma\tau^c$ for $\gamma \in \Gamma_0(2)$. For a discussion, see Taniguchi (1998).

Problem 5 *Does $\Gamma_0(2)$ play the role of a symmetry group in the study of IQHE transitions?*

7.26 Two Dimensional MITs

In this section we discuss two dimensional metal insulator transistions or MITs. The work of Abrahams, Anderson, Licciardello and Ramakrishnan (1979) predicted that all electron states in a disordered two dimensional electron system in a zero magnetic field are localized at zero temperature, i.e. there is no metal-insulator transition (MIT) in an infinite 2D sample; in particular, lowering temperature should make a 2DES more localized. However, the recent results of Kravchenko, Kravchenko and Furneaux (1994) have demonstrated possible metal-insulator transition at $B = 0$ in two dimensional systems. Their results used ultra high mobility silicon MOSFETs with very low electron density, n_s. At zero magnetic field weak localization is observed for $T \geq 1-2\,K$. For $n_s < n_{cr} \sim 10^{11}\,cm^{-2}$, resistivity monotonically increases as $T \to 0$ indicating an insulating or localized state. However, it was observed for $n_s > n_{cr}$ there is a strong decrease of ρ with decreasing temperature below $1-2\,K$. This absence of localization is consistent with Azbel's (1992) conjecture on the existence of a mobility edge in zero magnetic field for 2D systems. Other explanations of this result include the work of He and Xie (1999) who argue that there is a new liquid phase in 2DES in Si MOSFETs at low enough electron densities so that the MIT results as a "crossover from percolation transition of the liquid phase through the disorder landscape below the liquid-gas critical temperature."

7.26.1 Temperature Behavior

A system is defined to be insulating if its conductivities exhibit insulator like temperature behavior, i.e.

$$d\sigma_{xx}/dT, d\sigma_{xy}/dT > 0$$

with

$$\sigma_{xy} < \sigma_{xx} < e^2/h.$$

A system is defined to be metallic if

$$d\sigma_{xx}/dT, d\sigma_{xy}/dT < 0.$$

And the system is defined to be in a quantum Hall state if $\sigma_{xy} \to e^2/h$ and $\sigma_{xx} \to 0$ as the temperature is lowered.

7.26.2 Nonzero Magnetic Fields

Kravchenko, Mason, Furneaux, Caulfield, Singleton and Pudalov (1994) have also examined temperature induced transitions between insulator, metal and quantum Hall behaviors for transport coefficients in very dilute high mobility 2DES in silicon with a magnetic field corresponding to Landau level factor $\nu = 1$.

In the presence of a magnetic field, Kravchenko et al. have observed that for their samples for $T \geq 2.5\,K$ $\sigma_{xy} < \sigma_{xx} < e^2/h$ and both diminish with decreasing temperature. This phase is characteristic of an insulating state. Since σ_{xy} is a counter of the number of bands of extended states below E_F, there are no extended states below E_F which is also indicative of an insulating state in this phase.

For the region $1 < T < 2.5\,K$, both σ_{xx} and σ_{xy} increase with decreasing T, indicating metallic behavior. Both reach $e^2/2h$, the expected value for an extended state.

Below $T \sim 1\,K$, σ_{xx} tends to zero as $T \to 0$ while σ_{xy} approaches e^2/h, corresponding to one band of extended states below E_F; i.e. this is a quantum Hall state.

For a recent review of these results, see Abrahams (1999a, 1999b), Dobrosavljevic et al. (1997) and He and Xie (1997).

7.27 Localization in One Dimension

We have seen in the last section that real world disordered two dimensional devices may have MITs in contrast to the perhaps simplistic model as a disordered system. What about one dimensional disordered systems, which in the usual Anderson model have all states localized in the presence of disorder? Are they always localized? Are there conditions on the potential which provides for extended states? Hilke (1997) has considered a disordered one dimensional Anderson model with an underlying periodicity. Hilke's model is a tight binding model where

$$(V_l - E)\psi_l + \psi_{l+1} + \psi_{l-1} = 0.$$

The periodicity is introduced by using two lattices, one composed of random sites and one composed of deterministic sites. In the case that each random site is surrounded by at least one constant neighbor site, then there exist discrete resonance energies for which the states are extended, i.e. infinite localization length. Hilke finds the critical energies occur at the resonance energies noted in the work of Derrida and Gardner (1984). For any d they are roots of the equation

$$(1,0)T_d \begin{pmatrix} 1 \\ 0 \end{pmatrix} = 0$$

where

$$T_d = \prod_{n=1}^{d-1} \begin{pmatrix} \epsilon & -1 \\ 1 & 0 \end{pmatrix}.$$

In the case that one has multiple neighboring random sites, say $d = 3$ means every third site is non-random and the sites between are random, then the resonance energies still occur as described above; however, the localization lengths are no longer infinite but only enhanced at the resonance energies. Of course, as $d \to \infty$, one recovers the usual uncorrelated result.

Hilke uses a result of Felderhof (1986) which states that the transmission and reflection ratio is given by $R/T = (P(2,2) - 1)/2$ where

$$P = \Gamma_n G_n \Gamma_{n-1} G_{n-1} \cdots$$

In this formula

$$\Gamma_n = \begin{pmatrix} 1 - W_n^2 - 2iW_n & -W_n^2 - iW_n & -W_n^2 \\ 2W_n^2 + 2iW_n & 1 + 2W_n^2 & 2W_n^2 - 2iW_n \\ -W_n^2 & -W_n^2 + iW_n & 1 - W_n^2 + 2iW_n \end{pmatrix}$$

and

$$G_n = \begin{pmatrix} e^{2ik(X_n - X_{n-1})} & 0 & 0 \\ 0 & 1 & 0 \\ 0 & 0 & e^{-2ik(X_n - X_{n-1})} \end{pmatrix}$$

where $W_n = V_{dn}/2sin(k)$, $2cos(k) = E$, and X_n are the positions of the impurities with value V_{dn}. Note V_{dn} are random and V at the deterministic sites is taken to be zero. For Hilke's model for d even, $X_n - X_{n-1} = d$ and at the band center $G_n = I$ and $P(2,2) = 2(\sum_{n=1}^{N} W_n)^2 + 1$. So if the sum of impurities is zero, the reflection coefficient vanishes and we have total transmission.

Derrida and Gardner (1984) have examined the complex Lyapunov exponent γ for the Anderson model where the real part corresponds to the invers localization length; viz. they show $L_c^{-1} \sim Re(\gamma) \sim E^{2/3} < W^2 >^{1/3}$ and $\rho(E) \sim \partial Im(\gamma)/\partial E \sim E^{-1/3}$.

7.27.1   Delocalized States in Random Dimer Superlattices

The Anderson model normally considers a totally uncorrelated potential. The work of Dunlap, Wu and Phillips (1990) and Phillips and Wu (1990) showed that a discrete set of delocalized states appear if short range correlations are introduced in the random potential. See also Flores (1989). Izrailev and Krokhin (1999) have shown that in the tight-binding approximation with correlations, there is a relationship between the pair correlation and the localization length implying that there is a mobility edge in these one dimensional systems. Kuhl, Izrailev, Krokhin and Stöckmann (1999) have presented experimental observations of mobility edge in a waveguide with correlated disorder. They used a 2.15 meter waveguide with 100 micrometer screws which can be adjusted. Approximating the screws as delta scatterers, the wave equation is given by the Kronig-Penny model

$$\psi''(z) + E\psi(z) = \sum_{n=-\infty}^{\infty} EU_n\psi(z_n)\delta(z - nd).$$

Here the wave function ψ is associated with the electric field of the TE-mode and the energy is given by $E = k^2$ for wavenumber k. The discrete form of the wave equation in this case is

$$\psi_{n+1} + \psi_{n-1} = [2cos(kd) + U_nkdsin(kd)]\psi_n = 0$$

where $\psi_n = \psi(z_n)$, and the potential strength is split into two parts $U_n = \epsilon + \epsilon_n$ given by the mean value $\epsilon =< U_n >$ and the fluctuation component ϵ_n. This model can be re-expressed in terms of a two dimensional Hamiltonian map given by a linear oscillator with linear periodic delta kicks. Let μ denote the phase shift between the kicks. Then the dispersion relation for the Kronig-Penny model is

$$2cos\mu = 2cos(kd) + kd\epsilon sin(kd)$$

for $0 \leq \mu < \pi$.

Theorem 40 *(Izrailev and Krokhin) In the presence of correlated disorder the inverse localization length for the Kronig-Penny model is*

$$l^{-1}(E) = k^2\frac{< \epsilon_n^2 >}{8}\frac{sin^2(kd)}{sin^2\mu}\phi(\mu)$$

where

$$\phi(\mu) = 1 + 2\sum_{m=1}^{\infty} \xi_m cos(2\mu m)$$

with correlation described by $\xi_m =< \epsilon_{n+m}\epsilon_n > / < \epsilon_n^2 >$.

Let

$$\phi(\mu) = \begin{cases} C_0^2 & 0 < \mu_1 < \mu < \mu_2 < \pi/2 \\ 0 & \mu < \mu_1 \text{ and } \pi/2 > \mu > \mu_2 \end{cases}.$$

Here $C_0^2 = \pi/2(\mu_2-\mu_1)$ is a normalization constant so that $\xi_0 = 1$. From this result one can calculate ϵ_n (see Kuhl et al. (1999)). In this paper $\mu_1/\pi = 0.2m, \mu_2/\pi =$

$0.4, \sqrt{<\epsilon_n^2>} = 0.1$ and $\epsilon = -.1$ and their experimental work showed mobility edges near the points $kd/\pi = 0.38, 0.57$ and 0.76 which are roots of the dispersion relation for $\mu/\pi = 0.4, 0.6$ and 0.8. For the theoretical case of $N = 10^4$ scatterers their model shows two bands where the transmission coefficient is nearly one. For related microwave studies, see Stöckmann, Barth et al. (1999) and Kuhl and Stöckmann (1999).

Bellani et al. (1999) examined the photoluminescence spectra of GaAs-AlGaAs superlattices which have intentional correlated disorder. They have observed evidence of discrete extended states. This is based on earlier work of Diez, Sánchez and Domínguez-Adame (1995) where they modeled disordered quantum well based semiconductor superlattices with short range correlations. Their models are basically random dimer quantum well superlattices. The correlated disorder is introduced by taking quantum wells with two different average thicknesses, placed at random with the constraint that they appear in pairs. The authors numerically demonstrated the existence of a band of extended states in perfect correlated disordered superlattices, which gives rise to a strong enhancement of the finite temperature dc conductance when the Fermi level matches this band.

Tessieri and Izrailev (1999) have argued that one dimensional random lattices can have a continuum of extended states and mobility edges provided that the disorder exhibits appropriate long range correlations. The reader should also note the papers of Kottos, Izrailev and Politi (1998) and Izrailev, Ruffo and Tessieri (1998).

7.27.2 Delocalization in Continuous Disordered Systems

Hilke and Flores (1997) have studied delocalization in continuous disordered systems. Their model is a one dimensional sequence of scattering barriers centered at random position x_l, without overlap. Each barrier is symmetric with random support $2a_l$ and distance between the barriers is $x_{l+1} - x_l = d_l$, which is random. The Hamiltonian is given by $H = \frac{p^2}{2m} + \sum_l V_l(x - x_l)$. The wavefunction just after collision with the lth barrier is

$$\psi_l^+ = A_l e^{ik(x_l + a_l)} + B_l e^{-ik(x_l + a_l)}$$

and the condition for delocalization is

$$\psi_{l+1}^+ = \pm \psi_l^+.$$

Here k is the wave number where $E = \hbar^2 k^2 / 2m$. The amplitudes (A_{l+1}, B_{l+1}) satisfy the equation

$$\begin{pmatrix} A_{l+1} \\ B_{l+1} \end{pmatrix} = \begin{pmatrix} e^{-ikx_l} & 0 \\ 0 & e^{ikl_l} \end{pmatrix} \begin{pmatrix} \alpha_l & \beta_l \\ \beta_l^* & \alpha_l^* \end{pmatrix} \begin{pmatrix} e^{ikx_l} & 0 \\ 0 & e^{-ikl_l} \end{pmatrix} \begin{pmatrix} A_l \\ B_l \end{pmatrix}.$$

Hilke and Flores consider in one case a δ-barrier sequence where

$$V_l(x) = V_l \delta(x - x_l)$$

with $d_l = d$. Here V_l are random uncorrelated parameters and d is a constant lattice parameter. The scattering matrix is given by $\alpha_l = 1 + i\frac{V_l}{2k}$ and $\beta_l = i\frac{V_l}{2k}$. The delocalization condition described above is

$$e^{ikd} - \frac{V_l}{k}\sin(kd) = \pm 1.$$

If $k = n\pi/d$ for $n \in \mathbf{Z}^*$, we have a set of delocalized states with $|\psi_{l+1}| = |\psi_l|$ as noted earlier by Ishii (1973). What about the divergence of the localization length $L_c(E)$ near the critical energies $E_c = \frac{1}{2m}(\frac{\hbar n\pi}{d})^2$? They find the critical exponent ν where

$$L_c(E) \sim \frac{1}{|E - E_c|^\nu}$$

has the value 2/3 based on the work of Derrida and Gardner (1984).

Hilke and Flores also consider the case of a quantum well sequence. The extended states arise when

$$\beta_l = i\frac{(p_l^2 - k^2)\sin(2a_l p_l)}{2p_l k} = 0$$

where $p_l = \sqrt{|k^2 - V_l|}$. Thus,

$$E = k^2 = V_l + \frac{n_l^2\pi^2}{4a_l^2}$$

are the energies of the extended states for $n_l \in \mathbf{Z}$. Let $d_l = d$ and $a_l = a$. The critical exponent in this case is 2/3 or 2 depending on the well's parameters, i.e. if $k_c(d - 2a)/\pi$ is an integer or not.

The authors also consider the conductance and its fluctuations. For the delta doping case, conductance peaks at the critical energies with the form

$$G(\mu) \sim e^{-\gamma|\mu - E_n|^\nu}$$

where μ is the chemical potential and E_n are the critical energies, viz. $E_n = \frac{1}{2m}(\frac{\hbar n\pi}{d})^2$, and $\nu = 2/3$. Here γ depends on the size and the disorder of the system. The relative fluctuations of the conductance around these critical energies are vanishing. For further details, we direct the reader to the original paper.

7.28 Acoustics and Electromagnetics

Band edge localization for acoustic and electromagnetic wave in random media has been treated by Figotin and Klein (1994b) in the discrete case; Faris (1986, 1987) also treated the discrete case. The continuous case has been examined by Figotin and Klein (1996, 1997) for the case of Anderson like perturbations of periodic operators. Combes, Hislop and Tip (1999) and Stollmann (1999b) have treated both the acoustic and electromagnetic case. Making assumptions similar to (H1)-(H9) they have shown:

Theorem 41 *(Combes, Hislop and Tip) Assume (H1)-(H8); then there are constants* $E_{\pm}$ *with* $B_- \leq E_- \leq \tilde{B}_-$ *and* $\tilde{B}_+ \leq E_+ \leq B_+$ *such that the spectrum* $\Sigma \cap (E_-, E_+)$ *is pure point with exponentially decaying eigenfunctions.*

Theoretical work on photonic crystals has been pursued by Figotin and Kuchment (1996) and Figotin and Godin (1997) A theoretical discussion of the localization of light is given in Figotin and Klein (1999) and Klein (1999). For developments on the localization of light, see John (1988, 1991, 1993), Wiersma et al. (1999) and de Vries et al. (1998). This is a very interesting area of study. E.g., an optical Hall effect has been reported by van Tiggelen. Weak localization effects have been observed including enhanced backscattering of light from disordered media and universal conductance fluctuations have been shown in the transmission of light through disordered systems by Genack. For a review of these items, the reader is directed to de Vries et al. (1998).

Briefly, the theoretical approach is to note that one can re-write Maxwell's equations as a Schrödinger like equation

$$-i\frac{\partial}{\partial t}\Psi_t = \mathcal{M}(\varepsilon, \mu)\Psi_t$$

where ε is the dielectric constant and μ is the magnetic permeability. Here

$$\mathcal{M} = \begin{pmatrix} 0 & \frac{i}{\mu}\nabla^\times \\ \frac{i}{\varepsilon}\nabla^\times & 0 \end{pmatrix}$$

with $\nabla^\times \Psi = \nabla \times \Psi$. The Maxwell operator is defined by $\mathbf{M}_{\mu,\varepsilon} = \frac{1}{\sqrt{\mu}}\nabla^\times \frac{1}{\sqrt{\varepsilon}}\nabla^\times \frac{1}{\sqrt{\mu}}$. The eigenvalues and eigenfunctions of $\mathcal{M}$ and $\mathbf{M}$ are directly related and localized electromagnetic waves are eigenfunctions of $\mathbf{M}$.

A defect is a variation of a given medium in a bounded domain, i.e. if $\varepsilon(x) - \varepsilon_0(x)$ has compact support, one says that $\varepsilon(x)$ and $\varepsilon_0(x)$ differ by a defect. Figotin and Klein (1997) have shown that the essential spectra of Maxwell operators are not changed by defects:

Theorem 42 *(Figotin and Klein) Let* $\mathbf{M}_0$ *and* $\mathbf{M}$ *denote the Maxwell operators for two dielectric media which differ by a defect. Then*

$$\sigma_{ess}(\mathbf{M}) = \sigma_{ess}(\mathbf{M}_0).$$

And if (a, b) *is a gap in the spectrum of* $\mathbf{M}_0$*, the spectrum of* $\mathbf{M}$ *in* (a, b) *consists at most of isolated eigenvalues with finite multiplicity, with corresponding eigenfunctions decaying exponentially fast.*

Figotin and Klein show that one can define defects which create eigenvalues in any given subinterval of a spectral gap of $\mathbf{M}_0$; i.e., localized waves are allowed by Maxwell equations.

Theorem 43 *(Figotin and Klein) Let* $\mathbf{M}_{\mu_0,\varepsilon_0}$ *and* $\mathbf{M}_{\mu,\varepsilon}$ *be the Maxwell operators for two dielectric media which differ by a defect. Let* (a, b) *be a gap in the spectrum* of $\mathbf{M}_{\mu_0, \varepsilon_0}$. Select $\beta \in (a, b)$ and pick $\theta > 0$ such that the interval $[\beta - \theta, \beta + \theta]$ is a subset of (a, b). If $\mu(x) = \hat{\mu}$ and $\varepsilon(x) = \hat{\varepsilon}$ in a cube of side l with

$$l^2 \hat{\mu} \hat{\varepsilon} > \frac{79\beta}{\theta^2},$$

then the Maxwell operator $\mathbf{M}_{\mu, \varepsilon}$ has at least one eigenmode with corresponding eigenvalue inside the interval $[\beta - \theta, \beta + \theta]$.

Maxwell operators with spectral gaps arise from photonic crystals, which are crystals with periodic dielectric constant and magnetic permeability, i.e. $\varepsilon_0(x + \tau) = \varepsilon_0(x)$ and $\mu_0(x + \tau) = \mu_0(x)$ for τ in the lattice of periods, say $q\mathbf{Z}^3$. By the periodicity, Floquet-Bloch theory implies that the spectrum has band structure and if the spectral bands do not cover $[0, \infty)$ then there will be a spectral gap. As we have noted single defects in a periodic dielectric medium with a spectral gap can generate localized electromagnetic waves. In addition, one can show that a random array of such defects can create a similar effect. In particular let

$$\varepsilon_{g,\omega}(x) = \varepsilon_0(x)\gamma_{g,\omega}(x)$$

where

$$\gamma_{g,\omega}(x) = 1 + g \sum_{i \in \mathbf{Z}^3} \omega_i u_i(x).$$

Here $u_i(x) = u(x - i)$ are nonnegative, measurable, real valued functions with compact support such that

$$0 \leq U_- \leq U(x) = \sum_{i \in \mathbf{Z}^3} u_i(x) \leq U_+ < \infty.$$

And similarly for $\mu_{g,\omega}(x)$ with

$$0 \leq V_- \leq V(x) = \sum_{i \in \mathbf{Z}^3} v_i(x) \leq V_+ < \infty.$$

Let $\mathbf{M}_g = \mathbf{M}_{\mu_{g,\omega}, \varepsilon_{g,\omega}}$. One can show that there is a g_0 such that for $g < g_0$ there is a gap $(a(g), b(g))$ of the operator $\mathbf{M}_g$ inside the gap (a, b) of the unperturbed operator $\mathbf{M}_0$. Figotin and Klein have shown that there is Anderson localization near the edges of the gap:

Theorem 44 *(Figotin and Klein) If the probability density ρ of the random variables ω_i satisfies*

$$\int_{1-\gamma}^{1} \rho(t)dt \leq K\gamma^\eta$$

for $0 \leq \gamma \leq 1$, $K < \infty$ and $\eta > 3/2$, then for any $g < g_0$ there is a $\delta(g) > 0$ such that the random operator $\mathbf{M}_g$ exhibits exponential localization in the interval $[a(g) - \delta(g), a(g)]$.

The proof again depends on a Wegner type estimate for random Maxwell operators and multiscale analysis. For a proof, see Klein (1999). In particular the Wegner estimate for this problem states:

Theorem 45 *(Figotin and Klein) Let $R_{g,\omega_\Lambda}(E)$ denote the resolvent of the random operator $M_{g,\omega,\Lambda}$. Then there is a $Q < \infty$ such that*

$$\mathbf{P}\{\|R_{g,\omega,\Lambda}(E)\| \geq \frac{1}{\eta}\} \leq \frac{Q\|\rho\|_\infty \sqrt{|E|}}{g(1 - gmax\{U_+, V_+\})(U_- + V_-)} \eta |\Lambda|^2$$

for all $E > 0$, cubes Λ in $\mathbf{R}^3$ and all $0 < \eta < E$.

7.29 Birman-Schwinger Method

Exner and coworkers have used the Birman-Schwinger method to examine spectral properties of the problems with deformed quantum wave guides, viz. to show that the Dirichlet Laplacian on a suitable, deformed quantum waveguide has at least one isolated eigenvalue below the bottom of the essential spectrum. Figotin and Klein (1998) have recently utilized a modified Birman-Schwinger result to derive equations for the defect eigenmodes and midgap eigenvalues. We review the Birman-Schwinger methodology in this section.

Let $\mathbf{M}$ and $\mathbf{M}_0$ denote the operators with and without a defect. One cannot apply the standard Birman-Schwinger method since $\mathbf{M} - \mathbf{M}_0$ is not relatively compact with respect to $\mathbf{M}_0$. However, if one uses the resolvents $H = (\mathbf{M} + I)^{-1}$ and $H_0 = (\mathbf{M}_0 + I)^{-1}$, then one can show that $V = H - H_0$ is a Hilbert-Schmidt operator in dimension $d \leq 3$ (v., Figotin and Klein (1998).) Thus, Figotin and Klein use the Birman-Schwinger method for $H = H_0 + V$.

Let $H_g = H_0 + gV$ denote the Schrödinger operator where H_0 and V are self-adjoint operators. Assume that H_0 has a gap (a, b) in the spectrum and $V \geq 0$ is bounded and relatively compact with respect to H_0. Then $\sigma_{ess}(H_g) = \sigma_{ess}(H_0)$ (e.g., see Reed and Simon (1978)). So the spectrum of H_g in the gap (a, b) consists of isolated eigenvalues with finite multiplicity.

The eigenvalue problem $H_g\psi = \lambda\psi$ for $\lambda \in (a, b)$ can be rewritten as

$$\psi = -gR_0(\lambda)V\psi$$

where $R_0(\lambda) = (H_0 - \lambda I)^{-1}$. Set

$$\mathcal{R}(\lambda) = -\sqrt{V}R_0(\lambda)\sqrt{V}.$$

And one obtains the eigenvalue problem, which is equivalent to the original problem,

$$\mathcal{R}(\lambda)\phi = \frac{1}{g}\phi$$

where $\phi = \sqrt{V}\psi$.

The Birman-Schwinger operator $\mathcal{R}(\lambda)$ is a self-adjoint Hilbert-Schmidt operator. Klaus (1982) has shown:

Theorem 46 *(Klaus) For $g \neq 0$, then $\lambda \in (a, b)$ is an eigenvalue of H_g with multiplicity m if and only if $1/g$ is an eigenvalue of $\mathcal{R}(\lambda)$ with multiplicity m.*

For $\lambda \in (a,b)$ let $r_1^+(\lambda) \geq r_2^+(\lambda) \geq ... \geq 0$ denote the infinite sequence of positive eigenvalues of the Birman-Schwinger operator $\mathcal{R}(\lambda)$. And similarly for the negative eigenvalues $r_j^-(\lambda)$. Klaus has shown

Theorem 47 *(Klaus) The eigenvalues $r_n^\pm(\lambda)$ are monotonically decreasing, continuous functions of $\lambda \in (a,b)$ for each $n = 1,2,...$*

As Figotin and Klein observe, this gives a criterion for the absence of eigenvalues of H_g in the gap (a,b) for small g:

Theorem 48 *(Figotin and Klein) The following two statements are equivalent:*

- *the operator H_g has no eigenvalues in the gap (a,b) for small g;*
- $max_{\lambda \in (a,b)} \|\mathcal{R}(\lambda)\| = max\{r_1^+(a+0), -r_1^-(b-0)\} < \infty.$

Thus, the Birman-Schwinger method allows one to conclude:

Theorem 49 *(Figotin and Klein) Set $H = H_0 + V$, where H_0 and V are as above; if $V \geq 0$, then the only possible point of accumulation of $\sigma(H) \cap (a,b)$ is a; in this case the eigenvalues $\lambda_1 \geq \lambda_2 \geq ...$ of H in the gap (a,b) coincide with the set of solutions of the equations*

$$r_n^+(\lambda) = 1$$

for $n = 1,2,...$ where $r_n^+(\lambda)$ are positive eigenvalues of the operator $\mathcal{R}(\lambda)$; and if $\phi(\lambda)$ is a corresponding eigenfunction, then

$$\psi = (H_0 - \lambda I)^{-1}\sqrt{V}\phi(\lambda)$$

is an exponentially localized eigenfunction for the operator H with eigenvalue λ. And similarly for the case $V \leq 0$.

For the details of the proof, see Figotin and Klein (1998).

Using their Birman-Schwinger results Figotin and Klein have estimated the number of eigenvalues generated in a gap of a periodic medium, assuming a regularity condition of the resolvent in the region of the edges of the spectral gap (a,b) similar to that used by Klopp (1999). Based on this result, they show that if the defect is small, then there are no midgap eigenvalues.

Chapter 8

Selberg Trace Formula

8.1 Introduction

In the following chapter the topic is on the spectral theory of finite volume homogeneous graphs, which provide an excellent model for studying mesoscopic systems, extending the work of Exner and coworkers and others. As a form of background, we review certain Selberg conjectures in the finite volume manifold setting, which relate to the Selberg zeta function and the Selberg trace formula. The Selberg-Roelke and Phillips-Sarnak conjectures are outlined. The relationship of the Phillips-Sarnak dissolving conjecture and the Fermi Golden Rule is developed. The general background for this chapter is contained in Hejhal (1983) and Venkov (1990).

8.2 Automorphic Laplacian

Let $H = \{z \in \mathbf{C} | z = x + iy, y > 0\}$ be the hyperbolic upper half plane and let

$$\Delta = y^2 \left(\frac{\partial^2}{\partial x^2} + \frac{\partial^2}{\partial y^2} \right)$$

be the Laplacian associated with the Poincaré metric

$$ds^2 = \frac{dx^2 + dy^2}{y^2}.$$

Consider a subgroup Γ of the group of isometries acting on H and let χ be a one dimensional unitary representation of Γ. Let $\mathcal{H}(\Gamma)$ denote the Hilbert space of complex valued (Γ, χ)−automorphic functions f such that

$$f(\gamma z) = \chi(\gamma) f(z)$$

for all γ in Γ and z in H, with

$$\|f\|^2 = \int_F |f(z)|^2 d\mu(z) < \infty.$$

Here F is the fundamental domain of Γ in H and $d\mu$ is the invariant Riemannian measure given by

$$d\mu = \frac{dxdy}{y^2}.$$

The automorphic Laplacian $A(\Gamma, \chi)$ is defined by

$$A(\Gamma, \chi)f = -\Delta f$$

on smooth (Γ, χ)−automorphic functions $f \in L^2(F, d\mu)$. The closure of $A(\Gamma, \chi)$ in $\mathcal{H}(\Gamma)$ is a nonnegative, self-adjoint operator which is also denoted by $A(\Gamma, \chi)$.

8.3 Weyl's Law

If Γ is co-compact, then $A(\Gamma, \chi)$ has purely discrete spectrum consisting of eigenvalues with finite multiplicity. Define the spectral counting function $N(\lambda, \Gamma, \chi)$ by

$$N(\lambda, \Gamma, \chi) = |\{\lambda_j \leq \lambda\}|.$$

Weyl's law states that:

Theorem 1 *(Weyl)*

$$N(\lambda, \Gamma, \chi) \sim \frac{|F|}{4\pi}\lambda$$

as $\lambda \to \infty$.

8.4 Cofinite Subgroups

Let Γ be a cofinite discrete subgroup of $G = PSL(2, \mathbf{R})$ where G acts naturally on H; here cofinite means that $vol(\Gamma\backslash H) < \infty$. For a one dimensional unitary representation χ of Γ, χ is said to be singular or authentic at a cusp z_j of the fundamental domain F of Γ if $\chi(S_j) = 1$ where S_j is the generator of a parabolic group Γ_j which fixes the cusp z_j. The total degree of singularity of (χ, Γ) is the number of all pairwise inequivalent singular cusps of F, which we denote h_1.

If Γ is non-compact, cofinite and χ is singular, then $A(\Gamma, \chi)$ has an absolutely continuous spectrum $[1/4, \infty)$ of multiplicity h_1 and possibly discrete spectrum in $[0, \infty)$, a finite number of which lie in $[0, 1/4)$ and are called exceptional. The eigenfunctions corresponding to the finite or infinite sequence of eigenvalues λ_j in $[1/4, \infty)$ are called Maass cusp forms.

8.5 Selberg-Weyl Law

Let $N(\lambda, \Gamma, \chi)$ denote the spectral counting function

$$N(\lambda, \Gamma, \chi) = |\{\lambda_k \leq \lambda\}|,$$

i.e., the number of Maass cusp forms with eigenvalues less than λ. Let $M(\lambda, \Gamma, \chi)$ denote the number of poles of the determinant of the scattering matrix with modulus less than $\sqrt{\lambda}$. Selberg (1955) showed the following generalization of Weyl's law:

Theorem 2 *(Selberg) For any* Γ

$$N(\lambda, \Gamma, \chi) + M(\lambda, \Gamma, \chi) \sim \frac{|F|}{4\pi}\lambda$$

as $\lambda \to \infty$ *where*

$$M(\lambda, \Gamma, \chi) = -\frac{1}{4\pi}\int_{-T}^{T}\frac{\phi'}{\phi}(\frac{1}{2} + it)dt$$

where $\lambda = \frac{1}{4} + T^2$, $T > 0$, $\phi(s) = det\,\Phi(s, \Gamma, \chi)$ *and* $\Phi(s, \Gamma, \chi)$ *is the scattering matrix.*

8.6 Essentially Cuspidal Groups

Let

$$\Gamma(m) = ker(SL(2, \mathbf{Z}) \to SL(2, \mathbf{Z}/m\mathbf{Z})) = \{g \in SL(2, \mathbf{Z}) | g \equiv \begin{pmatrix} 1 & 0 \\ 0 & 1 \end{pmatrix} mod\, m\}.$$

If Γ is a subgroup of finite index of $\Gamma(1)$ and it contains $\Gamma(m)$, then Γ is called a congruence subgroup.

In the case that Γ is the modular group $\Gamma(1) = PSL(2, \mathbf{Z})$ or a congruence subgroup $\Gamma \subset \Gamma(1)$, then Selberg showed that

$$\frac{M(\lambda, \Gamma, \chi)}{\lambda} \sim 0.$$

A group Γ with this property is called essentially cuspidal. In other words, a group for which the set of eigenvalues $\{\lambda_j\}$ of $A(\Gamma)$ satisfies Weyl's law

$$N(\lambda, \Gamma, \chi) \sim \frac{|F|}{4\pi}\lambda$$

as $\lambda \to \infty$ is said to be essentially cuspidal.

8.7 Primitive Elements

Consider a subgroup $\Gamma \subset SL(2, \mathbf{R})$. An element γ in Γ is called primitive if γ is not a positive power of another element. Let $\mathcal{P}_\Gamma$ denote the conjugacy classes of primitive elements in Γ. For $p \in \mathcal{P}_\Gamma$ let $N(p) = \xi^2$ where ξ is the maximum eigenvalue of a representative of p. One notes that if $l(p)$ is the length of the unique prime closed geodesic whose homotopy class corresponds to p, then

$$N(p) = exp(l(p)).$$

8.8   Dimension of Cusps

Let $\Gamma(N)$ denote the principal congruence group of level N. $\Gamma(N)$ is a normal subgroup of $\Gamma(1) = SL(2, \mathbf{Z})$ of index

$$\mu = [\Gamma(1) : \Gamma(N)] = N^3 \prod_{p|N}(1 - p^{-2}).$$

The number of inequivalent cusps in this case is given by

$$\mu/N = N^2 \prod_{p|N}(1 - p^{-2}).$$

One notes that there are no elliptic elements in $\Gamma(N)$ if $N \geq 3$.

For the case of the Hecke congruence group of level N, $\Gamma_0(N)$, this group has finite index

$$\mu = [\Gamma_0(1) : \Gamma_0(N)] = N \prod_{p|N}(1 + \frac{1}{p})$$

and the number of inequivalent cusps is given by

$$h = \sum_{ab=N} \phi((a, b)).$$

In case N is prime, then there are two inequivalent cusps for $\Gamma_0(N)$ at ∞ and 0.

8.9   Eisenstein Series

For a cusp z_α, consider the parabolic subgroup $\Gamma_\alpha \subset \Gamma$ where

$$\Gamma_\alpha = \{\gamma \in \Gamma | \gamma z_\alpha = z_\alpha\}.$$

Here Γ_α is an infinite cyclic subgroup of Γ, generated by a parabolic generator S_α. There is an element g_α in $PSL(2, \mathbf{R})$ such that $g_\alpha \infty = z_\alpha$ and $g_\alpha^{-1} S_\alpha g_\alpha z = S_\infty z = z + 1$ for all $z \in H$.

Let $y(z) = Im(z)$. Then the Eisenstein-Maass series is given by

$$E_\alpha(z, s, \Gamma, \chi) = \sum_{\gamma \in \Gamma_\alpha \backslash \Gamma} y^s(g_\alpha^{-1} \gamma z) \bar{\chi}(\gamma).$$

Here $\bar{\chi}$ is the complex conjugate of χ. The series is absolutely convergent for $Re(s) > 1$ and there is an analytic continuation to the whole complex plane as a meromorphic function of s.

Since $p(z) = y^s$ is an eigenfunction of Δ with eigenvalue $\lambda = s(1 - s)$, then so is $E_\alpha(z, s)$. The Fourier series expansion for the Eisenstein series is given by:

Theorem 3

$$E_\alpha(g_\beta z, s) = \delta_{\alpha\beta} y^s + \phi_{\alpha\beta} y^{1-s} + \sum_{n\neq 0} \phi_{\alpha\beta}(n, s) W_s(nz)$$

where $W_s(z)$ is the Whittaker function.

As an example consider the case that Γ is the modular group which has a single cusp at $\alpha = \infty$; then as shown by Selberg and Chowla (1967)

$$\phi_{\alpha\alpha}(s) = \pi^{1/2} \frac{\Gamma(s - 1/2)}{\Gamma(s)} \frac{\zeta(2s - 1)}{\zeta(2s)}.$$

And in this case there is only one simple pole in the half-plane $Re(s) \geq 1/2$, viz. at $s = 1$.

8.10 Scattering Matrix

The scattering matrix is defined by $\Phi(s) = (\phi_{ab}(s))$. Let $\mathcal{E}(z, s)$ denote the column vector of Eisenstein series $E_\alpha(z, s)$ where α ranges over the inequivalent cusps. Then, Selberg has shown:

Theorem 4 *The following functional equations hold:*

$$\mathcal{E}(z, s) = \Phi(s)\mathcal{E}(z, 1 - s)$$

$$\Phi(s)\Phi(1 - s) = I.$$

For s with $Re(s) = 1/2$, Φ is unitary.
The scattering matrix elements $\phi_{\alpha\beta}(s)$ are holomorphic in $Re(s) \geq 1/2$ except for a finite number of simple poles in the segment $(1/2, 1]$. And if $s = s_j$ is a pole of $\phi_{\alpha\beta}(s)$, it is also a pole of $\phi_{\alpha\alpha}(s)$.
The poles of $E_\alpha(z, s)$ in $Re(s) > 1/2$ are among the poles of $\phi_{\alpha\alpha}(s)$ and they are simple. The residues of these poles are Maass forms, i.e. square integrable on the fundamental domain and orthogonal to cusp forms.

8.11 Scattering Determinant

The scattering determinant $\phi(s) = det\ \Phi(s)$ has the properties:

$$\phi(s, \Gamma, \chi)\phi(1 - s, \Gamma, \chi) = 1$$

and

$$|\phi(\frac{1}{2} + it, \Gamma, \chi)| = 1$$

for $t \in \mathbf{R}$. ϕ is holomorphic for $Re(s) = \sigma \geq 1/2$ except for finitely many poles σ_j in the interval $1/2 < s \leq 1$ of order less than or equal to h_1. One can show that $\phi(s, \Gamma, \chi)$ is a meromorphic function of order ≤ 2.

$\phi(s, \Gamma, \chi)$ has infinitely many poles in $\{Re(s) < 1/2\}$ and they correspond to zeros of $\phi(s, \Gamma, \chi)$ in the half-plane $\{Re(s) > 1/2\}$. Moreover, if $\rho = \sigma + i\tau$ is a pole of ϕ, then $\bar{\rho} = \sigma - i\tau$ is also a pole of ϕ and the numbers $1 - \rho$ and $1 - \bar{\rho}$ are zeros of ϕ.

ϕ has trivial poles (related to $\Gamma(s - 1/2)^{h_1}$) at the points $s = -1/2, -3/2, \ldots$. For a fixed group Γ, all nontrivial poles of $\phi(s, \Gamma, \chi)$ in the half-plane $\{Re(s) < 1/2\}$

lie in the strip $\{-a < Re(s) < 1/2\}$ for some $a = a(\Gamma, \chi) > 0$. Balslev and Venkov (1997) have extended Selberg's asymptotic formula regarding the distribution of poles of ϕ as follows:

Theorem 5 *(Balslev and Venkov) Define the counting function for poles of ϕ as*

$$P(\lambda, \Gamma, \chi) = |\{\rho = \sigma + i\tau | 0 \leq \tau \leq T, \sigma < 1/2\}|.$$

Then

$$N(\lambda, \Gamma, \chi) + P(\lambda, \Gamma, \chi) = \frac{|F|}{4\pi}T^2 - \frac{h_1}{\pi}T\log T + O(T)$$

as $T \to \infty$.

8.12   Example – $\Gamma_0(2)$

Hejhal (1983) computed the Eisenstein series for $\Gamma_0(2)$. He showed that

$$\phi_{11}(s) = \frac{\sqrt{\pi}}{(2^{2s} - 1)} \frac{\Gamma(s - 1/2)}{\Gamma(s)} \frac{\zeta(2s - 1)}{\zeta(2s)},$$

$$\phi_{22}(s) = \phi_{11}(s),$$

and

$$\phi_{12}(s) = \phi_{21}(s) = (2^s - 2^{1-s})\phi_{11}.$$

One can prove that $\phi_{11}(s)$ has no pole in $1/2 < s < 1$, i.e. no pole for $s(1-s) < 1/4$.

8.13   Kloosterman Sums and Scattering Matrices

Consider the case of a cofinite subgroup $\Gamma \subset PSL(2, \mathbf{R})$. In terms of Kloosterman sums $S_{\alpha\beta}(m, n, c, \Gamma, \chi)$ where

$$S_{\alpha\beta}(m, n, c, \Gamma, \chi) = \sum_{d \bmod c} \bar{\chi} \begin{pmatrix} a & b \\ c & d \end{pmatrix} exp\, 2\pi i \frac{ma + nd}{c}$$

the scattering matrices have the form

$$\phi_{\alpha\beta}(s, \Gamma, \chi) = \sqrt{\pi} \frac{\Gamma(s - 1/2)}{\Gamma(s)} \sum_{c>0} S_{\alpha\beta}(0, 0, c) c^{-2s}$$

$$\phi_{\alpha\beta}(n, s, \Gamma, \chi) = \frac{2\pi^s}{\Gamma(s)} |n|^{s-1/2} \sum_{c>0} S_{\alpha\beta}(0, n, c) c^{-2s}.$$

The case for which $\Gamma = \Gamma_0(8)$ with the primitive character $\chi(\gamma) = \chi(d)$ where

$$\chi(n) = \chi_8(n) = \begin{cases} 1 & n = \pm 1 \,(mod\, 8) \\ -1 & n = \pm 3 \,(mod\, 8) \\ 0 & \text{otherwise} \end{cases}$$

has been evaluated by Balslev and Venkov (1999). Note that $\chi_8(n) = \chi_8(-n)$ and $\chi_8(n) = \left(\frac{8}{n}\right)$. In this case there are four cusps, only $z_1 = 0$ and $z_3 = \infty$ are left open, in the sense to be discussed below. They show that the scattering matrix in this case is

$$\Phi(s, \Gamma, \chi) = \begin{pmatrix} 0 & \phi_{13}(s) \\ \phi_{31}(s) & 0 \end{pmatrix}$$

where

$$\phi_{13}(s) = \phi_{31}(s) = \sqrt{\pi}\frac{\Gamma(s - \frac{1}{2})}{\Gamma(s)}\frac{1}{8^s}\frac{L_8(2s - 1)}{L_8(2s)}$$

with the Dirichlet series

$$L_8(s) = \sum_{n=1}^{\infty}\frac{\chi_8(n)}{n^s}$$

for $Re(s) > 1$.

8.14   Selberg Trace Formula

The Selberg-Weyl law follows from the Selberg trace formula which is stated in the following theorem. First, assume h is a $\mathbf{C}$-valued function with the properties: (1) $h(r) = h(-r)$, (2) h is analytic in the strip $\{z||Imr| < 1/2 + \epsilon\}$ for some $\epsilon > 0$ and (3) $h(r) = O((1 + |r|^2)^{-1-\epsilon})$ in the strip. Then,

Theorem 6 *(Selberg)*

$$\sum_j h(r_j) = \frac{|F|}{2\pi}\int_{-\infty}^{\infty} r\tanh(\pi r)h(r)dr+$$

$$2\sum_{p\in\mathcal{P}_\Gamma}\sum_{k=1}^{\infty}\frac{\chi^k(p)logN(p)}{N(p)^{k/2} - N(p)^{-k/2}}g(klogN(p)) + \frac{1}{2\pi}\int h(r)\frac{\phi'}{\phi}(\frac{1}{2} + ir, \Gamma, \chi)dr-$$

$$\frac{h_1}{\pi}\int_{-\infty}^{\infty} h(r)\frac{\Gamma'}{\Gamma}(1 + ir)dr + \frac{1}{2}(h_1 - tr\Phi(1/2, \Gamma, \chi))h(0)-$$

$$2(h_1 log2 + \sum_{j=h_1}^{h_2} log|1 - \chi(S_j)|)g(0)$$

where $g(u) = \frac{1}{2\pi}\int_{-\infty}^{\infty} e^{-iru}h(r)dr$, h_1 is the total degree of singularity of χ relative to Γ and $\mathcal{P}_\Gamma$ runs through all primitive conjugacy classes in Γ. The sum on the left hand side is over all eigenvalues of the discrete spectrum of $A(\Gamma, \chi)$. h_2 is the number of essential cusps of the domain F and $h_1 \leq h_2$ is the total degree of singularity of χ with respect to Γ.

8.15   Selberg Zeta Function

The Selberg zeta function is defined by

$$Z(s, \Gamma, \chi) = \prod_{k=0}^{\infty} \prod_{p \in \mathcal{P}_\Gamma} det(I_n - \chi(p)N(p)^{-(s+k)})$$

where $\mathcal{P}_\Gamma$ runs over the set of all primitive hyperbolic conjugacy classes and $\chi :$ $\Gamma \to U(n)$ is a unitary representation.

Let $s_j = \frac{1}{2} + ir_j$ where $\lambda_j = \frac{1}{4} + r_j^2$ are eigenvalues of $A(\Gamma, \chi)$. Since $\lambda = \frac{1}{4} + r^2 \geq 0$, it follows that $r \in i[-\frac{1}{2}, \frac{1}{2}] \cup \mathbf{R}$ or $s = \frac{1}{2} + ir \in [0, 1] \cup \{\frac{1}{2} + i\mathbf{R}\}$.

Selberg (1956) showed that $Z(s, \Gamma, \chi)$ can be extended homomorphically to $\mathbf{C}$ and the zeros of $Z(s, \Gamma, \chi)$ are $0, -1, -2, ...$ (the trivial zeros) and zeros at $s = \frac{1}{2} + ir$ where $\lambda = \frac{1}{4} + r^2$, i.e. at

$$\frac{1}{2}(1 \pm (1 - 4\lambda_n(\chi))^{1/2}$$

where $n = 0, 1, 2, ...$ and $\{\lambda_n(\chi)\}$ are the eigenvalues of the Laplacian associated to χ; and zeros at the poles of the Eisenstein series, which are all in $Re(s) < 1/2$, (except for a finite number in $(1/2, 1]$).

8.16   Riemann Hypothesis for the Selberg Zeta Function

One notes the analogue of the Riemann hypothesis for $Z(s, \Gamma, \chi)$; viz., all zeros whose real parts are in $(0, 1)$ lie on the line $Re(s) = 1/2$ except for a finite number that lie on the real line, corresponding to the so-called exceptional eigenvalues $\lambda_n(\chi) \in (0, 1/4)$.

8.17   Properties of the Selberg Zeta Function

The functional equation for the Selberg zeta function in the finite volume case has the form

$$Z(1 - s, \Gamma, \chi) = Z(s, \Gamma, \chi)\phi(s, \Gamma, \chi)\Psi(s, \Gamma, \chi).$$

For details, see Venkov (1990) and Hejhal (1983) for a discussion and the definition of $\phi(s, \Gamma, \chi)$ and $\Psi(s, \Gamma, \chi)$.

If χ_1, χ_2 are two finite dimensional unitary representations of Γ, then

$$Z(s, \Gamma, \chi_1 \oplus \chi_2) = Z(s, \Gamma, \chi_1)Z(s, \Gamma, \chi_2).$$

If U^χ is a finite dimensional unitary representation of Γ induced by a finite dimensional representation χ of a subgroup Γ_1 of finite index in Γ, then

$$Z(s, \Gamma_1, \chi) = Z(s, \Gamma, U^\chi).$$

From this relationship, the Venkov-Zograf (1983) factorization follows:

Theorem 7 *(Venkov-Zograf) If* Γ_1 *is a normal subgroup of finite index in* Γ, *then*

$$Z(s,\Gamma,1) = \prod_{\chi \in \widehat{\Gamma_1 \backslash \Gamma}} Z(s,\Gamma,\chi)^{dim\chi}$$

where χ *runs over the set of all finite dimensional irreducible unitary representations of the factor group* $\Gamma_1 \backslash \Gamma$.

In other words, there is a relationship

$$L^2(\Gamma_1 \backslash X) = \oplus L^2(\Gamma \backslash X, \chi)^{deg(\chi)}$$

where Γ_1 is a normal subgroup of finite index in Γ and χ runs over the irreducible representations of $\Gamma_1 \backslash \Gamma$.

8.18 Selberg's Full Density Result

Let Γ be an arithmetic group, say $\Gamma = SL(2, \mathbf{Z})$, and let Γ_1 be a congruence subgroup of Γ, say $\Gamma_1 = \Gamma(l)$, the principal congruence subgroup of level l. Then Selberg showed that

$$N(\lambda, \Gamma_1) \sim [\Gamma : \Gamma_1] N(\lambda, \Gamma)$$

as $\lambda \to \infty$. Thus, the discrete spectra occur in all $L^2(\Gamma \backslash X, \chi)$ with full density.

8.19 The Selberg-Roelke Conjecture

The Selberg-Roelke conjecture states that for every cofinite discrete subgroup Γ of $SL(2, \mathbf{R})$ the corresponding upper half plane has infinitely many cusp forms associated to $A(\Gamma, \chi)$. That is, Selberg and Roelke conjectured that all discrete subgroups are essentially cuspidal.

This conjecture has not been proven and more recent work seems to indicate that the conjecture is not true. Phillips and Sarnak (1985) showed that under certain conditions $M(\lambda, \Gamma, \chi) \geq c(\epsilon)\lambda^{1-\epsilon}$ for every $\epsilon > 0$, generically for Γ in the Teichmüller space of a given congruence group Γ_0. Wolpert (1994) and Judge and Phillips (1997) considered the singular perturbation of the spectrum of $A(\Gamma_0)$ for certain families of groups Γ_0. Under specific assumptions (viz., the multiplicity one conjecture), Wolpert showed that the generic member of each family has at most a finite number of cusp forms with certain symmetries.

8.20 Dissolving Eigenvalues

What is happening can be understood as follows. The Maass cusp form eigenvalues are dissolving into the continuous spectrum so the asymptotic contribution of $M(\lambda, \Gamma, \chi)$ increases. For the physicist this is interesting because it means that resonances are created in the region $Re(s) < 1/2$ where $\lambda = s(1-s)$ is the usual spectral parameter.

Specifically, Sarnak (1986) has shown that for discrete subgroups of $SO(n, 1)$ the nonvanishing of certain Rankin-Selberg L-series at special points on their critical line implies the disappearance of discrete spectra under deformations of the type $L^2(\Gamma\backslash X, \chi_\theta)$.

As an example, consider the following deformation for $\Gamma(2)$, which is freely generated by

$$A = \begin{pmatrix} 1 & 2 \\ 0 & 1 \end{pmatrix}$$

and

$$B = \begin{pmatrix} 1 & 0 \\ -2 & 1 \end{pmatrix}.$$

For $0 < \theta < 1$, set $\chi_\theta(A) = 1$ and $\chi_\theta(B) = exp(2\pi i\theta)$. Then $\Gamma(2)\backslash H$ has three cusps $\{0, 1, \infty\}$, of which only ∞ is authentic. As shown in Phillips and Sarnak (1985) there are no cusp forms in this case for $0 < \theta < 1$ with eigenvalues in the interval $[0, \frac{1}{4}]$. For further discussion of this model see Phillips and Sarnak (1992).

8.21 Dissolving Functional

Phillips and Sarnak (1985a) considered the dissolving functional

$$< \dot{\Delta}\phi, E > = \int_{\Gamma\backslash H} E \dot{\Delta}\phi \, dx dy / y^2$$

where E is an Eisenstein series for Γ, ϕ is a Maass cusp form and $\dot{\Delta}$ is the first derivative of the perturbed Laplacian. By selecting ϕ to be a Hecke eigenform, they found that for generic $\Gamma \in \mathbf{T}(\Gamma)$, the Teichmüller space, and for every $\epsilon > 0$,

$$M(\lambda, \Gamma, \chi) \sim \lambda^{1-\epsilon}.$$

8.22 Deformation of Discrete Groups

Phillips and Sarnak (1985b) considered the case that $\Gamma_0(q)$ is the Hecke congruence subgroup of level q, i.e.

$$\Gamma_0(q) = \{ \begin{pmatrix} a & b \\ c & d \end{pmatrix} \in SL(2, \mathbf{Z}), q|c \}$$

where q is prime. Let $Q(z)$ be a holomorphic cusp form of weight 4 for $\Gamma_0(q)$ which is also a Hecke eigenform. Let $\{u_j\}$ be an orthonormal basis for Maass cusp forms for $\Gamma_0(q)$, which are also Hecke eigenforms. Let $\lambda_j = \frac{1}{4} + r_j^2$ be the Maass eigenvalue of u_j. Let $L(Q \otimes u_j, s)$ denote the Rankin-Selberg L−function. L is proportional to the dissolving functional:

$$< \dot{\Delta}\phi, E > = c(\lambda)L(Q \otimes \phi, \frac{1}{2} + ir)$$

where ϕ is a Maass cusp form with eigenvalue $\lambda = \frac{1}{4} + r^2$, and $c(\lambda) > 0$. So the disappearance of the cusp form is equivalent to the nonvanishing of the special value of the Rankin-Selberg L–function.

For a real number r, let $m(r)$ denote the dimension of the space of cusp forms of $\Gamma_0(q)$ with eigenparameter r. One says that $\Gamma_0(q)$ has a cusp form degeneracy of order β if

$$m(r) << r^\beta.$$

For the modular group it is conjectured that all the eigenvalues are simple, i.e. $m(r) \leq 1$.

The extended Lindelöf hypothesis states that for $\epsilon > 0$

$$L(Q \otimes u_j, \frac{1}{2} + it) << [(1 + |t|)(1 + |r_j|)]^\epsilon (\cosh \pi r_j)^{1/2}.$$

Phillips and Sarnak (1985b) have shown:

Theorem 8 *(Phillips and Sarnak) If the extended Lindelöf hypothesis holds for $L(Q \otimes u_j, s)$ and the cusp form degeneracy of $\Gamma_0(q)$ is of order β, then for generic $\Gamma \in T(\Gamma_0(q))$*

$$M(\lambda, \Gamma, \chi) >> \lambda^{1-\beta-\epsilon}$$

for all $\epsilon > 0$.

For generic Γ, it is expected that β will be arbitrarily small, which leads to the above cited result. The proof of this theorem hinges on the result of Phillips and Sarnak (1985a) that a cusp form v associated with eigenparameter r is dissolved under perturbation if $L(Q \otimes v, \frac{1}{2} + ir) \neq 0$. Numerical work of Phillips and Sarnak (1985) showed that $L(Q \otimes u_j, \frac{1}{2} + ir_j)$ differed from zero for the first nine eigenfunctions which they tested. Later, Deshouillers and Iwaniec (1986) proved that on average these L–function values are not zero and thus many cusp forms are transformed into resonances.

8.23 Fermi's Golden Rule

The singular set, $\sigma(\Gamma) \subset \mathbf{C}$, is defined as follows. As above, let $\lambda = \frac{1}{4} + r^2$. For $Im(r) \leq 0, r \neq 0$, let $m(r)$ denote the multiplicity of λ and for $Im(r) > 0$, let $m(r)$ denote the order of the pole of ϕ at $s = \frac{1}{2} + ir$ plus the multiplicity of λ. For $r = 0$, $m(r)$ is twice the multiplicity of cusp forms at $\lambda = 1/4$ plus $(Tr(\Phi(\frac{1}{2})) + n)/2$, where n is the number of cusps. Then the singular set $\sigma(\Gamma)$ is the set of r's in $\mathbf{C}$ such that $m(r) > 0$, counted with their multiplicities. Thus, $m(r) \geq 0$ and $m(r)$ is integer valued. For $r \in \mathbf{R}, r \neq 0$, $m(r)$ is the dimension of the cusp forms with eigenvalue λ.

Let $\mathbf{T}(\Gamma)$ be the Teichmüller space and let Γ vary on a real algebraic curve Γ_t in $\mathbf{T}(\Gamma_0)$. Phillips and Sarnak (1992) show that in this case $\sigma(\Gamma_t)$ has at most algebraic singularities. In addition, Phillips and Sarnak (1992) developed an explicit formula for

$$I = \frac{d^2}{dt^2} Im \, r_j(t)|_{t=0}$$

where $0 \neq r_j(0) \in \mathbf{R}$ corresponds to a simple cusp form for $\Gamma_t|_{t=0}$. As with Fermi's Golden Rule in physics $I \neq 0$ gives a sufficient condition for a point $r(0)$ in $\sigma(\Gamma_0)$ to dissolve into a resonance, i.e. a pole of the Eisenstein series, $r(t)$ with $Im\, r(t) > 0$. More precisely, they showed:

Theorem 9 *(Phillips and Sarnak) If $r(0) \in \sigma(\Gamma_0), Im\, r(0) = 0, r(0) \neq 0$ and $m(r(0)) = 1$ and ψ_0 is the corresponding cusp form; then if $r(t)$ is the corresponding trajectory in $\sigma(\Gamma_t)$, then*

$$\frac{d^2}{dt^2} Im\, r(t)|_{t=0} = \frac{1}{2} \sum_k |(E_k(.,s_0), \dot{\Delta}\psi_0)|^2$$

where $\dot{\Delta}$ is the first order variation of Δ_t at $t = 0$ and $E_k(z, s_0)$ is the Eisenstein series for the kth cusp at $s_0 = 1/2 + ir(0)$.

Here $(E_k(.,s_0), \dot{\Delta}\psi_0)$ is the value of the Rankin-Selberg L−function $L(\psi_0 \otimes Q, s_0)$, where s_0 is on the critical line of the L-function, Γ_0 is a congruence subgroup of $SL(2, \mathbf{Z})$ and the curve Γ_t is generated by a holomorphic cusp form Q of weight 4. As noted above, Phillips and Sarnak (1985) have showed that an L^2 cusp form in $\Gamma\backslash H$ is dissolved under deformation if the Rankin-Selberg L−function does not vanish at a special point on its critical line.

Wolpert (1989) has shown that $I \neq 0$ is both necessary and sufficient to first order for the destruction of a cusp form. For further details, see Wolpert (1989) and Phillips and Sarnak (1992).

8.24   Sarnak's Conjecture

Sarnak (1986) has proposed the following conjecture: Γ is essentially cuspidal if and only if Γ is arithmetic. Phillips and Sarnak (1991) modified this conjecture to allow for certain non-congruence arithmetic groups to be non-cuspidal (v.i.).

8.24.1   Higher Rank Case

Sarnak (1986) considered also the higher rank symmetric space M of non-compact type; viz., his conjecture if $rank(M) > 1$ is that Γ is essentially cuspidal. In support of this Efrat (1983) has shown

Theorem 10 *(Efrat) Let Γ be a nonuniform (irreducible) lattice in $(SL(2, \mathbf{R}))^n$; then Γ is essentially cuspidal and*

$$N_{cusp}(R) = C_\Gamma R^{2n} + O(R^{2n-1}/logR).$$

For the higher rank spaces, one can apply the Selberg trace formula due to the super-rigidity results of Selberg and Margulis. In this case Γ is commensurable with a Hilbert modular group, in fact Γ must contain a congruence subgroup of the Hilbert modular group by the Serre's (1970) theorem. We also note that Γ is arithmetic, fitting Sarnak's conjecture.

For a discussion of scattering theory and automorphic functions on a symmetric space of rank greater than one, see Phillips (1996), where he treats the case of the Hilbert modular group.

8.25 Multiplicity Hypothesis

The dissolving functional approach posits the multiplicity hypothesis that the eigenvalue of the to-be-dissolved cusp form is simple. More precisely, it states that the newform eigenfunctions are assumed to lie in simple eigenspaces.

For $K > 0$, let $N_K(\lambda)$ denote the number of Maass cusp form eigenvalues λ of $\Delta(0)$ with multiplicity $m(\lambda) \leq K$.

Judge and Phillips (1997) have shown:

Theorem 11 *(Judge - Phillips) Let χ_θ denote the family of characters on $\Gamma(2)$ considered in Phillips and Sarnak (1994). If for some $K > 0$*

$$lim\, inf_{\lambda \to \infty} \frac{N_K(\lambda)}{\lambda} > 0 \quad (*)$$

then the spectrum of $(\Gamma(2), \chi_\theta)$ is not essentially cuspidal for the generic θ.

Under the collar deformation (see Wolpert (1994)) if (*) holds, then a positive percentage of Maass forms dissolve under perturbations for $SL(2, \mathbf{Z})$, v. Judge and Phillips (1996).

8.26 Judge's Theorem

Judge (1997) has considered perturbation theory of the Laplacian of hyperbolic surfaces with conical singularities belonging to a fixed conformal class. Let $\mathbf{T}_{g,n,m}$ denote the Teichmüller space of hyperbolic metrics for a surface of genus g, with n conical singularities, and m cusps. He has shown:

Theorem 12 *(Judge) If there is a metric in $\mathbf{T}_{g,n-1,m}$, $m > 0$, whose Laplacian has only simple embedded eigenvalues, then the Laplacian of a generic point in $\mathbf{T}_{g,n,m}$ has no Maass cusp forms. In particular the Laplacian of the generic point has only finitely many L^2 eigenfunctions.*

In this paper Judge looks at the Hecke triangle groups G_k. Hejhal (1992) has performed numerical studies on the spaces H/G_k. Hejhal found no even eigenfunctions for the non-arithmetic Hecke groups; and the generalized eigenfunctions which span the continuous spectrum are even (v., Venkov (1990)). Recall that G_k are arithmetic for $k = 3, 4, 6, \infty$ and these are expected to be essentially cuspidal. For $k > 6$, G_k is not arithmetic and one expects G_k to have no associated discrete spectrum. Judge has shown:

Theorem 13 *(Judge) If the eigenvalues associated to the group $SL(2, \mathbf{Z})$ and new-form eigenvalues associated to $\Gamma_0(2)$ are simple and these sets of eigenvalues are disjoint, then at most a countable number of metrics in the Hecke triangle deformation $H(t)$ have even cusp forms.*

8.27   Phillips-Sarnak Conjecture

Theorem 14 *(Phillips-Sarnak) Consider the operators $A(\Gamma(2), \chi_\theta)$ for characters $\chi_\theta, 0 < \theta < 1$. Then for at most countably many $\theta \in (0, 1)$ the pair $(\Gamma(2), \chi_\theta)$ is essentially cuspidal.*

The Phillips-Sarnak conjecture for $(\Gamma(2), \theta)$ is that $(\Gamma(2), \theta)$ is essentially cuspidal only for $\theta = n/d = j/8$ for $0 \leq j \leq 7$.

More generally Phillips and Sarnak have conjectured that there exist non-congruence groups which admit no cusp forms except for those which they must inherit from a congruence overgroup.

8.28   Balslev-Venkov Theorem

Consider the principal congruence subgroup of $PSL(2, \mathbf{Z})$ of level 2. $\Gamma(2)$ is generated by $A = \begin{pmatrix} 1 & 2 \\ 0 & 1 \end{pmatrix}$ and $B = \begin{pmatrix} 1 & 0 \\ -2 & 1 \end{pmatrix}$. Each element of $\Gamma(2)$ can be written uniquely as

$$\gamma = A^{n_1} B^{m_1} ... A^{n_k} B^{m_k}$$

for $n_j, m_j \in \mathbf{Z}, 1 \leq j \leq k$. Set $P_A(\gamma) = n_1 + ... + n_k$ and define Γ_d by

$$\Gamma_d = \{\gamma \in \Gamma(2) | P_A(\gamma) \equiv 0 \, mod \, d\}.$$

Balslev and Venkov (1997) have shown:

Theorem 15 *(Balslev-Venkov) Γ_d is a congruence subgroup of $PSL(2, \mathbf{Z})$ only for $d = 1, 2, 4, 8$.*

A curve Ω_{T_0} is said to separate resonances of $A(\Gamma(2), \chi_{n,d})$ from the spectrum for $\epsilon_1 \leq n/d \leq \epsilon_2$ and starting from height T_0 if Ω_{T_0} satisfies: (1) Ω_{T_0} lies in the half-plane $Re(s) < \frac{1}{2}$; (2) Ω_{T_0} contains two components, which are symmetric with respect to $\{Im(s) = 0\}$; (3) the upper component of Ω_{T_0} is a continuous curve without self-intersections starting at the point iT_0 and going to $i\infty$; and (4) the domain between the line $Re(s) = 1/2$ and Ω_{T_0} does not contain any poles of $\phi(s, \Gamma(2), \chi_{n,d})$ for any $\frac{n}{d} \in [\epsilon_1, \epsilon_2]$.

For $d = 1, 2, 4, 8$ the scattering determinant $\phi(s, \Gamma_d)$ is a meromorphic function of order one and Weyl's law hold for Γ_d. Balslev and Venkov have developed a sufficient condition in order for $A(\Gamma_d)$ to satisfy Weyl's law:

Theorem 16 *(Balslev-Venkov) If the operator $A(\Gamma(1), \chi_2)$ has a simple spectrum and for $[\epsilon_1, \epsilon_2] \subset (0, 1/4)$ there exists a curve separating from some height $T_0 = T_0(\epsilon_1, \epsilon_2)$ the resonances of $A(\Gamma(2), \chi_{n,d})$ from the spectrum $\frac{1}{2} + ir$ for all (n, d) with*

$$\epsilon_1 \leq n/d \leq \epsilon_2,$$

then Γ_d are essentially cuspidal for only finitely many d.

8.29 Riemann Manifold with Cusps

Consider the case of (M, g) where M is a complete surface with finite area and hyperbolic ends. M has the form $M = M_0 \cup Y_1 \cup ... \cup Y_h$ where M_0 is a compact Riemannian manifold with boundary and Y_j are a finite number of cusps or ends. The spectrum of the Laplace operator on M consists of a sequence of discrete eigenvalues $0 = \lambda_0 < \lambda_1 \leq \lambda_2...$ and an absolute continuous spectrum which is the half line $[\frac{(d-1)^2}{4}, \infty) = [1/4, \infty)$ where $d = dim(M)$. Let $N(\lambda)$ denote the number of eigenvalues of Δ, counted with multiplicity, which are less than λ. The analogue of $N(\lambda)$ for the continuous spectrum is given by the winding number

$$M(r) = -\frac{1}{4\pi} \int_r^r Tr(\Phi'(\frac{N}{2} + i\lambda)\Phi(\frac{N}{2} - i\lambda))d\lambda$$

where $N = d - 1$ and Φ is the $h \times h$ scattering matrix. Let

$$\phi(s) = det(\Phi(s))$$

and poles of this meromorphic function are called resonances. For the case at hand Müller (1986) has shown that

$$|M(r)| \leq C_1(1 + r^d).$$

Weyl's law is then

$$N(\lambda) + M(r) \sim (4\pi)^{-d/2} \frac{Vol(M)}{\Gamma(\frac{d}{2} + 1)} \lambda^{d/2}$$

as $\lambda \to \infty$ and $\lambda = \frac{N^2}{4} + r^2$.

As we know, eigenvalues which are embedded in the continuous spectrum are very unstable. Colin de Verdiere (1983) has shown that for a generic metric on M, the Laplace operator has only finitely many eigenvalues and all of them lie below the continuous spectrum. More specifically he has shown that a generic compactly supported conformal deformation of the metric will destroy all embedded eigenvalues and convert them into poles of the scattering determinant $\phi(s)$. The work of Colin de Verdiere provided impetus to the conjectures of Sarnak and Phillips.

Let $N_p(T)$ denote the number of poles ρ of $\phi(s)$, counted with order, satisfying $|\rho| < T$. Based on Colin de Verdiere's result, Müller (1992) has shown in this case

Theorem 17 *(Müller) For a generic metric on M*

$$N_p(T) = \frac{Vol(M)}{2\pi} T^2 + o(T^2)$$

as $T \to \infty$.

8.30   Hecke Operators

Define the operators T_n by

$$T_n f(z) = \frac{1}{\sqrt{n}} \sum_{ad=n, b \bmod d} f(\frac{az+b}{d})$$

and

$$T_\infty f(z) = -\Delta f(z) = -y^2 (\partial_x^2 + \partial_y^2) f$$

for $n = 1, 2, \ldots$ The family $\{T_n\}$ forms a commutative family of self-adjoint operators and $\mathcal{H} = L^2(\Gamma \backslash H)$ decomposes into Hecke invariant subspaces:

$$\mathcal{H} = \mathcal{R} \oplus \mathcal{E} \oplus \mathcal{C}.$$

Here $\mathcal{R}$ is the constant functions, $\mathcal{E}$ is spanned by the Eisenstein functions and $\mathcal{C}$ is orthogonal to these spaces and consists of nonconstant discrete eigenfunctions, the cuspidal forms. In other words, $\mathcal{R} \oplus \mathcal{C}$ and $\mathcal{E}$ are the subspaces spanned by discrete and continuous eigenfunctions, respectively.

On $\mathcal{C}$ there is a simultaneous orthonormal basis of $\{T_n\}$ say $\psi_j(z)$ where

$$T_p \psi_j = \rho_j(p) \psi_j$$

and

$$T_\infty \psi_j = (\frac{1}{4} + r_j^2) \psi_j = \lambda_j \psi_j.$$

8.31   Ramanujan Conjecture

The Ramanujan conjecture for $\rho_j(p)$ is

$$|\rho_j(p)| \leq 2$$

for all j and prime p. One notes that the Ramanujan conjecture holds for Eisenstein series $E(z, \frac{1}{2} + it)$. The inequality above, is equivalent to

$$|<T_p f, f>| \leq 2 <f, f>$$

for all $f \in L^2(\Gamma \backslash H)$ for which $<f, 1> = 0$. Thus, the Ramanujan conjecture means that

$$Spec(T_p|\{1\}^\perp) \subset [-2, 2]$$

where $Spec(T)$ is the spectrum of T.

8.32   Sato-Tate Conjecture

A sequence $\{x_j\}$ in a topological space X is said to be μ−equidistributed if for all $f \in C_c(X)$

$$lim_{N \to \infty} \frac{1}{N} \sum_{j \leq N} f(x_j) \to \int_X f(x) d\mu(x)$$

where μ is a Radon measure.

The Sato-Tate conjecture states that for fixed j, $\rho_j(p)$ is $\mu-$ equidistributed for the semi-circle distribution.

8.33 Sarnak's Distribution Theorems

The Sato-Tate conjecture for $\rho_j(p)$ looks at the distribution of $\rho_j(p)$ for fixed j. Sarnak (1987) considered the distribution of $\rho_j(p)$ as a function of both j and p. First, for fixed p and $\alpha \geq 2$, set

$$N(\alpha, K) = |\{j | r_j \leq K, |\rho_j(p)| \geq \alpha\}|,$$

where $\lambda_j = \frac{1}{4} + r_j^2$.

Theorem 18 *(Sarnak)*

$$N(\alpha, K) << K^{2 - \frac{\log \alpha/2}{\log p}}.$$

In other words, almost all $\rho_j(p)$ lie in $[-2, 2]$.

Now, consider variations in both j, p and set

$$x_j = (\rho_j(2), \rho_j(3), \rho_j(5), ...)$$

$$x_j \in X = \prod_p [-n(p), n(p)]$$

where $n(p) = \|T_p\| = p^{1/2} + p^{-1/2}$.

Theorem 19 *(Sarnak) $\{x_j\}$ is $\mu-$ equidistributed on X where $\mu = \prod_p \mu_p$ where*

$$d\mu_p(x) = \begin{cases} \frac{(1+p)}{(n(p)^2 - x^2)} \mu_{ST} & \text{if } |x| < 2 \\ 0 & \text{otherwise} \end{cases}$$

where $\mu_{ST} = \frac{1}{\pi} \sqrt{1 - \frac{x^2}{4}} dx$ is the Sato-Tate measure.

8.34 Phillips-Sarnak Conjecture for $\bar{\Gamma}_0(N)$

Balslev and Venkov (1999) have proven the following results related to the Phillips-Sarnak conjecture regarding the disappearance of embedded eigenvalues for the automorphic Laplacian $A(\bar{\Gamma}_0(N), \chi)$ for the case of projective Hecke groups $\Gamma = \bar{\Gamma}_0(N) \subset PSL(2, \mathbf{R})$ where χ is a primitive character mod N. More precisely, χ denotes the real primitive Dirichlet character mod N:

$$\chi(\gamma) = \chi_N(n)$$

where $\gamma = \begin{pmatrix} a & b \\ Nc & n \end{pmatrix} \in \bar{\Gamma}_0(N)$ with $an - bcN = 1$. For $N = |d|$ these are identical to the symbols $\left(\frac{d}{n}\right)$. For details, see Balslev and Venkov (1999). The

Riemann surfaces associated with $\bar{\Gamma}_0(N)$ have $d(N)$ cusps z_d where $d(N)$ is the number of divsors of N. Let $\bar{\Gamma}_d = \{\gamma \in \bar{\Gamma}_0(N) | \gamma z_d = z_d\}$. Let S_d denote the generator of $\bar{\Gamma}_d$. Three cases are considered by Balslev and Venkov: (1) N_1 any positive square-free integer with $N_1 \equiv 1 \bmod 4$, (2) N_2 any square-free positive integer with $N_2 \equiv 3 \bmod 4$, and (3) N_3 any square-free positive integer with $N_3 \equiv 2 \bmod 4$. In these cases the inequivalent cusps are identified, and which are open and closed. E.g., in case (2) Balslev and Venkov show:

Theorem 20 *(Balslev and Venkov) There is a complete set of inequivalent cusps given by $z_d = 1/d$ where $d|4N_2, d > 0$. The set of all parabolic generators, S_d, is given by*

$$S_d = \begin{pmatrix} 1 - 4N_2 & 4N_2/d \\ -4N_2 d & 1 + 4N_2 \end{pmatrix}$$

where $d|N_2, d > 0$,

$$S_d = \begin{pmatrix} 1 - 2N_2 & N_2/d_1 \\ -4N_2 d_1 & 1 + 2N_2 \end{pmatrix}$$

where $d_1|N_2, d_1 > 0, d = 2d_1$,

$$S_d = \begin{pmatrix} 1 - 4N_2 & 4N_2/d_2 \\ -16N_2 d_2 & 1 + 4N_2 \end{pmatrix}$$

where $d_2|N_2, d_2 > 0, d = 4d_2$. The character χ_{4N_2} is singular for the group $\bar{\Gamma}_0(4N_2)$. The open cusps are z_d with $d|N_2, d > 0$ or $d = 4d_2$ where $d_2|N_2, d_2 > 0$ and $\chi(S_d) = 1$; the closed cusps are z_d for $d = 2d_1, d_1 > 0, d_1|N_2$ and $\chi(S_d) = -1$. So the primitive character χ closes one third of the cusps in this case.

We note that Huxley (1984) has shown for $\bar{\Gamma}_1(N)$ that the determinant of the scattering matrix is given by

$$det\ \Phi(s, \bar{\Gamma}_1(N), 1) = (-1)^{(k-k_0)/2} \left(\frac{\Gamma(1-s)}{\Gamma(s)} \right)^k \left(\frac{\mathcal{A}}{\pi^k} \right)^{1-2s} \prod_\chi \frac{L(2 - 2s, \bar{\chi})}{L(2s, \chi)}$$

where k is the number of cusps, $-k_0 = tr\phi(1/2, \bar{\Gamma}_1(N), 1)$ and $\mathcal{A}$ is a positive integer composed of the primes dividing N and the product has k terms, in each of which χ is a Dirichlet character to some modulus dividing N and $L(s, \chi)$ is the corresponding Dirichlet L−series. In particular, one sees that

Theorem 21 *(Huxley) The scattering determinant $det\ \Phi(s, \bar{\Gamma}_1(N), 1)$ is a meromorphic function of order one.*

Using the factorization formula for the Selberg zeta function and the Huxley explicit formula just described allows Balslev and Venkov to prove that $A(\bar{\Gamma}_0(N), \chi)$ has an infinite sequence of eigenvalues which satisfies Weyl's law.

Theorem 22 *(Balslev and Venkov) For $\bar{\Gamma}_0(N)$ with real primitive character χ mod N the Weyl law holds.*

Problem 1 *Does the whole spectrum of $A(\bar{\Gamma}_0(N), \chi)$ belong to $[1/4, \infty)$?*

Let $\mathcal{H}(\Gamma) = L^2(F, d\mu)$ denote the Hilbert space with inner product

$$< v_j, v_k >= \int_F v_j(z)\overline{v_k(z)}d\mu(z)$$

where F is the fundamental domain for Γ where $d\mu(z) = dxdy/y^2$.

Let $T(n)$ denote the Hecke operator

$$T(n)f(z) = \frac{1}{\sqrt{n}} \sum_{ad=n} \chi(a) \sum_{b \bmod d} f(\frac{az+b}{d})$$

where f is a $(\bar{\Gamma}_0(N), \chi)$–automorphic function. Let $U(q) = T(q)$ for $q|N$ and let $T(p)$ be reserved for $p \nmid N$. $T(n)$ acts in the subspace of cusp forms of weight zero $\mathcal{C} \subset \mathcal{H}(\Gamma) = L^2(F, d\mu)$. These operators are bounded and $\chi(n)$ hermitian

$$< T(n)f, g >= \chi(n) < f, T(n)g > .$$

All $T(n)$ commute with each other and with the Laplacian $A(\bar{\Gamma}_0(N), \chi)$. One can take in $\mathcal{C}$ a common basis of eigenfunctions $\{v_j(z)\}$ for all $T(n)$ and A, viz.

$$Av_j = \lambda_j v_j$$

$$T(m)v_j = \rho_j(m)v_j$$

where $(m, N) = 1$.

8.34.1   Newforms and Old

We continue with the case $\bar{\Gamma}_0(N), \chi$. For χ mod M and $v(z) \in \mathcal{C}$, then $v(dz) \in \mathcal{C}$ whenever $dM|N$. The set of old forms $\mathcal{C}^{old}$ is given by the subspace of $\mathcal{C}$ spanned by all forms $v(dz)$ where $v(z)$ is defined for $\Gamma_0(M)$ with character χ mod $M, M < N, dM|N$ and v is a common eigenfunction of $T(m)$ for $(m, M) = 1$. The space of newforms $\mathcal{C}^{new}$ is the orthogonal complement so that

$$\mathcal{C} = \mathcal{C}^{old} \oplus \mathcal{C}^{new}.$$

One notes that there are no old forms for $(\bar{\Gamma}_0(N_1), \chi_{N_1}), (\bar{\Gamma}_0(4N_2), \chi_{4N_2})$, and $(\bar{\Gamma}_0(4N_3), \chi_{4N_3})$.

One can show that $U(q)$ are unitary and have only the eigenvalues ± 1 for the case $A(\bar{\Gamma}_0(N), \chi)$ with χ not trivial. In the trivial case the operators $U(q)$ are not normal, or they are normal but not unitary in the space of newforms.

8.34.2   Multiplicity One Theorem

Let v_j be a cusp form, i.e. an eigenfunction of $A(\Gamma, \chi)$ with $Av_j = \lambda_j v_j$ where $\lambda_j = s_j(1 - s_j)$. Then one also has $T(n)v_j(z) = \rho_j(n)v_j(z)$ for $(n, N) = 1$. One can show:

Theorem 23 *(Balslev-Venkov) There exists a unique basis of eigenfunctions for all operators $A(\Gamma_0(N), \chi), T(n), T^*(n), n \geq 1$ in the space of cusp forms; each eigenfunction $v_j(z)$ for this basis can be taken with the normalization $\rho_j(1) = 1$ and is uniquely determined by the eigenvalues $\lambda_j, \rho_j(n), (n, N) = 1$.*

8.34.3   Non-vanishing Hecke L-functions

Any eigenfunction v_j of A has the Fourier decomposition

$$v_j(z, \Gamma, \chi) = \sqrt{y} \sum_{n \neq 0} \rho_j(n) K_{s_j - \frac{1}{2}}(2\pi |n| y) e^{2\pi i n x}.$$

Attached to v_j are the L–functions given by the Dirchlet series

$$L(s, v_j) = \sum_{n=1}^{\infty} \frac{\rho_j(n)}{n^s}$$

and

$$L(s, \hat{v}_j) = \sum_{n=1}^{\infty} \frac{\overline{\rho_j(n)}}{n^s}$$

which are absolutely convergent for $Re(s) > 1$. One can show that

Theorem 24 *(Balslev and Venkov) There is an Euler product representation for* $L(s, v_j)$, *viz.*

$$L(s, v_j) = \prod_p (1 - \rho_j(p) p^{-s} + \chi(p) p^{-2s})^{-1}$$

where the product is over all primes p.

Based on general theory of L-series (v., Murty and Murty (1997)) one can show the following non-vanishing result:

Theorem 25 *(Balslev and Venkov) The L–functions $L(s, v_j)$ and $L(s, \hat{v}_j)$ are regular for $s = 1 + it, s = it, t \in \mathbf{R}$ and $L(1 + it, v_j) \neq 0, L(it, v_j) \neq 0$ and similarly for $\hat{v}_j$, for $j = 1, 2, \dots$.*

8.34.4   Perturbations

Set $L = A(\Gamma, \chi)$ for $\Gamma = \bar{\Gamma}_0(N)$. Let $\omega(z)$ be a holomorphic modular form of weight 2 for Γ; so $\omega(\gamma z) = (cz + d)^2 \omega(z)$ for $\gamma = \begin{pmatrix} a & b \\ c & d \end{pmatrix}$. Define

$$\chi_\alpha(\gamma) = exp(2\pi i \alpha Re \int_{z_0}^{\gamma z_0} \omega(t) dt)$$

for $\alpha \in \mathbf{R}$ and $z \in H$. This defines a family of unitary characters for Γ, which is independent of the choice of z_0. Consider the family of self-adjoint operators $A_\alpha = A(\Gamma, \chi \cdot \chi_\alpha)$. One can show that A_α is unitarily equivalent to the operator

$$L(\alpha) = L + \alpha M + \alpha N$$

where

$$M = -4\pi i y^2 (\omega_1 \frac{\partial}{\partial x} - \omega_2 \frac{\partial}{\partial y})$$

and

$$N = 4\pi^2 y^2 (\omega_1^2 + \omega_2^2)$$

with $\omega = \omega_1 + i\omega_2$.

Functions are said to be odd if $f(-x + iy) = -f(x + iy)$ and functions are said to be even if $f(-x + iy) = f(x + iy)$. Automorphic functions f, where $f(\gamma z) = \chi(\gamma)f(z)$, are allowed to be either even or odd. One can check that the operator M maps odd functions to even and even to odd. In addition, M and N map $(\Gamma, \chi^{(\alpha)})$-functions into $(\Gamma, \chi^{(\alpha)})$-functions.

The approach of Balslev and Venkov is to construct the form ω from the holomorphic Eisenstein series of weight 2,

$$P(z) = E_2(z) = 1 - 24 \sum_{n=1}^{\infty} \sigma(n)e^{2\pi inz},$$

for the modular group so that the corresponding perturbed operator leaves the same cusps open and closed which are already open and closed by the primitive character χ. That is, one looks for a holomorphic modular form $\omega(z)$ of weight 2 for the group Γ which is exponentially small at the open cusps and behaves like $j^2(g_\beta, z)$ at each closed cusp β, where $j(\gamma, z) = (cz + d)$. Recall, one says that there is a closing of a cusp if the continuous spectrum associated with the cusp disappears. The expression for ω for case N_2 has the form

$$\omega(z) = \sum_{d|4N_2,d>0} P(dz)\alpha_d$$

with real coefficients α_d.

In the cases (2) and (3) cited above, $\bar{\Gamma}_0(4N_2)$ with χ_{4N_2} and $\bar{\Gamma}_0(4N_3)$ with χ_{4N_3} Balslev and Venkov have shown that there exist forms $\omega(z)$ with the desired properties.

8.34.5   Phillips-Sarnak Integral

For any odd eigenfunction which corresponds to an embedded eigenvalue $\lambda_j > 1/4$, define the integral

$$I_j(s) = \int_{\mathcal{F}} (Mv_j)(z)E_\infty(z, s)d\mu(z)$$

where $E_\infty(z, s)$ is the Eisenstein series

$$E_\infty(z, s) = \sum_{\Gamma_\infty \backslash \Gamma} y^s(\gamma z)\chi(\gamma).$$

If the integral $I_k(s_j)$ is nonzero where $\lambda_j = s_j(1 - s_j)$, for at least one of the Eisenstein series $E_k(s_j)$, then the eigenvalue λ_j disappears under the perturbation $\alpha M + \alpha^2 N$ for small $\alpha \neq 0$ and becomes a resonance; the corresponding eigenfunction ϕ_j becomes a resonance function with resonance $\lambda_j(\alpha)$. This follows from the proof that the sum over k of $|I_k(s_j)|^2$ is a constant times $Im(\lambda_j''(0))$, which is known as Fermi's Golden Rule. The proof that $I_j(s_j) \neq 0$ for odd eigenfunctions v_j with embedded eigenvalue $s_j(1 - s_j)$ is based on the non-vanishing of L−functions $L(s, v_j)$ on the line $\{s|Re(s) = 1\}$. Of course, $I_j(s_j) = 0$ for even eigenfunctions due to symmetry.

Theorem 26 *(Balslev-Venkov) For $\omega(z)$ described above, $I_j(s_j) \neq 0$ if $s_j = 1/2 + ir_j$ does not belong to any of the following sequence $s_n = 1/2 + ir_n$ with $r_n = n\pi/\log q$, where q is prime, $q|N, n \in \mathbf{Z}$.*

To show that the Phillips-Sarnak integral is nonzero, one reduces the integral to a product of L-functions and applies the nonvanishing result described above. For details, see Balslev and Venkov (1999).

8.34.6   Fermi's Golden Rule, Part II

Assume that there are h cusps $z_1, ..., z_h$ and under character χ_α the cusps $z_1, ..., z_k$ are open and $z_{k+1}, ..., z_h$ are closed. Let $R(s, \alpha) = (L(\alpha) - s(1 - s))^{-1}$ denote the resolvent for $L(\alpha) = L + \alpha M + \alpha^2 N$. $R(s, \alpha)$ has an analytic continuation $\tilde{R}(s, \alpha)$ to $\{s|0 < Re(s) < 2\}$ for any $\alpha \in (-1/2, 1/2)$ Here one dimensional eigenvalues λ_j continue analytically as functions $\lambda_j(\alpha)$ and multidimensional eigenvalues in general split up into Puiseux cycles or order $p \geq 1$. Here $\lambda_j(\alpha)$ are poles of $\tilde{R}(s, \alpha)$. If $p = 1$, the corresponding $\lambda_j(\alpha)$ is analytic for $|\alpha| < \epsilon$. There two cases (1) $\lambda_1(\alpha)$ is real for all real α and $\lambda_i(\alpha)$ is an embedded eigenvalue of $L(\alpha)$ for $|\alpha| < \epsilon$. Or (2) $\lambda_i(\alpha) = \lambda_0 + a_1\alpha + ... + a_{2l-1}\alpha^{2l-1} + a_{2l}\alpha^{2l} + \sum_{m \geq 2l+1} a_m\alpha^m$ where $a_1, ..., a_{2l-1}$ are real, $Im(a_{2l}) > 0$ for $s_0 = \frac{1}{2} + it_0$ and $Im(a_{2l}) > 0$ for $s_0 = \frac{1}{2} - it_0$ for $t_0 > 0$. If $p \geq 2$, then the functions $\lambda_{j1(\alpha)}, ..., \lambda_{jp(\alpha)}$ have expansions of the form

$$\lambda_{jl}(\alpha) = \lambda_0 + b_1\alpha + ... + b_{2m-1}\alpha^{2m-1} + b_{2m}\alpha^{2m} + b_{2m+1}\omega^l\alpha^{(2m+1)/p} + ...$$

where $l = 1, .., p, b_1, ..., b_{2m-1}$ are real, $Im(b_{2m}) > 0$ for $s_0 = \frac{1}{2} - it_0$ and $Im(b_{2m}) > 0$ for $s_0 = \frac{1}{2} - it_0$ for $t_0 > 0$.

Howland's (1974) result has been extended by Balslev and Venkov to the following:

Theorem 27 *(Balslev and Venkov) Let $\lambda_0 = \frac{1}{4} + it_0^2 > 1/4$ be an eigenvalue of $L = A(\bar{\Gamma}_0(N), \chi)$ with eigenspace $\mathcal{N}$ of dimension m. Let $\mathcal{K} \subset \mathcal{N}$ denote the subspace of odd eigenfunctions with $dim\mathcal{K} = k, 0 < k \leq m$. Under the perturbation $\alpha M + \alpha^2 N$ of L corresponding to the character $\chi(\alpha)$ all the eigenfunctions ϕ in $\mathcal{K}$ are transformed into resonance functions for $0 < |\alpha| < \epsilon$. The eigenvalue λ_0 splits into at most m eigenvalues and resonances of total multiplicity m. Eigenfunctions in $\mathcal{K}$ give rise to resonance functions associated with resonances $\lambda_i(\alpha), i = 1, ...k$. The functions $\lambda_i(\alpha)$ are either analytic or form branches of a Puiseux series of order $p \geq 2$. In both cases they have the form*

$$\lambda_i(\alpha) = \lambda_0 + a_2\alpha^2 + o(\alpha^2)$$

where the Fermi golden rule is

$$Im(\lambda_i''(0)) = Im(a_2) = \pm \frac{1}{4t_0} \sum_{j=1}^{h} |<M\phi_i|E_j(\frac{1}{2}-it_0)>|^2$$

for $t_0 > 0$ resp. $t_0 < 0$ and $\lambda_0 = s_0(1-s_0)$ with $s_0 = 1/2 + it_0$ and $\phi_i \in \mathcal{K}$.

Eigenfunctions in $\mathcal{N} \ominus \mathcal{K}$ give rise to poles $\lambda_j(\alpha), j = k+1, ..., m$ where the functions $\lambda_j(\alpha)$ are either analytic and remain embedded eigenvalues or $\lambda_j(\alpha)$ becomes a resonance for small $\alpha \neq 0$ and is of the form

$$\lambda_j(\alpha) = \lambda_0 + a_2\alpha^2 + ... + a_{2n}\alpha^{2m} + o(\alpha^{2m})$$

where $a_2, ..., a_{2n-1}$ are real, $n \geq 2$, $Im(a_{2n}) \neq 0$ and $\lambda_j(\alpha)$ is either analytic or a branch of a Puiseux series of order $p \geq 2$.

The embedded eigenvalue λ_j can only move to the second sheet (v., Balslev and Venkov (1999)). As noted above, the resonances $\lambda_i(\alpha)$ are poles of the analytic continuation of the resolvent $R(s, \alpha)$ from the upper half plane across $(1/4, \infty)$ to the second sheet. In the s−plane this corresponds to $\{s = \sigma + i\tau | \sigma < 1/2, t < 0\}$.

Assuming boundedness, i.e. the dimension of the eigenspaces, say $m(\lambda_i) \leq m$, then a positive portion of the odd eigenfunctions leave as resonance functions and this part of the Phillips-Sarnak conjecture is true.

8.35 Selberg Resonances

Selberg (1989) has noted the presence of resonances, or poles of the S−matrix, of $A(\Gamma, \alpha)$ for $\alpha \neq 0$ which condense at every point of the continuous spectrum of A as $\alpha \to 0$. Selberg addressed the case $\Gamma = \Gamma(2)$ with a singular character perturbation closing two cusps. That is, these resonances arise from the continuous spectrum of the cusps for the cusps which are closed by the perturbations. For further discussion of Selberg resonances see Balslev and Venkov (1999) for the more general case of $\Gamma_0(p)$.

8.36 Time Dependent Resonance Theory

In this section we review related work to the Phillips-Sarnak results which use the Fermi golden rule. Soffer and Weinstein (1998) have studied the problem of perturbations of systems with embedded eigenvalues in their continuous spectrum, in particular they present a time-dependent theory of these quantum resonances. Their work makes two basic assumptions: (1) non-vanishing of the Fermi golden rule and (2) certain local decay estimates. Assume that H_0 is a self-adjoint operator in a Hilbert space $\mathcal{H} = L^2(\mathbf{R}^n)$ and assume that H_0 has a simple eigenvalue λ_0 which is embedded in the continuous spectrum:

$$H_0\psi_0 = \lambda_0\psi_0$$

with $\|\psi_0\| = 1$. The perturbed self-adjoint Hamiltonian is

$$H = H_0 + W.$$

Soffer and Weinstein show that if W is a "small" perturbation, then the embedded eigenvalue moves off the real axis and becomes a resonance pole. They also show that in the neighborhood of this embedded eigenvalue there are no new embedded eigenvalues which appear.

Let g_Δ be a smooth characteristic function for the open interval Δ. Let P_0 denote the projection on ψ_0; i.e. $P_0 f = (\psi_0, f)\psi_0$. Let P_{1b} denote the spectral projection on $\mathcal{H}_{pp} \cap \{\psi_0\}^\perp$, i.e. the pure point spectral part of H_0 orthogonal to ψ_0. Let Δ_* denote the union of intervals disjoint from Δ containing all thresholds of H_0 and a neighborhood of infinity; set $P_1 = P_{1b} + g_{\Delta_*}$ where $g_{\Delta_*} = g_{\Delta_*}(H_0)$ is a smoothed characteristic function of the set Δ_*.

Let $< x >^2 = 1 + |x|^2$ and $P_c^\# = I - P_0 - P_1$. One sees that $P_c^\#$ is the smoothed spectral projection of the set $T^\#$, where

$$T^\# = \sigma(H_0) - \{\text{eigenvalues, real neighborhoods of thresholds and infinity}\}.$$

The basic assumptions are as follows:

(H1) H_0 is a self-adjoint operator with dense domain $\mathcal{D}$ in $\mathcal{H}$;

(H2) λ_0 is a simple embedded eigenvalue of H_0 with eigenfunction ψ_0;

(H3) there is an open interval Δ containing λ_0 and no other eigenvalue of H_0;

(H4) the local decay estimate: for $r \geq 2 + \epsilon$ for $\epsilon > 0$, there is a $\sigma > 0$ such that if $< x >^\sigma f \in L^2$, then

$$\| < x >^{-\sigma} e^{itH_0 t} P_c^\# f\|_2 \leq C < t >^{-r} \| < x >^\sigma f\|_2;$$

(H5) by appropriate choice of c, the L^2 operator norm of

$$< x >^\sigma (H_0 + c)^{-1} < x >^{-\sigma}$$

can be made sufficiently small;

And the assumptions on W are:

(W1) W is symmetric and $H = H_0 + W$ is self adjoint on $\mathcal{D}$ and there exists a $c \in \mathbf{R}$ such that c lies in the resolvent sets of H_0 and H;

(W2) for some σ

$$\|\|W\|\| = \| < x >^{2\sigma} W g_\delta(H_0)\| + \| < x >^\sigma W g_\delta(H_0) < x >^\sigma \|$$

$$+ \| < x >^\sigma W (H_0 + c)^{-1} < x >^{-\sigma} \| < \infty$$

and

$$\| < x >^\sigma W (H_0 + c)^{-1} < x >^\sigma \| < \infty;$$

(W3) there is non vanishing of the Fermi golden rule:

$$\Gamma = \pi(W\psi_0, \delta(H_0 - \tilde{\omega})(I - P_0)W\psi_0) \neq 0$$

for $\tilde{\omega}$ near λ_0 and

$$\Gamma \geq \delta_0 \|\|W\|\|^2$$

for some $\delta_0 > 0$;

(W4) $\|\|W\|\| < \theta|\Delta|$ for some $\theta > 0$.

Let $\mathcal{F}_c^{H_0}$ denote the generalized Fourier transform with respect to the continuous spectral part of H_0. Then the Fermi golden rule can be restated as

$$\Gamma = \pi |\mathcal{F}_c^{H_0}[W\psi_0](\lambda)|^2 > 0.$$

Theorem 28 *(Soffer and Weinstein) Let H_0 and W satisfy the conditions above. Then (1) $H = H_0 + W$ has no eigenvalues in Δ; (2) the spectrum of H in Δ is purely absolutely continuous; (3) local decay estimates hold for $e^{-iHt}g_\Delta(H)$; viz., for ϕ_0 with $<x>^\sigma \phi_0 \in L^2$:*

$$\| <x>^{-\sigma} e^{-iHt}g_\Delta(H)\phi_0\|_2 = \mathcal{O}(<t>^{-r+1}).$$

For ϕ_0 in the range of $g_\Delta(H)$ one has

$$e^{-iHt}\phi_0 = (I + A_W)(e^{-i\omega_* t}a(0)\psi_0 + e^{-iH_0 t}\phi_d(0)) + R(t)$$

where $a(0)$ is a complex number, $\phi_d(0)$ is a complex function in the range of $P_c^{\#}$; the complex frequency ω_* is given by

$$\omega_* = \omega - \Lambda - i\Gamma + \mathcal{O}(\|\|W\|\|^3)$$

where

$$\omega = \lambda + (\psi_0, W\psi_0),$$
$$\Lambda = (W\psi_0, P.V.(H_0 - \omega)^{-1}W\psi_0)$$

and

$$\Gamma = \pi(W\psi_0, \delta(H_0 - \omega)(I - P_0)W\psi_0).$$

The result of Soffer and Weinstein can be extended to the case where λ_0 is not simple, or the case where λ_0 is a threshold. They note that one approach to verify the local decay hypothesis is by using the Mourre estimates.

8.36.1 Direct Sum

Let $H_0 = \begin{pmatrix} -\Delta_x & 0 \\ 0 & -\Delta_x + q(x) \end{pmatrix}$ act on $\mathbf{C}^2 \otimes L^2(\mathbf{R}^n)$. Here $q(x)$ is a potential having some positive discrete eigenvalues; e.g. $q(x) = P(x)$ where $P(x)$ is a polynomial which is bounded below. The spectrum of $-\Delta_x + P(x)$ is then $\lambda_1 < \lambda_2 < ...$ and the spectrum of H is {eigenvalues of $-\Delta_x + P(x)$}$\cup [0,\infty)$. Thus the operator H_0 has nonnegative eigenvalues embedded in the continuous spectrum.

Let $W = \begin{pmatrix} 0 & W(x) \\ W(x) & 0 \end{pmatrix}$ where $W(x)$ satisfies the conditions above. Then Soffer and Weinstein have shown for the direct sum case:

Theorem 29 *(Soffer and Weinstein) If H_0 and W satisfy the conditions above, and if for some strictly positive simple eigenvalue $\lambda > 0$ the resonance condition, i.e. the Fermi golden rule, holds, then in an interval Δ around λ, the spectrum of H is absolutely continuous and the conclusions of Theorem 28 holds. And if $n > 4$, these results hold even if $\lambda = 0$.*

8.36.2   Tensor Product

Consider the case $H_0 = 1 \otimes h_1 + h_2 \otimes 1$ acting on $L^2(\mathbf{R}^n) \otimes L^2(\mathbf{R}^n)$ where $h_1 = -\Delta_{x_1}$ and $h_2 = -\Delta_{x_2} + q(x_2)$; then $\sigma(H_0) = \{\lambda = \lambda_1 + \lambda_2 | \lambda_1 \in \sigma(-\Delta_{x_1}), \lambda_2 \in \sigma(-\Delta_{x_2} + q(x_2))\}$. Let $W(x_1, x_2)$ act on $L^2 \otimes L^2$ and satisfy the assumptions (W1)-(W4) with $< x >^2 = 1 + |x_1|^2 + |x_2|^2$; then Soffer and Weinstein have shown:

Theorem 30 *(Soffer and Weinstein) The embedded eigenvalues of H_0 are unstable and the conclusions of Theorem 28 hold.*

8.36.3   Threshold Eigenvalues

Let $H_0 = -\Delta + V(x) \in L^2(\mathbf{R}^n)$ with $n > 4$. Assume V is smooth and rapidly decaying. If H_0 has local decay and L^∞ decay with a rate $r > 2$ (see Jensen and Kato (1979) and Journe, Soffe and Sogge (1990)), and if H_0 has a threshold eigenvalue $\lambda = 0$, then it is unstable with respect to small, generic perturbations W.

8.36.4   Stark Effect

Consider the Stark Hamiltonian for an atom in an uniform electric field

$$H = -\Delta + V(x) + E \cdot x$$

acting on $\mathcal{H}$. If $V(x)$ is real valued, not too singular, then for $E \neq 0$, the continuous spectrum of H is $(-\infty, \infty)$; and if H has an eigenvalue, it is necessarily embedded in the continuous spectrum. By the results above, any embedded eigenvalue is generically unstable if the Fermi golden rule holds and it perturbs to a resonance. For a discussion of how an eigenvalue becomes a complex resonance for the acoustic waveguide problem, see Aslanyan, Parnovski and Vassiliev (1999).

8.37   Hecke Operators and Scarring

Hecke operators have played a role in better understanding quantum chaos. We briefly review two results in this area. Jakobson and Zelditch (1998) have looked at the question of Hecke operators on the sphere. In general let (M, g) denote a Riemannian manifold with universal cover $(\tilde{M}, \tilde{g})$ and deck transformation group Γ. The Hecke operator on $L^2(M)$ is defined by the finite Radon transform

$$T_C f(x) = \frac{1}{2N} \sum_{j=1}^{N} f(C_j x) + f(C_j^{-1} x)$$

where C_j are isometries of the universal cover. The Hecke operator T_C is assumed to commute with Γ. Since the C_j's are isometries, it follows that $[\Delta, T_C] = 0$ and there is an orthonormal basis of joint eigenfunctions

$$\Delta \phi_j = \lambda_j \phi_j$$

and
$$T_C \phi_j = \mu_j(C)\phi_j.$$

One studies the weak* limits of linear functionals $\Phi_j(A) = (A\phi_j, \phi_j)$ for the algebra of bounded pseudodifferential operators. One version of the scarring problem is as follows:

Problem 2 *Does there exist a sparse subsequence of eigenfunctions concentrating on a closed geodesic, $(A\phi_{j_k}, \phi_{j_k}) \to \int_\gamma \sigma_A$ for observable A with classical limit σ_A?*

For more details on the problem of scarring, see Zelditch (1999) and Jakobson and Zelditch (1998). Jakobson and Zelditch have shown that joint eigenfunctions of Δ and a single Hecke operator on S^n cannot scar on a single closed geodesic:

Theorem 31 *(Jakobson and Zelditch) If T_C is a Hecke operator on S^n defined as above such that (1) any closed geodesic γ on S^n is fixed by at most two isometries C_j, C_j^{-1} and by at most two words of length two in the free group $\mathcal{F}$ generated by the C_j and (2) if γ is a closed geodesic fixed by some word $W \in \mathcal{F}$ of length at most two and if W is not a power of another element of $\mathcal{F}$, then the only words in $\mathcal{F}$ of length at most four fixing γ are those which reduce to the powers of W; then there is no sequence ϕ_j of joint $T_C - \Delta$-eigenfunctions such that the corresponding Φ_j's converge to δ_γ where δ_γ is the delta measure on a single closed geodesic.*

Let $\mathbf{H}^2$ denote the upper half-plane and let $\Gamma \subset G = SL(2, \mathbf{R})$ denote a discrete subgroup such that $\mathbf{H}^2/\Gamma$ has finite volume. An element $g \in G$ is said to be in the commensurator of Γ if the index $[\Gamma, \Gamma \cap g^{-1}\Gamma g]$ is finite. The subgroup Γ is arithmetic if its commensurator is dense in G.

As we have seen above, Hecke operators arise naturally in the study of discrete arithmetic subgroups of $SL(2, \mathbf{R})$. In this case Rudnick and Sarnak (1994) have shown:

Theorem 32 *(Rudnick and Sarnak) Joint Hecke-Laplace eigenfunctions on arithmetic hyperbolic surfaces cannot singularly concentrate on a finite union of closed geodesics.*

8.38 Sunada's Theorem

In the next chapter we will examine the question of isometric and isospectral graphs. What about isospectral surfaces? Sunada (1985) developed the following method to study this problem:

Theorem 33 *(Sunada) Let G be a finite group and let H_1, H_2 be two subgroups of G which satisfy the following condition: for all $g \in G$,*

$$|([g] \cap H_1)| = |([g] \cap H_2)|.$$

Then for any homomorphism $\phi : \pi_1(S) \to G$ which is onto, the two surfaces S^{H_1} and S^{H_2} corresponding to $\phi^{-1}(H_1)$ and $\phi^{-1}(H_2)$ are isospectral. Here $[g] = \{\sigma g \sigma^{-1} | \sigma \in G\}$ denotes the conjugacy class of $g \in G$.

As shown by McKean (1972) for the case of Riemannian surfaces, the spectrum of the Laplacian determines and is determined by the set of lengths of closed geodesics including multiplicities. Let γ be any closed path in S. Then γ determines a conjugacy class $[\gamma]$ in $\pi_1(S)$ and the number of lifts of γ which close up is given by $|([\gamma] \cap H_1)|$ and similarly for H_2. Hence, the length spectra of S^{H_1} and S^{H_2} agree and by McKean's result they are isospectral.

Brooks and Tse (1987) used Sunada's method to show:

Theorem 34 *(Brooks and Tse) For all $g \geq 4$ there exist Riemannian surfaces S_1 and S_2 of genus g which are isospectral but not isometric.*

The standard approach here (for $g \neq 5$) is to use (G, H_1, H_2) with $G = PSL(3, \mathbf{Z}/2)$,

$$H_1 = \begin{pmatrix} * & * & * \\ 0 & * & * \\ 0 & * & * \end{pmatrix}$$

and

$$H_2 = \begin{pmatrix} * & 0 & 0 \\ * & * & * \\ * & * & * \end{pmatrix}.$$

The following problems are discussed by Brooks (1999):

Problem 3 *To what extent does the Sunada method account for all isospectral Riemannian surfaces?*

Let $N(g)$ denote the largest number of distinct Riemannian surfaces $S_1, \ldots S_{N(g)}$ of genus g such that they are all pairwise isospectral.

Problem 4 *What is the growth rate of $N(g)$?*

Using the approach of Gordan and Wilson on isospectral deformation of nilpotent manifolds, Brooks and coworkers have shown

$$N(g) > g^{C \log g}.$$

Problem 5 *Is there a Buser-type upper bound, i.e.*

$$N(g) \leq C_1 g^{C_2 \log g}?$$

For further discussion we direct the reader to Brooks (1999) and Buser (1992).

Chapter 9

STF and Finite Volume Graphs

9.1 Introduction

The interrelationship between graph theory and differential geometry has played an important role in these two areas of research over the last ten years. Applications of this work have appeared in communication networks, the work on expanders, superconcentrators and Ramanujan graphs (v., Lubotzky (1994)) and in work on coding theory (v., e.g. the papers of Tillich and Zemor (1997) and Lafferty and Rockmore (1997)). In this review the focus is on the recent work on zeta functions and the Selberg trace formula for finite graphs and the extensions of these results to infinite graphs. The general theme is spectral theory on finite volume graphs.

Graph theoretic models have been utilized in understanding the the quantum physics of mesoscopic systems. E.g., in the work of Exner and coworkers they have examined graph systems as limits of quantum wires. For finite diameter systems, Exner and Seba (1989) have shown that any branched (star shaped) system of infinitely long waveguides with Dirichlet boundary conditions has at least one bound state. Exner (1996) has shown that the existence of a bound state is preserved in the zero diameter limit with appropriate coupling constants. However, Exner's approach does not allow for multiple bound states, i.e. greater than two. The models presented in this review provide a much richer class for the study of mesoscopic systems. In related work, Cayley graphs have been introduced by mathematical physicists in the study of the quantum conductance and the quantum Hall effect, v. Shapiro (1983). The recent work of Carey, Hannabuss and Mathai (1998) and Marcolli and Mathai (1999) also examines a discrete, graph theoretic model of the quantum Hall effect on the hyperbolic plane. Here their discrete model views the electrons as hopping along the vertices of the Cayley graph, based on a subgroup $\Gamma \subset SL(2, \mathbf{Z})$; that is, the motion is along the geodesics joining the vertices.

The Ihara zeta function and the Selberg trace formula for compact homoge-

203

neous graphs are presented in this chapter and results on the spectral theory and
Selberg conjectures for finite volume graphs are outlined.

The model for mesoscopic systems arises in this case since the finite volume
graphs have the form of a disjoint union of a finite graph and a finite number of half
lines, which are called cusps. Although the details of the Ihara zeta functions have
not been completely worked out in this case, we review what is known, specifically
recent results of Scheja (1998a) on the evaluation of the Ihara zeta function for the
finite graphs portion of the finite volume graphs. Scheja (1998b) has noted that
in the finite volume case, the L−functions are still rational, where the numerator
is determined by the number of cusps, or infinite half lines. In analogy with the
Ihara-Hashimoto-Bass zeta and L−functions for finite graphs, the L−function is
the determinant of a Laplacian. And the L−functions still have the functorial
properties of zeta and L−functions of finite graphs, which are discussed below. In
this chapter, only Scheja's work on the zeta function for the finite portion of finite
volume graphs is reviewed. Several examples are developed which are natural
models to study in mesoscopic systems. In the last sections of this chapter we
cover recent work of Shirai (1997) on a discrete analogue of the Gutzwiller trace
formula for graphs and the recent development of Nagoshi (1998) of the Selberg
trace formula for finite volume graphs for the case of the principal congruence
subgroups.

9.2   Graphs

A graph $X = (V(X), E(X), o, t, -)$ is given by two sets $V(X)$, the vertices, and
$E(X)$, the edges, and the maps, the origin and the terminus, $o, t : E(X) \to V(X)$
and an orientation reversal, $- : E(X) \to E(X) : y \to \bar{y}$, where $\bar{\bar{y}} = y$, $\bar{y} \neq y$ for
every $y \in E(X)$, $o(y) = t(\bar{y})$ and $t(y) = o(\bar{y})$. The size of X, or $|X|$, is given by
$|V(X)|$. The cardinality of $\{y | o(y) = x\}$ is called the degree or valency of x. A
graph is said to be regular of valence $q + 1$ if for every $x \in V(X)$, the set of edges
of origin x has $q + 1$ elements. We assume that X is locally finite, i.e. the degree
of each $x \in VX$ is finite.

A path of length r in X is a set $y = (y_1, ..., y_r)$ of r edges $y_i \in E(X)$ such that
$t(y_i) = o(y_{i+1})$ for $1 \leq i < r$. A path is said to be closed if $o(y_1) = t(y_r)$.

A path is said to be without backtracking if $y_{i+1} \neq \bar{y}_i$ for $1 \leq i < r$. A path y
is a circuit if it is closed, without backtracking and if $y_r \neq \bar{y}_1$. A circuit Cir_n is a
graph with vertices $\mathbf{Z}/n\mathbf{Z}$. A circuit is said to be primitive if it is not equal to z^s
for $s > 1$ and every circuit can be written as a power of a primitive circuit.

A graph is said to be connected if for any two vertices, x, x_1 there is a path in
X from x to x_1.

9.3   Uniform Tree Lattices

Let X be a locally finite tree and let Γ be a group acting on X with the properties:
(1) Γ acts without inversion, i.e. $\Gamma e \neq \Gamma \bar{e}$ for any edge e of X; (2) Γ acts discretely,

i.e. $\Gamma_x = \{\gamma \in \Gamma | \gamma(x) = x\}$ is finite for all $x \in X$; and (3) Γ acts uniformly, in which case $\Gamma \backslash X$ is a finite graph.

More generally, let $G = Aut(X)$ and let Γ be a subgroup of G acting without inversions on X. Γ is said to be discrete if Γ is a discrete subgroup of G, i.e. if Γ_x is finite for all $x \in X$. Γ is said to be a lattice if

$$Vol(\Gamma \backslash X) = \sum_{x \in \Gamma \backslash X} \frac{1}{|\Gamma_x|} < \infty.$$

And Γ is said to be a uniform lattice if $\Gamma \backslash X$ is a finite graph.

Bass and Kulkarni (1990) have characterized uniform lattices as follows:

Theorem 1 *(Bass and Kulkarni) X admits a uniform lattice if and only if $G \backslash X$ is finite and G is unimodular. Or more generally, $\Gamma \subset G$ contains a uniform lattice if and only if $\Gamma \backslash X$ is finite and the closure $\bar{\Gamma}$ of Γ is unimodular.*

Such a locally finite tree is called a uniform tree.

Theorem 2 *(Bass and Kulkarni) X is a uniform tree if and only if any of the following equivalent conditions hold: (a) X covers a finite graph; (b) $G \backslash X$ is finite and G is unimodular; (c) G contains a free uniform lattice.*

Bass and Kulkarni present examples of locally finite trees for which $G \backslash X$ is finite and yet X is not uniform, viz. failing to be unimodular. They also address other interesting issues relating to commensurability and finiteness properties of lattices. E.g., their commensurability theorem is equivalent to Leighton's (1982) common covering theorem:

Theorem 3 *(Leighton) If two finite connected graphs have a common covering, then they have a common finite covering.*

Leighton's theorem has also been used by Brooks in his study of the isospectral problem for graphs, which is discussed below.

9.4 Adjacency Matrix

Associated to every graph, X, is the adjacency matrix A. It is a square matrix of size $n = |VX|$ where the i, jth element is equal to the number of edges joining vertices v_i, v_j of X.

For a regular graph the adjacency matrix, A, is related to the Laplacian by

$$\Delta = kI - A.$$

Δ is a nonnegative, self-ajoint operator and the constant function on a connected component of X is the eigenfunction of Δ with eigenvalue zero. The number of connected components of X is the multiplicity of this eigenvalue.

A is a real, symmetric matrix. Let $\phi_X(z)$ denote the characteristic polynomial

$$\phi_X(z) = det(zI_n - A).$$

The set of roots, counted with multiplicities, of $\phi_X(z)$, i.e. the spectrum of A, is called the spectrum of X, or $Spec(X)$. One notes that $Spec(X) \subset [-k, k]$ for a k-regular graph and $Spec(\Delta) \subset [0, 2k]$. For a k-regular graph X, define

$$\lambda(X) = max\{|\lambda_i| : |\lambda_i| \neq k\}.$$

As an example, the circuit graph $C_n = Cir_n$ has the eigenvalues $2cos(2k\pi/n)$ for $k = 0, 1, ..., n - 1$.

The spectral density of A with spectrum $\lambda_1, ..., \lambda_n$, is defined by

$$\frac{1}{n} \sum_{i=1}^{n} \delta(x - \lambda_i)$$

9.5   Spectra of Infinite Graphs

The spectra of finite and infinite graphs has been reviewed in several places; see Mohar (1982, 1988), Mohar and Woess (1989); in particular Mohar has considered the spectra of product graphs, etc.

For a locally finite graph X, the adjacency operator A is a linear operator on l^2. The spectrum of A consists of point, continuous and residual components, which are denoted $Spec(X)_p, Spec(X)_c$ and $Spec(X)_r$. Since A is a closed operator, $Spec(X)$ is a closed subset of $\mathbf{C}$ with $Spec(X)_p \subset \mathbf{R}, Spec(X)_c \subset \mathbf{R}$ and $Spec(X)_r = \emptyset$ or $\mathbf{C} \backslash \mathbf{R} \subset Spec(X)_r$. Mohar (1982) has shown:

Theorem 4 *(Mohar) A is bounded if and only if there is an $M < \infty$ such that $deg(v) \leq M$ for all $v \in V(X)$. In this case $\|A\| \leq M$ and $Spec(X) \subset [-M, M]$. And, the adjacency operator A is compact if and only if X has only finitely many edges.*

Recall that the essential spectrum consists of the continuous spectrum and all eigenvalues with infinite multiplicity.

Theorem 5 *(Mohar) If one adds or deletes finitely many edges in a graph X, then its essential spectrum remains the same.*

If X_i are two locally finite graphs with adjacency operators A_i, then

Theorem 6 *(Mohar) If $X = X_1 \times X_2$, then*

$$Spec(X) = \{\lambda | \lambda = \mu_1 + \mu_2 | \mu_i \in Spec(A_i)\}$$

and

$$Spec(X)_p = \{\lambda | \lambda = \mu_1 + \mu_2 | \mu_i \in Spec(A_i)_p\}.$$

In particular if P_∞ is the one-way infinite path, then

$$Spec(P_\infty) = Spec(P_\infty)_c = [-2, 2]$$

and $Spec(P_\infty)_p = \emptyset$; and if C_∞ is the infinite circuit graph, then

$$Spec(C_\infty) = Spec(C_\infty)_c = [-2, 2]$$

and $Spec(C_\infty)_p = \emptyset$. And one can easily calculate the spectrum of $P_\infty \times P_\infty$.

If X is the two way infinite path, then the spectrum of X is given by $2cos(\alpha)$ for $\alpha \in [0, 2\pi)$. One notes that the spectral density in this case is given by

$$\phi(x) = \frac{1}{\pi} \frac{1}{\sqrt{4 - x^2}}.$$

Simon (1996) has developed examples of graphs whose Laplacian has singular continuous spectra; one example is a ladder with missing rungs and in another case Simon produces a regular graph whose Laplacian has singular spectrum.

9.6 Bipartite Graphs

A graph is said to be bipartite if the set V can be partitioned into disjoint sets $V = V_1 \cup V_2$ such that all edges have one end point in V_1 and the other in V_2.

Theorem 7 *A $k-$regular graph is bipartite if and only if $2k$ is an eigenvalue of Δ; in this case, the spectrum of Δ has the symmetry property $E \to 2k - E$.*

9.7 Diameter of Graphs

Let X denote a graph with n vertices. As noted above the distance $d(u, v)$ between any two vertices u, v is the length of the shortest path between u, v. The diameter $diam(X)$ is the maximum distance over all pairs of vertices. The diameter and the first eigenvalue are related. E.g. Chung (1989) has shown:

Theorem 8 *(Chung) For a connected $k-$regular graph X (except for the complete graph $K(n)$) with n vertices which is not bipartite, then*

$$diam(X) \leq \frac{log(n - 1)}{log(k) - log(\lambda_1)} + 1.$$

So if λ_1 is bounded away from zero, then the diameter is small and the boundary of a subset is large. For sparse graphs like $k - regular$ graphs, $1 - \lambda_1$ cannot be too small, in fact $1 - \lambda_1 \geq 1/\sqrt{k}$.

As will be discussed below, improvements on this estimate are possible based on the trace formula for graphs.

9.8 Isoperimetric Dimension

A graph X is said to have isoperimetric dimension δ with isoperimetric constant c_δ if for every subset $U \subset V(X)$ the number of edges between U and its complement $\bar{U}$, which we denote $|E(U, \bar{U})|$, satisfies

$$|E(U, \bar{U})| \geq c_\delta (vol(U))^{\frac{\delta - 1}{\delta}}$$

where $vol(U) \leq vol(\bar{U})$. Here $vol(U) = \sum_{x \in U} d(x)$, where $d(x)$ is the degree of vertex x.

The Cheeger constant h_X of a graph is given by $h_X = min_U h_X(U)$ where

$$h_X(U) = \frac{|E(U, \bar{U})|}{min(\sum_{x \in U} d(x), \sum_{y \in \bar{U}} d(y))}.$$

We see that the Cheeger constant can be viewed as a special case of the isoperimetric constant c_δ with $\delta = \infty$.

Let $0 = \lambda_0 \leq \lambda_1 \leq ... \leq \lambda_{n-1}$ denote the eigenvalues of the Laplacian of a graph X with n vertices. The Cheeger inequality states that

$$2h_X \geq \lambda_1 \geq \frac{h_X^2}{2}.$$

Mohar (1988) has extended these results to locally finite graphs. In particular,

Theorem 9 *(Mohar) For an arbitrary locally finite graph X with bounded degree*

$$h_X \leq \sqrt{\lambda_1(X)(2\Delta(X) - \lambda_1(X))}$$

where $\Delta(X)$ is the maximal degree of X.

Chung has shown the following which result from a discrete version of Sobolev inequalities:

Theorem 10 *(Chung)*

$$\sum_{i \neq 0} e^{-\lambda_i t} \leq c_1(\delta) vol(X)/t^{\delta/2}$$

and

$$\lambda_k \geq c_2(\delta)(\frac{k}{vol(X)})^{2/\delta}.$$

The latter result resembles an analogue of Polya's conjecture for Dirichlet eigenvalues of regular domains M in $\mathbf{R}^n$ that

$$\lambda_k \geq \frac{2\pi}{w_n}(\frac{k}{vol(M)})^{2/n}$$

where w_n is the volume of the unit disk in $\mathbf{R}^n$; see Polya and Szego (1951).

9.9 Trees

A nonempty, connected (nondirected) graph without circuits is a tree. A path without backtracking in a tree is said to be a geodesic. For any two vertices x, x' in a tree X, there is a unique geodesic from x to x'. The length of the geodesic from x to x' is called the distance, $d(x, x')$.

A homogeneous tree of order $q + 1$ is a locally finite tree where each vertex has valence $q + 1$.

An automorphism of a tree $X = (V, E)$ is a bijective map $h : V \to V$ which preserves edges, i.e. $h(x), h(x')$ are adjacent if and only if x, x' are adjacent.

Theorem 11 *h is an automorphism of $X = (V, E)$ if and only if h is an isometry of (V, d_V).*

An automorphism h of $X = (V, E)$ is said to be a rotation about x if h stabilizes some vertex x, i.e. there is an $x \in V$ such that $h(x) = x$. h is said to be an inversion about an edge e if h stabilizes e but exchanges its endpoints. Finally, h is a translation of size j along a geodesic y if there is a geodesic $y = (...x_{-2}, x_{-1}, x_0, x_1, x_2, ...)$ such that $h(n) = x_{n+j}$ for all n.

Theorem 12 *For all automorphisms h of a tree $X = (V, E)$ h is a rotation, an inversion or a translation.*

If an automorphism acts without inversions on X, then h is said to be hyperbolic if it has no fixed vertices. A group Γ of automorphisms of X is said to be strictly hyperbolic if every $\gamma \in \Gamma$ for $\gamma \neq id_\Gamma$ is hyperbolic. If h is a hyperbolic automorphism of a tree X, define the degree of h to be

$$deg(h) = min_{x \in V(X)} d(x, hx).$$

The following properties hold:

Theorem 13

$$deg(h) = deg(\{h\}),$$

where $\{h\}$ is the conjugacy class in the group $Aut(X)$ with representative h;

$$deg(h^m) = m deg(h);$$

and

$$deg(h^{-1}) = deg(h).$$

Theorem 14 *If X is a $(q+1)$-regular infinite tree and Γ is a strictly hyperbolic free group of isometric automorphisms, then $\Gamma \backslash X$ is a finite $(q+1)$-regular graph.*

9.10 Conjugacy Classes

Let Γ_γ denote the centralizer of $\gamma \in \Gamma$:

$$\Gamma_\gamma = \{g \in \Gamma | g\gamma = \gamma g\}.$$

Γ_γ is a free cyclic group. Any element $\gamma \neq e$ in Γ or a conjugacy class $\{\gamma\} \neq \{e\}$ in Γ is said to be primitive if γ generates Γ_γ in Γ.

9.11 The Group of Tree Automorphisms

The group $G = Aut(X)$ of automorphisms of a tree is a locally compact group and is unimodular. The irreducible unitary representations of $Aut(X)$ have been studied by Figà-Talamanca and Nebbia (1991), Ol'shianskii (1977) and Webster (1996).

9.12   Tits' Theorem

The elements of G can be classified as: (1) the identity; (2) hyperbolic elements which have no fixed vertices on X and two fixed on the boundary ∂X; (3) elliptic elements which have fixed vertices on X and no fixed points on ∂X; (4) parabolic elements which have fixed vertices on X and a fixed point on ∂X; and (5) split hyperbolic elements which have fixed vertices on X and two fixed points on ∂X.

Theorem 15 *(Tits) If* $g \in G$ *is hyperbolic and* $T_g = \{v \in V(X)|d(v, gv) = deg(g)\}$, *then (1)* T_g *is the vertex set of an infinite path in* X; *(2)* g *induces a shift by distance* $deg(g)$ *on* T_g; *(3) is a vertex* u *is of distance* d *from* T_g, *then* $d(u, gu) = deg(g) + 2d$.

For details, see Serre (1977) and Venkov and Nikitin (1994).

9.13 $\mathcal{L}_\infty(X, \mathcal{V})$

Let $\mathcal{V}$ be a complex vector space, with $dim_\mathbb{C}\mathcal{V} < \infty$. $\mathcal{L}_\infty(X, \mathcal{V})$ is the linear space of all $\mathcal{V}-$ valued uniformly bounded functions in $V(X)$, i.e. $\phi \in \mathcal{L}_\infty(X, \mathcal{V})$ if and only if $|\phi(x)|_\mathcal{V} < C_\phi$ for any $x \in V(X)$ where C_ϕ is a constant. $\mathcal{L}_\infty(X, \mathcal{V})$ is a normed vector space with norm

$$\|\phi\| = sup_{x \in V(X)}|\phi(x)|.$$

Similarly, let $\mathcal{L}_2(X, \mathcal{V})$ denote the Hilbert space of functions $\phi : V(X) \to \mathcal{V}$ such that

$$\|\phi\| = \{ \sum_{x \in V(X)} < \phi(x), \phi(x) >_\mathcal{V}\}^{1/2} < \infty.$$

9.14   Hecke Operators on Graphs

The Hecke operators T_m on graphs are defined by

$$(T_m\phi)(x) = \sum_{x' \in V(X), d(x,x')=m} \phi(x')$$

for $\phi \in \mathcal{L}_\infty(X, \mathcal{V})$. The operators $\{T_m\}$ satisfy the relationships:

$$T_1^2 = T_2 + (q + 1)T_0$$

$$T_1T_m = T_{m+1} + qT_{m-1}$$

for $m \geq 2$. Let Θ_r denote the polynomials in $T = T_1$:

$$\Theta_0 = 1,$$

$$\Theta_1 = T,$$

$$\Theta_2 = T^2 - (q + 1),$$

or generally:

$$T\Theta_r = \Theta_r + \begin{cases} q+1 & \text{for } r = 1 \\ q\Theta_{r-1} & \text{for } r > 1. \end{cases}$$

Theorem 16 *There is a generating series for Θ_r given by*

$$\sum_{r=0}^{\infty} \Theta_r t^r = \frac{1-t^2}{1-tT+qt^2}.$$

The trace of Θ_r, $Tr\,\Theta_r = f_r$, is the number of closed paths without backtracking of length r. If c_r, resp. c_r^0, is the number of circuits, resp. primitive circuits, of length r, then

$$c_r = \sum_{s|r} c_s^0$$

and

$$f_r - c_r = \sum_{1 \le i < r/2} (q-1)q^{i-1}c_{r-2i}.$$

Theorem 17

$$Tr\,\Theta_r = c_r + \sum_{1 \le i < r/2} (q-1)q^{i-1}c_{r-2i}.$$

9.15 Parameterization of Eigenvalues of T_1

There is a natural parameterization of eigenvalues of T_1 which arises as follows. For the continuous spectrum, the eigenfunctions and eigenvalues of T_1 have the form:

$$T_1\Phi(v,s) = \sqrt{q}(q^{s-\frac{1}{2}} + q^{\frac{1}{2}-s})\Phi(v,s)$$

where

$$\Phi(v,x) = c(s)q^{-sd(v,v_0)} + c(1-s)q^{(s-1)d(v,v_0)}$$

and

$$c(s) = \frac{1}{q+1}\frac{q^{1-s}-q^{s-1}}{q^{-s}-q^{s-1}}.$$

E.g., see Figà-Talamanca and Nebbia (1991) or Venkov and Nikitin (1994). This results in the parameterization of the eigenvalues of T_1 as:

$$\lambda_j(\Gamma,\chi) = \sqrt{q}[z_j(\Gamma,\chi) + z_j^{-1}(\Gamma,\chi)].$$

9.16 Automorphic Functions on Graphs

A function $\phi \in \mathcal{L}_\infty(X,\mathcal{V})$ is called (Γ,χ)–automorphic if

$$\phi(\gamma x) = \chi(\gamma)\phi(x)$$

for every $\gamma \in \Gamma$ and $x \in V(X)$. Let $\mathcal{A}(\Gamma,\chi)$ denote the space of automorphic functions.

Theorem 18
$$dim_{\mathbb{C}}\mathcal{A}(\Gamma,\chi) = |\Gamma\backslash X|dim_{\mathbb{C}}\mathcal{V}.$$

Theorem 19 $\|T_m\| \leq (q+1)q^{m-1}, m = 1, 2, \ldots$

Since T_m commutes with the action of Γ, let $T_m(\Gamma,\chi)$ denote the restriction of T_m to the space $\mathcal{A}(\Gamma,\chi)$. The operators $T_m(\Gamma,\chi)$ form a family of self-adjoint operators with finitely many eigenvalues: $\lambda_1^m(\Gamma,\chi), \ldots \lambda_M^m(\Gamma,\chi)$ where $M = dim_{\mathbb{C}}\mathcal{A}(\Gamma,\chi)$. Set $\lambda_j(\Gamma,\chi) = \lambda_j^{(1)}(\Gamma,\chi)$.

Theorem 20 *The eigenvalues* $\lambda_j(\Gamma,\chi)$ *are real and*

$$|\lambda_j(\Gamma,\chi)| \leq q + 1.$$

9.17   Ihara Zeta Function

Given a homogeneous tree X and a strictly hyperbolic group of isometric automorphisms, the Ihara (1966) zeta function is defined by

$$Z(\Gamma, u, \chi) = \prod_{P \in \mathcal{P}r} det(I - \chi(P)u^{deg P})^{-1}$$

for $|u| < q^{-1}, u \in \mathbb{C}$. The product is taken over the primitive conjugacy classes in Γ, different from $\{e\}$.

9.18   Properties of the Ihara Zeta Function

The properties of the Ihara zeta function are analogous to the properties of the Selberg zeta functions discussed in Chapter 8:

Theorem 21 *The following properties hold:*

$$Z(\Gamma, 0, \chi) = 1;$$

the zeta function $Z(\Gamma, u, \chi)$ *admits a meromorphic extension to the entire complex* $u-plane$; $Z(\Gamma, u, \chi)$ *satisfies the functional equation:*

$$Z(\Gamma, u\sqrt{q}, \chi) = u^{-(q+1)/M}(\frac{qu^2 - 1}{q - u^2})^{g_\chi} Z(\Gamma, u^{-1}/\sqrt{q}, \chi).$$

$$Z(\Gamma, u, \chi_1 \oplus \chi_2) = Z(\Gamma, u, \chi_1)Z(\Gamma, u, \chi_2)$$

where $M = dim_{\mathbb{C}}\mathcal{A}(\Gamma,\chi)$, $g_\chi = r - 1$ *and* $r = rank\, H^1(X, \mathbb{Z})$.

Ihara showed the following rationality result:

Theorem 22 *(Ihara) If X is a k-regular graph and $k = q + 1$, then the Ihara zeta function $Z(\Gamma, u, \chi)$ has the expression*

$$Z(\Gamma, u, \chi) = (1 - u^2)^{g_\chi} det(I - T_1(\Gamma, \chi)u + qu^2)^{-1}$$

or

$$Z^{-1}(\Gamma, u, \chi) = (1 - u^2)^{-g_\chi} \prod_{\lambda \in Spec(X)} (1 - \lambda u + qu^2)$$

where $g_\chi = r - 1, r = rank\ H^1(X, \mathbf{Z})$ and $u = q^{-s}$. Thus, $Z(\Gamma, u, \chi)$ has trivial poles at $u = \pm 1$ with multiplicity $g_\chi = \frac{1}{2}(q - 1)M$ and nontrivial poles at points where $det(I - T_1(\Gamma, \chi)u + qu^2) = 0$.

9.19 Functional Equations

If X is a connected $(q + 1)$-regular graph, with n vertices, then one has the following functional equations:

Theorem 23

$$\Lambda_X(u) = (1 + u^2)^{n/2 + r - 1}(1 - q^2 u^2) Z_X(u) = (-1)^n \Lambda_X(1/qu);$$

$$\xi_X(u) = (1 + u)^{r-1}(1 - u)^{r-1+n}(1 - qu)^n Z_X(u) = \xi_X(1/qu);$$

$$\Xi_X(u) = (1 - u^2)^{r-1}(1 + qu^2) Z_X(u) = \Xi_X(1/qu).$$

For a proof see Stark and Terras (1996).

9.20 Riemann Hypothesis

The analogue of the Riemann hypothesis for the Ihara zeta function $Z(\Gamma, u, \chi)$ is if $|\lambda_j(\Gamma, \chi)| < q + 1$, then $|\lambda_j(\Gamma, \chi)| \leq 2\sqrt{q}$. Set $Y = \Gamma \backslash X$ and let $\mu(Y) = max_j\{|\lambda_j(\Gamma, 1)| < q+1\}$. Then, a finite $(q+1)$-regular quotient graph Y is called Ramanujan if $\mu(Y) < 2\sqrt{q}$. That is, a k-regular graph X is a Ramanujan graph if for every eigenvalue λ of the adjacency matrix A_X either $\lambda = \pm k$ or $|\lambda| \leq 2\sqrt{k-1}$.

As a corollary to Ihara's theorem discussed above, we have:

Theorem 24 *If X is a k-regular graph, then X is a Ramanujan graph if and only if $Z_X(s)$ satisfies the Riemann hypothesis, i.e. all poles of $Z_X(s)$ in the region $0 < Re(s) < 1$ lie on the line $Re(s) = 1/2$.*

In terms of the parameters $z_j(\Gamma, \chi)$ the Riemann hypothesis implies that if

$$\sqrt{|(z_j(\Gamma, \chi) + z_j^{-1}(\Gamma, \chi)|} < q + 1$$

then $|z_j(\Gamma, \chi)| = 1$. In this case the $z_j(\Gamma, \chi)$ can be parameterized by

$$z_j(\Gamma, \chi) = q^{i\theta_j(\Gamma, \chi)}$$

or
$$\lambda_j(\Gamma, \chi) = 2\sqrt{q}\cos(\theta_j(\Gamma, \chi)lnq).$$

For further discussion on Ramanujan graphs, see Lubotzky (1994), Terras (1992, 1997), Lubotzky, Phillips and Sarnak (1988), Margulis (1988), Venkov and Nikitin (1994), Hashimoto (1989), Morgenstern (1994) and Li (1995, 1998).

9.21 Finite Irregular Graphs

Bass (1992) showed that Ihara's theorem holds for finite irregular graphs. The approach of Bass differs from that of Hashimoto (1989) and Hashimoto and Hori (1989), which was based on geometric edges and required the space to be bipartite. Bass' method provides a simpler, more general approach although it does require the use of noncommutative determinants. For the details the reader is directed to Bass (1992). We only state the final result on the zeta function for uniform tree lattices. Let $Q_X = Q$ denote the $n \times n$ diagonal matrix with jth diagonal element q_j where $q_j + 1$ is the degree of the jth vertex of X and $n = |V(X)|$. It still holds that $r - 1 = \frac{1}{2}Tr(Q - I)$ where r is the rank of the fundamental group of X.

Theorem 25 *(Bass) The Ihara zeta function of X is given by*

$$Z_X(u)^{-1} = (1 - u^2)^{r-1}det(I - Au + Qu^2).$$

For a simplified proof see Stark and Terras (1996) who follow Sunada (1986). For related work see Foata and Zeilberger (1998) and Northshield (1998).

9.22 Equidistribution of Eigenvalues

Let $T' = T/\sqrt{q}$ and similarly $\Theta'_r = \Theta_r/q^{r/2}$. Let X_m be a family of finite nonempty regular graphs of valence $q + 1$. Let $c_{r,m}$, resp. $c^0_{r,m}$ denote the number of circuits, resp. primitive circuits of X_m of length r. The eigenvalues of T are real and of absolute value $\leq q + 1$. Similarly, the eigenvalues of T'_m are in the interval $\Omega_q = [-\omega_q, \omega_q]$ where $\omega_q = q^{1/2} + q^{-1/2}$. Serre (1998) has shown:

Theorem 26 *(Serre) The following are equivalent:*
(1) there is a measure μ on Ω_q such that the eigenvalues of T_m have the property that they are $\mu-$equidistributed.
(2) for every $r \geq 1$, $c_{r,m}/|X_m|$ has a limit a $m \to \infty$.

For a $(q + 1)-$regular graph X with eigenvalues $\{\lambda_i\}$ define the measure

$$\mu_X = \frac{1}{|X|} \sum_{\lambda_i} \delta_{\lambda_i/\sqrt{q}}$$

where δ is the Dirac measure.

Earlier, McKay (1981) had shown that

Theorem 27 *(McKay) The eigenvalues of T'_m are equidistributed with respect to μ_q if and only if $\gamma_r = 0$ for all $r \geq 1$; i.e.,*

$$lim_{m \to \infty} c_{r,m} / |X_m| = 0$$

for $r = 1, 2, \ldots$ In particular, if $\{X_m\}$ be a family of connected undirected $(q + 1)-$regular graphs with $|X_m| \to \infty$. Assume

$$c_{r,m} / |X_m| \to 0$$

as $m \to \infty$ for each r. Then the sequence of measures $\{\mu_{X_m}\}$ converges weakly to the measure μ supported on $[-2, 2]$ given by

$$\mu = \frac{q+1}{M^2 - x^2} \mu_{ST}$$

where $M = \sqrt{q} + 1/\sqrt{q}$ and μ_{ST} is the Sato-Tate measure

$$\mu_{ST} = \frac{1}{\pi} \sqrt{1 - \frac{x^2}{4}} dx.$$

Furthermore, for the Ihara zeta function, which is given by

$$Z_{X_m}(t) = exp(\sum_{r=1}^{\infty} c_{r,m} t^r / t) = 1 / \prod_{r=1}^{\infty} (1 - t^r)^{c_{r,m}^0 / r},$$

Serre showed:

Theorem 28 *(Serre) The graphs X_m satisfy conditions (1) and (2) above if and only if the formal series $Z_{X_m}(t)^{1/|X_m|}$ has a limit in $\mathbf{R}[[t]]$:*

$$z(t) = lim_{m \to \infty} Z_{X_m}(t)^{1/|X_m|} = 1 / \prod_{r=1}^{\infty} (1 - t^r)^{\gamma_r^0 / r}$$

where $\gamma_r^0 = lim \, c_{r,m}^0 / |X_m|$. The series $z(t)$ determines the measure μ and conversely.

9.23 Level Spacing Distribution

Jakobson, Miller, Rivin and Rudnick (1998) have studied the level spacing distribution of a generic $k-$regular graph. Recall that the statistical behavior of the eigenvalue spacing for Laplacian operators on manifolds falls into two main classes, Poisson (related to integrable systems) and GOE/GUE (related to chaotic systems). Viz., if the spacings are normalized to have unity mean, then in the Poisson case the density of spacings has the form e^{-x} and in the GOE case it can be approximated by the Wigner surmise $\frac{\pi x}{2} e^{-\pi x^2 / 4}$. Here GOE stands for the

Gaussian orthogonal ensemble of random matrix theory, which for $M \times M$ real symmetric matrices has a limiting density given by Wigner's semicircle law:

$$R_1(x) = \begin{cases} \frac{1}{\pi}\sqrt{2M - x^2} & |x| \le \sqrt{2M} \\ 0 & |x| > \sqrt{2M}. \end{cases}$$

The numerical work of Schmit (1991) indicates that the spacing distribution for $SL(2, \mathbf{Z})\backslash H$ is Poisson; see also the work of Luo and Sarnak (1994) and Rudnick and Sarnak (1994) for the more general case of arithmetic groups and Katz and Sarnak (1996) and Sarnak (1993).

Jakobson et al. computed the eigenvalues of generic adjacency matrices and unfolded the spectrum using McKay's law; they computed the level spacing distribution and compared it with the GOE. Viz., let $\bar{N}(\lambda)$ denote the expected number of levels below λ, then one unfolds the spectra by setting

$$\hat{\lambda}_j = \bar{N}(\lambda_j).$$

This gives rise to a sequence $\hat{\lambda}_j$ with mean unity spacing

$$s_j = \hat{\lambda}_{j+1} - \hat{\lambda}_j \sim 1.$$

The level spacing distribution is given by

$$p_N(s) = \frac{1}{N} \sum_{j=1}^{N} \delta(s - s_j)$$

Jakobson et al. showed statistically that for a generic k–regular graph the level spacing distribution approaches that of the GOE as the number of vertices goes to infinity.

The reader should also note the relationship to the work of Shirai (1997, 1997) in this connection, which is discussed below, and the recent work of Kottos and Smilansky (1998).

9.24   Spectra of Cayley Graphs

Lafferty and Rockmore (1993, 1997) have examined the eigenvalue spacings for adjacency matrices for 4-regular (2-generator) Cayley graphs for three families of groups: $SL(2, \mathbf{F}_p)$, the symmetric groups, S_m for $m = 10$, and large cyclic groups; in addition they examined the eigenvalue spacings for random 4-regular graphs (which are not Cayley graphs). The generators for $SL(2, \mathbf{F}_p)$ included the Selberg set

$$\left\{ \begin{pmatrix} 1 & 2 \\ 0 & 1 \end{pmatrix}, \begin{pmatrix} 1 & 0 \\ 2 & 1 \end{pmatrix} \right\},$$

a non-Selberg set and a set due to Shalom. Their numerical work shows strong agreement in the 4-regular Cayley case with Poisson distribution and in the random case the behavior appears to follow GOE distribution, which is consistent with the work of Jakobson et al. discussed above. Based on their numerical work and their earlier work (Lafferty and Rockmore (1997)), they have conjectured:

1) if $\{G_n\}$ denote any of the three families of groups, the symmetry group S_n, the cyclic group, $\mathbf{Z}/n\mathbf{Z}$ or $SL(2, \mathbf{F}_{p_n})$ (for p_n any increasing sequence of primes), then as $n \to \infty$ for $\epsilon > 0$ the probability of choosing a pair of generators S for G_n with $\|P(S) - (1 - e^{-S})\| \geq \epsilon$ goes to zero. Here $P(S)$ is the cumulative distribution function for eigenvalue spacings of the adjacency operator:

$$P(S) = \frac{1}{N} \sum_{j=1}^{N} [\lambda_{j-1} - \lambda_j \leq S]$$

where $[a \leq b]$ is one if $a \leq b$ and zero otherwise; and it is assumed that the eigenvalues have been normalized so that the spacings have mean one:

$$\frac{1}{N} \sum_{j=1}^{N} (\lambda_{j-1} - \lambda_j) = 1.$$

2) if the set

$$X_p = X(SL(2, \mathbf{F}_q), S)$$

is a family of k-regular Cayley graphs for $SL(2, \mathbf{F}_q)$, then asymptotically as $p \to \infty$ the distributions of eigenvalue spacings for X_p and its individual Fourier transforms $\hat{\delta}_S(\rho)$ are Poisson.

9.25 Abelian Ramanujan Graphs

Li (1992) has given a construction of Ramanujan graphs based on abelian groups. They are shown to be Ramanujan graphs based on Deligne's (1977) estimate on generalized Kloosterman sums. Let G be a finite (additive) abelian group and let S be a k-element subset of G. The sum and difference graphs are defined as follows. The neighbors of a vertex x of $X_s(G, S)$ are $-x + s$ for $s \in S$ and the neighbors of $X_d(G, S)$ are $x + s$. For each character ψ of G, set $e(\psi, S) = \sum_{s \in S} \psi(s)$. One can show that each character ψ of G is an eigenfunction of the Hecke operator A on $X_d(G, S)$ with eigenvalue $e(\psi, S)$. The graphs $X_s(G, S)$ and $X_d(G, S)$ are k-regular graphs.

Let ψ be a character of G. If we view ψ as a function on $X_d(G, S)$, then

$$A\psi(x) = \sum_{s \in S} \psi(x + s) = \psi(x) \sum_{s \in S} \psi(s) = \psi(x)e(\psi, S).$$

Theorem 29 *(Li) Each character of G is an eigenfunction of the Hecke operator A on $X_d(G, S)$ with eigenvalue $e(\psi, S)$.*

If $e(\psi, S) = 0$, then ψ and ψ^{-1} are both eigenfunctions with eigenvalue 0 of the Hecke operator A on $X_s(G, S)$. If $e(\psi, S) \neq 0$, then $|e(\psi, S)|\psi \pm e(\psi, S)\psi^{-1}$ are two eigenfunctions of A on $X_s(G, S)$ with eigenvalues $\pm|e(\psi, S)|$.

It follows that

Theorem 30 *(Li) The absolute value of the eigenvalues of the Hecke operator on* $X_s(G, S)$ *and* $X_d(G, S)$ *are the same, viz.* $|\sum_{s \in S} \psi(s)|$ *where* ψ *runs through all characters of* G. *The Hecke operators on* $X_s(G, S)$ *and* $X_d(G, S)$ *are diagonalizable by orthonormal bases of functions on* G.

Let F be a finite field with q elements and let F_n be a degree n field extension of F. If ψ is a nontrivial additive character of F, then by Pontryagin duality all character of F_n are given by $\Psi^a(x) = \psi(T(ax))$ for $a \in F_N$ and T the trace from F_n to F. Let N denote the norm from F_n to F. For all nonzero a in F_n, Deligne (1977) showed that

$$|\sum_{N(x)=1} \Psi^a(x+c)| = |\sum_x \psi(T(ax))\psi(T(ac))| = |\sum_{N(y)=b} \psi(T(y))| \leq nq^{n-1/2}$$

where $b = N(a)$.

From these results we have:

Theorem 31 *(Li) Let* F *be a finite field with* q *elements and* F_n *a degree* n *field extension of* F. *Let* N_n *be the kernel of the norm map from* F_n *to* F; *it contains* $d_n = (q^n - 1)/(q - 1)$ *elements. Let* c *be an element in* F_n. *Then except for the trivial eigenvalue* d_n, *all the eigenvalues* λ *of the Hecke operators on* $X_s(F_n, N_n + c)$ *and* $X_d(F_n, N_n + c)$ *satisfy* $|\lambda| \leq nq^{(n-1)/2}$. *And the graphs* $X_s(F_2, N_2 + c)$ *and* $X_d(F_2, N_2 + c)$ *are* $(q + 1)-$ *regular Ramanujan graphs with* q^2 *vertices.*

9.26   Multigraphs

A multigraph X is a triple (VX, EX, ϵ) consisting of the set of vertices, VX, the set of edges, EX, and the incident map

$$\epsilon : EX \to VX \times VX$$

given by $\epsilon(y) = (o(y), t(y))$. Here $o(y)$ is the origin of y and $t(y)$ is the terminus of y. In addition there is an involution map

$$i_X : EX \to EX$$

denoted by $y \to \bar{y}(y \neq \bar{y})$ where $o(\bar{y}) = t(y)$ and $t(\bar{y}) = o(y)$. A pair $e = \{y, \bar{y}\}$ is a non-oriented edge and set $\epsilon(e) = \{o(y), t(y)\}$. An edge $e \in EX$ is incident to $P, Q \in VX$ if $\epsilon(e) = \{P, Q\}$. If P, Q are joined by e, then P, Q are said to be adjacent. If $\epsilon(e) = \{P, P\}$, then e is said to be a loop. If more than one non-oriented edge has the same pair $\{P, Q\}$ as their ends, then X is said to have multiple edges and is called a multigraph. If X has no loops and no multiple edges, then X is a graph.

A path C of length n on a multigraph X is a sequence

$$C = (P_o, y_1, P_1, y_2, ... P_{n-1}, y_n, P_n)$$

of $n+1$ vertices and n oriented edges where $\epsilon(y_i) = (P_{i-1}, P_i)$. The length of C is denoted by $|C|$ and one calls $P_0 = o(C)$ and $P_n = t(C)$. The inverse path C^{-1} is given by

$$C^{-1} = (P_n, \bar{y}_n, P_{n-1}, ..., \bar{y}_1, P_0).$$

The composition $C.C'$ of two paths is defined if $t(C) = o(C')$. If C is such that $y_i \neq \bar{y}_{i+1}$ for $1 \leq i \leq n-1$, the C is called a proper path or has no back tracking. A path is closed if $o(C) = t(C)$.

A closed proper path $C = (y_1, ..., y_n)$ on X is called reduced if either $n = 1$ or $y_1 \neq \bar{y}_n$. A cycle or reduced closed path C of length n is called primitive if and only if C is not of the form $C = D^m$, for $m > 1$.

9.27 Adjacency Matrix for Multigraphs

Let X be a finite multigraph with vertices $VX = \{P_1, ..., P_n\}$ and edges $EX = \{e_1, ..., e_m\}$. The $n \times n$ matrix $A(X)$ defined by

$$a_{ij} = |\{y \in EX | o(y) = P_i, t(y) = P_j\}|$$

is called the adjacency matrix. $A(X)$ is a real symmetric matrix and the characteristic polynomial is defined by

$$\phi_X = det(zI_n - A).$$

The set of roots of ϕ_X is the spectrum of X or $Spec(X)$.

9.28 Valency for Multigraphs

Let X be a multigraph as in the last section. If $P_i \in VX$, then $\sum_{j=1}^n a_{ij}$ is the number of edges incident to P_i. It is called the valency of P_i. If the valency is the same for all P_i, say k, then X is said to be regular of valency k.

9.29 Ihara's Theorem for Multigraphs

Hashimoto (1989) has generalized Ihara's theorem for multigraphs:

Theorem 32 *(Hashimoto-Ihara) If X is a regular multigraph with valency $q + 1$ and adjacency matrix A, then*

$$Z_X(u)^{-1} = (1 - u^2)^{r-1} det(I_n - Au + qu^2)$$

where $r = (q-1)n/2$ is the rank of $\Gamma = \pi_1(X, P_0)$ and $n = |VX|$.

9.30 Bipartite Multigraphs

If X is a bipartite graph, $Z_X(u,\chi)^{-1}$ is an even polynomial in u, i.e.

$$Z_X(-u,\chi) = Z_X(u,\chi).$$

In this case we define

$$Z_{X,b}(u,\chi) = Z_X(u^{1/2},\chi).$$

Theorem 33 *Assume X is a connected k-valency multigraph; then k belongs to $Spec(X)$ with multiplicity one, k is the eigenvalue of $A(X)$ with maximum absolute value, and $-k \in Spec(X)$ if and only if X is bipartite.*

Theorem 34 *(Hashimoto) If X is a bipartite multigraph and $VX = V_1 \cup V_2$ with $|V_i| = n_i, n_2 \geq n_1$; then*

$$Spec(X) = \{\pm\lambda_1, ..., \pm\lambda_{n_1}, 0, ..., 0\,(n_2 - n_1)\ times\}$$

where $\lambda_1 \geq \lambda_2, ... \geq \lambda_{n_1} \geq 0$.

An s-partite multigraph is called semiregular of valency $(k_1, ..., k_s)$ if each $P \in V_i$ is incident to exactly k_i edges for $i = 1, ..., s$.

Theorem 35 *(Hashimoto-Ihara) If X is a connected semiregular bipartite multigraph with valency $(q_1 + 1, q_2 + 1)$, where $|VX_i| = n_i$ and $q_1 > q_2$, then*

$$Z_{X,b}(u)^{-1} = (1-u)^{r-1}(1 + q_2 u)^{(n_2 - n_1)} det(I_{n_1} - (A^1 - q_2 + 1)u + q_1 q_2 u^2)$$

where $A^1 = A(X_1)$; and if $Spec(X) = \{\pm\lambda_1, \pm\lambda_2, ..., \pm\lambda_{n_1}, 0, ..., 0\}$, then

$$det(I_{n_1} - (A^1 - q_2 + 1)u + q_1 q_2 u^2) = \prod_{j=1}^{n_1} \{1 - (\lambda_1^2 - q_1 - q_2)u + q_1 q_2 u^2)\}.$$

9.31 Complementary Graphs

If X is a finite connected graph with n vertices, then there exists an injective morphism $i : X \to K(n)$ where $K(n)$ is the complete graph with n vertices; i.e. any two vertices in $K(n)$ are joined by an edge. If one removes from $K(n)$ all edges of X, then one obtains a graph, X^c, with n vertices VX called the complementary graph.

Theorem 36 *(Hashimoto) If X is a connected regular graph with valency k, and $Spec(X) = \{\lambda_1 = k, \lambda_2, ..., \lambda_n\}$. Then the complementary graph X^c is regular of valency $n - 1 - k$ and*

$$Spec(X^c) = \{n - 1 - k, -\lambda_2 - 1, ..., -\lambda_n - 1\}.$$

And if X^c is connected, then

$$Z_{X^c}(u)^{-1} = (1 - u^2)^{n(n-3-k)/2}(1-u)[1 - (n - 1 - k)u]\times$$

$$\prod_{j=2}^{n}[1 + (\lambda_j + 1)u + (n - 2 - k)u^2].$$

9.32 Cartesian Product Graphs

The cartesian product of two graphs X_1, X_2 is a graph $X = X_1 \times X_2$ with vertices $VX_1 \times VX_2$ and two vertices (P_1, P_2) and (Q_1, Q_2) are adjacent if either $P_1 = Q_1$ and P_2, Q_2 are adjacent or (2) $P_2 = Q_2$ and Q_1, P_1 are adjacent. If X_i are connected, then $X_1 \times X_2$ is connected.

Theorem 37 *(Hashimoto) If X_i are connected graphs, with spectra*

$$Spec(X_1) = \{\lambda_j | 1 \leq j \leq n_1\}$$

and

$$Spec(X_2) = \{\mu_j | 1 \leq j \leq n_2\},$$

then

$$Spec(X_1 \times X_2) = \{\lambda_i + \mu_j, 1 \leq i \leq n_1, 1 \leq j \leq n_2\}.$$

If X_i is regular of valency k_i, then $X_1 \times X_2$ has valency $q_1 + q_2$ and

$$Z_{X_1 \times X_2}(u)^{-1} = (1 - u^2)^{q - n_1 n_2/2} \prod_{i=1}^{n_1} \prod_{j=1}^{n_2} (1 - (\lambda_i + \mu_j)u + qu^2)$$

where $q = k_1 + k_2 - 1$.

9.33 Conjunction Graphs

The conjunction graph $X_1 \wedge X_2$ of two graphs X_1, X_2 has the set of vertices $VX_1 \times VX_2$ and (P_1, Q_1) and (P_2, Q_2) are adjacent if and only if both pairs $\{P_1, P_2\}$ and $\{Q_1, Q_2\}$ are adjacent. If X_i are connected, then X is connected.

Theorem 38 *(Hashimoto) The spectrum of the conjunction graph is given by*

$$Spec(X_1 \wedge X_2) = \{\lambda_i \mu_j, 1 \leq i \leq n_1, 1 \leq j \leq n_2\}.$$

If X_i is regular of valency k_i, then $X_1 \wedge X_2$ is regular of valency $k_1 \cdot k_2$ and

$$Z_{X_1 \wedge X_2}(u)^{-1} = (1 - u^2)^{(q-1)n_1 n_2/2} \prod_{i=1}^{n_1} \prod_{j=1}^{n_2} (1 - (\lambda_i \mu_j)u + qu^2)$$

where $q = k_1 k_2 - 1$.

9.34 Circuit Graph C_l

The circuit $C_l = Cir_l$ is the graph given by a closed proper path of length l. In this case,

$$Z_{C_l}(u)^{-1} = (1 - u^l)^2.$$

One can show that if $X = Cir_4$, then $r = 1$ and

$$Z_X(u)^{-1} = (1 - u)^2(1 + u)^2(1 + u^2)^2.$$

9.35   Complete Graph $K(l+2)$

The complete graph $K(l+2)$ is a regular graph with valency $l+1$; and for $l \geq 1$

$$Z_{K(l+2)}(u)^{-1} = (1-u)^r (1+u)^r (1-qu)(1+u+qu^2)^{q+1}$$

where $r = q(q+1)/2$.

9.36   Cayley Graphs

Let G be a group and S a symmetric set of generators, i.e. $S = S^{-1}$ and $G = < S >$. Assume 1 is not in S. The Cayley graph $X = X(G, S)$ is defined by $VX = G$ and g, h are adjacent if $g = hs$ for some $s \in S$. In this case X is a connected, regular graph of valency $|S|$.

The adjacency matrix for X is

$$A(G, f) = (f(g^{-1}h))_{g,h \in G} \in M(n, \mathbf{C})$$

where $n = |G|$. More generally, for any set of generators S, not necessarily symmetric, as an operator

$$Af(x) = \sum_{y \sim x} f(y) = \sum_{s \in S \cup S^{-1}} f(xs) = \sum_{s \in S \cup S^{-1}} (\rho(s)f)(x)$$

where ρ is the right regular representation. So we write

$$A(G, S) = \sum_{s \in S \cup S^{-1}} \rho(s).$$

Of course the right hand side is just the Fourier transform of the characteristic function for $S \cup S^{-1}$.

In terms of representation theory, one knows that

$$\rho(s) \sim diag(B_1(s), ..., B_h(s))$$

and

$$B_i(s) = diag(\rho_i(s), ..., \rho_i(s))$$

where $\rho_1, ..., \rho_h$ form a complete set of irreducible unitary representations of G. Here $B_i(s)$ has $deg(\rho_i)$ copies of $\rho_i(s)$ along the diagonal.

Theorem 39 *(Lafferty-Rockmore)*

$$Spec(X(G, S)) = \cup_{i=1}^{h} Spec(\sum_{s \in S \cup S^{-1}} \rho_i(s)).$$

For a proof, see Lafferty and Rockmore (1997).

For the case of a finite abelian group, this simplifies to:

Theorem 40 *If $X = X(G, S)$ is a Cayley graph for a finite abelian group, then*

$$Spec(X) = \{\sum_{s \in S} \chi(s), \chi \in \hat{G}\}$$

and

$$Z_X(u)^{-1} = (1 - u^2)^{(q-1)n/2} \prod_{\chi \in \hat{G}} [1 - (\sum_{s \in S} \chi(s))u + qu^2].$$

For a proof, see Terras (1992).

9.37 Inverse Spectral Problems

Can one hear the shape of a Cayley graph? Kesten (1959), Day and others have studied problems of this nature; for a review see Valette (1994). Define

$$h = \frac{A}{|S|} = \frac{1}{|S|} \sum_{s \in S} \rho(s),$$

which is called the transition operator. h is a selfadjoint operator with norm less than or equal to one. The inverse spectral problem is, knowing the spectrum, $Spec(h)$, what can be said about (G, S)? de la Harpe, Robertson and Valette (1993) have shown:

Theorem 41 *If $Spec(h) = C_n$, the group of nth roots of unit, then $G \simeq Z/nZ$ and $|S| = 1$. And if $Spec(h) = S^1$, then $G \simeq Z$ and $|S| = 1$.*

For further results in this direction, see Valette (1994).

9.38 Sunada's Theorem for Graphs

Brooks (1997) has examined the question of when two graphs are isospectral in terms of a Sunada-like condition; for the manifold case, see Sunada (1985). Let X be a k-regular graph and let G be a group of automorphisms of X. Let H_i be two subgroups of G which act freely on X. Then (X, G, H_1, H_2) is said to satisfy the Sunada condition if for all $g \in G$

$$|(\{g\} \cap H_1)| = |(\{g\} \cap H_2)|$$

where $\{g\}$ denotes the conjugacy class of g in G. For $n > 0$, let $H_i^{(n)}$ be the set of elements h of H_i such that there is a path of length less than or equal to n in X whose endpoints differ by multiplication by h. Then (X, G, H_1, H_2) is said to satisfy the Sunada condition up to length n if

$$|(\{g\} \cap H_1^{(n)})| = |(\{g\} \cap H_2^{(n)})|.$$

Brooks (1997) has shown:

Theorem 42 *(Brooks) Suppose* (X, G, H_1, H_2) *satisfies the condition of Sunada. Then the graphs* $X_1 = X/H_1$ *and* $X_2 = X/H_2$ *are isospectral. If* (X, G, H_1, H_2) *satisfies the Sunada condition up to length* N*, where*

$$N \geq max(|X_1|, |X_2|),$$

then X_1 *and* X_2 *are isospectral.*

One notes that two $k-$regular graphs are isospectral if and only if for any l, the number of closed paths of length l is the same for both graphs.

Brooks proves the following converse to the last theorem using the above mentioned Leighton's covering theorem.

Theorem 43 *(Brooks) Let* X_1 *and* X_2 *be two* $k-$*regular isospectral graphs. Then, for any* n*, there is a graph* $X^{(n)}$ *which covers both* X_1 *and* X_2*, a group of graph automorphisms* G^n *and two subgroups* $H_1^{(n)}$ *and* $H_1^{(n)}$ *which act freely on* $X^{(n)}$ *with*

$$X_i = X^{(n)}/H_i^{(n)}$$

for $i = 1, 2$ *so that* $(X^{(n)}, G^n, H_1^{(n)}, H_2^{(n)})$ *satisfies the Sunada condition up to length* n*.*

Brooks (1997) cites the following result due to Quenell:

Theorem 44 *(Quenell) For* $n \to \infty$*, there are sets of isospectral* $k-$*regular graphs with* n *vertices, where* $k \to \infty$ *with* n*, which grow in size like* n^{Cn} *for some constant* C*.*

9.39   Selberg Trace Formula for Graphs

Let X be a $(q+1)-$regular infinite tree and let Γ be a strictly hyperbolic group of isometric automorphisms of X and assume $|\Gamma \backslash X| < \infty$. The Selberg trace formula in this case was proven by Ahumada (1987):

Theorem 45 *(Ahumada) The eigenvalues of* $T_1(\Gamma, \chi)$ *are real and*

$$|\lambda_j(\Gamma, \chi)| \leq (q + 1)$$

where $T_1(\Gamma, \chi)$ *is the adjacency matrix of the finite* $(q+1)-$*regular quotient graph* $\Gamma \backslash X$*.*

For any sequence $\{h(n)\}_{n \in \mathbf{Z}}$ *of complex numbers such that*

$$\sum |h(n)||q|^{|n|/2} < \infty$$

define the Fourier transform

$$\hat{h}(z) = \sum_{n \in \mathbf{Z}} h(n) z^n.$$

Then the Selberg trace formula states that

$$\sum_{j=1}^{M} \hat{h}(z_j(\Gamma, \chi)) = |\Gamma \backslash V(X)| dim_{\mathbb{C}} V q \int_{|\lambda|=1} \hat{h}(\lambda) \frac{1-\lambda^2}{q-\lambda^2} d^x \lambda +$$

$$\sum_{\{P\} \in \mathcal{P}_r} \sum_{l=1}^{\infty} \frac{Tr\chi(P^l) deg P}{q^{l deg P/2}} h(l deg P)$$

where the left hand side is really a sum over the eigenvalues of T_1 and $M = dim_{\mathbb{C}} \mathcal{A}(\Gamma, \chi)$.

As Venkov and Nikitin (1994) have noted one can relate the Ihara zeta function and the Selberg trace formula as follows:

Theorem 46 (Venkov-Nikitin) The Ihara formula

$$Z(\Gamma, u, \chi) = (1 - u^2)^{-g \chi} det(I - T_1(\Gamma, \chi)u + qu^2)^{-1}$$

implies and is implied by the Selberg trace formula. In addition,

$$TrT_1(\Gamma, \chi) = \frac{d}{du} log Z(\Gamma, u, \chi)|_{u=0}.$$

9.40   Star Graph

Let Y denote the graph with only one vertex and $2r$ branches, $a_1, \bar{a}_1, ..., a_r, \bar{a}_r$. The group Π is isomorphic to the free group with r generators, $\{a_1, ..., a_r\}$, and X is the Cayley graph of Π, where $Y = \Pi \backslash X$. The Hecke operator in this case is $T_1(\chi) = \sum_{i=1}^{r} \chi(a_i) + \chi(\bar{a}_i)$. For the case of the trivial representation χ, then Ahumada (1987) has shown that

$$Z(u, \chi)^{-1} = (1 - u^2)^{(q-1)/2}(1 + qu^2 - (q+1)u)$$

and

$$\prod_{\{p\}} (1 - u^{l\{p\}})^{-1} = \frac{1}{(1 - u^2)^{(q-1)/2}(1 + qu^2 - (q+1)u)}.$$

9.41   Spectral Determinant

Akkermans et al. (1999) have examined the spectral determinant $S(\gamma)$ of the Laplacian on finite graphs. Viz., if E_n are the eigenvalues of $-\Delta$ and $\mathcal{Z}(t)$ is the partition function

$$\mathcal{Z}(t) = \sum_{n} e^{-E_n t},$$

then the spectral determinant is given by

$$S(\gamma) = det(-\Delta + \gamma) = \prod_{n} (\gamma + E_n).$$

The spectral determinant is related to $\mathcal{Z}(t)$ by

$$\int_0^\infty dt \mathcal{Z}(t) e^{-\gamma t} = \frac{\partial}{\partial \gamma} \ln S(\gamma).$$

For these models, the weak-localization correction, $< \Delta\sigma >$, and the variance in conductance, $< \delta\sigma^2 >$, can be expressed in terms of the partition function or the spectral determinant. Viz., Pascaud and Montambaux (1999) have shown that

$$< \Delta\sigma > = -\frac{2e^2 D}{\pi\Omega} \int_0^\infty dt \mathcal{Z}(t) e^{-\gamma t} = -\frac{2e^2 D}{\pi\Omega} \frac{\partial}{\partial\gamma} \ln S(\gamma)$$

and

$$< \delta\sigma^2 > = \frac{12e^4 D^2}{\beta\pi^2\Omega^2} \int_0^\infty dt \, t \mathcal{Z}(t) e^{-\gamma t} = -\frac{12e^4 D^2}{\beta\pi^2\Omega^2} \frac{\partial^2}{\partial\gamma^2} \ln S(\gamma).$$

Here D is the diffusion constant and Ω is the volume of the system.

For the case of a finite graph with V vertices and B edges, Akkermans et al. (1999) show that

$$S(\gamma) = (\frac{\sqrt{\gamma}}{2\pi})^{V-B} \prod_{(\alpha\beta)} sinh(\sqrt{\gamma} l_{\alpha\beta}) det(M)$$

where M is the $V \times V$ matrix given by

$$M_{\alpha\alpha} = \sum_{i=1}^{m_\alpha} coth(\sqrt{\gamma} l_{\alpha\beta_i})$$

and

$$M_{\alpha\beta} = \begin{cases} -\frac{1}{sinh(\sqrt{\gamma} l_{\alpha\beta})} & \text{if } (\alpha\beta) \text{ is a bond} \\ 0 & \text{otherwise} \end{cases}.$$

Here the sum is over the m_α neighboring sites of the vertex α and $l_{\alpha\beta}$ is the length of edge $(\alpha\beta)$.

Akkermans et al. show the following:

Theorem 47 *(Akkermans et al.) Let $\tilde{C}$ denote the primitive paths in a graph. Then*

$$S(\gamma) = \gamma^{\frac{V-B}{2}} e^{\sqrt{\gamma}L} (\prod_\alpha m_\alpha) 2^{-B} \prod_{\tilde{C}} (1 - \alpha(\tilde{C}) e^{-\sqrt{\gamma} l(\tilde{C})}).$$

And Roth's (1983) formula follows:

$$\mathcal{Z}(t) = \frac{L}{2\sqrt{\pi t}} + \frac{V-B}{2} + \frac{1}{2\sqrt{\pi t}} \sum_C l(\tilde{C}) \alpha(C) e^{-\frac{l(C)^2}{4t}}.$$

Here L is the total length $L = \sum_{\alpha\beta} l_{\alpha\beta}$. For the characterization of the weights $\alpha(C)$ and $l(\tilde{C})$, see Akkermans et al.

The authors relate their work to the developments of Kottos and Smilansky (1997, 1999). Akkermans et al. extend their results naturally to the case which includes a magnetic flux. E.g., for the case of a ring of perimeter l and an attached arm of length b with a flux ϕ, let $\theta = 2\pi\phi/\phi_0$ where $\phi_0 = h/e$ is the flux quantum. They show in this case

$$S(\gamma) = sinh(\sqrt{\gamma}b)sinh(\sqrt{\gamma}l) + 2[cosh(\sqrt{\gamma}l) - cos(\theta)]cosh(\sqrt{\gamma}b).$$

They develop the spectral determinant for the complete graph K_n and they also treat the case of a graph connected to an infinite lead for which they calculate the scattering matrix phase shift; we direct the readers to the paper for the details.

9.42 Spherical Functions

Let X be a $k-$regular graph and let Δ denote the Laplacian on $L^2(X)$ given by

$$\Delta f(x) = \frac{1}{k}\sum_{y \sim x}(f(x) - f(y)).$$

The universal cover of X is a $k-$tree T_k. Let $l(x, y)$ denote the number of edges in the unique path from x to y in T_k.

Theorem 48 *There is a unique function, called the spherical function S_λ, with eigenvalue λ such that (1) $S_\lambda(0) = 1$ and (2) $\Delta(S_\lambda) = \lambda S_\lambda$.*

Brooks (1991) has explicitly calculated S_λ.

9.43 Brooks' Trace Formula

Let $K(l)$ be a sufficiently rapidly decreasing function of l and set

$$K_X(x, y) = \sum_{y'} K(l(x', y'))$$

where x' is a fixed lift of vertex x to T_k and y' runs over all lifts of y to T_k. The pre-trace formula of Brooks states:

Theorem 49 *(Brooks)*

$$Tr(K_X) = \sum_{l} K(l)[k(k - 1)^{l-1}]S_{\lambda_i}(l).$$

In particular if

$$K_l(l') = \begin{cases} 1 & \text{if } l' = l \\ 0 & \text{otherwise,} \end{cases}$$

then

$$Tr(K_l) = k(k - 1)^{l-1}S_{\lambda_i}(l).$$

Here $Tr(K_l)$ counts the number of closed loops on X of length l.

Brooks' trace formulas follows from the pre-trace formula:

Theorem 50 *(Brooks) If* $K = K_l$, *then*

$$\sum_{\gamma'} F(\frac{l - l'}{2})l' = \sum_i k(k-1)^{l-1}S_{\lambda_i}(l)$$

where

$$F(\frac{l - l'}{2}) = \begin{cases} k(k-1)^{\frac{l-l'}{2}-1} & \text{if } l - l' \text{ is positive and even} \\ 1 & \text{if } l = l' \\ 0 & \text{otherwise} \end{cases}$$

The sum on the left is over all closed geodesics γ' *and only those of length less than or equal to* l *make a nonzero contribution. The sum on the right is over the spectrum of* Δ.

9.44   Lattice-Point Counting Theorem

Using his trace formula Brooks studied the function $N(X, l)$, the number of closed loops of length l in X. First, one notes that

$$lN(X, l) = \sum_{x \in X} K_l(x, x).$$

Define the generating function $f_X(x)$ by

$$f_X(s) = \sum lN(X, l)x^l.$$

Then in analogy with the Selberg zeta function, Brooks showed:

Theorem 51 *(Brooks) Let* X *be a* $k-$*regular graph. Then* $f_X(x)$ *is a rational function of* x. *If* β *is a pole of* f, *then so is* $1/(k-1)\beta$. *The eigenvalues of* X *and the poles* β *of the generating function are related by*

$$\lambda = 1 - \frac{(k-1)\beta^2 + 1}{k\beta}.$$

The β*'s are either purely real or of absolute values* $1/\sqrt{k-1}$. *And the poles are simple, except for* $\beta = \pm\frac{1}{\sqrt{k-1}}$ *which occur as double poles.*

As Brooks has noted, if one replaces x by q^{-s} where $q = k - 1$, then X is Ramanujan graph if and only if the poles of $f_X(q^{-s})$ occur only for $Re(s) = \frac{1}{2}$.

Using the trace formula, Brooks is also able to improve estimates on the diameter of a graph; e.g., Brooks shows:

Theorem 52 *(Brooks) If* X *is Ramanujan and bipartite, then*

$$diam(X) \leq \frac{2arccosh(\frac{|X|-2}{2}) + 1}{log(k-1)}.$$

For other results in this direction see Brooks (1991) and Quenell (1994).

9.45 Selberg's 3/16 Theorem for Graphs

Using his trace formula, Brooks has developed the following analogy with Selberg's 3/16 theorem:

Theorem 53 *(Selberg) If Γ is the congruence group, then*

$$\lambda_1(\Gamma) = \lambda_1(H/\Gamma) \geq 3/16.$$

The bound 3/16 has been improved to $\lambda_1(\Gamma) > 3/16$ by Gelbart and Jacquet (1978) and then to $\lambda_1(\Gamma) \geq 21/100$ by Luo, Rudnick and Sarnak (1995). In addition, Selberg conjectured that $\lambda_1(H/\Gamma) \geq 1/4$.

Brooks' theorem for graphs states:

Theorem 54 *(Brooks) Let X_p be a family of k−regular graphs and let $c_1, ..., c_4$ be constants independent of p such that (1) $c_1 p^3 \leq |X_p| \leq c_2 p^3$, (2) $\lambda_1(X_p)$ occurs with multiplicity at least $c_3 p$, (i.e., the graphs have interesting symmetries) and (3) the number of closed geodesics of length less than or equal $6\log(p)$ is at most $c_4 p^3$ (i.e., the graphs are short and fat). Then*

$$\liminf{}_p \lambda_1(X_p) \geq 1 - \frac{(k-1)^{1/6} + (k-1)^{5/6}}{k}.$$

Brooks looked at the case $S_p = H/\Gamma(p)$ where

$$\Gamma(p) = \{ \begin{pmatrix} a & b \\ c & d \end{pmatrix} \mid \begin{pmatrix} a & b \\ c & d \end{pmatrix} \equiv \pm \begin{pmatrix} 1 & 0 \\ 0 & 1 \end{pmatrix} \bmod p \}.$$

The analogue of Brooks graph theoretic result for manifolds gives the estimate

$$\liminf{}_p \lambda_1(S_p) \geq 5/36.$$

For more details see Brooks (1996).

9.46 Bruhat-Tits Buildings for $PGL(2, \mathbf{Q}_p)$

Let $V = \mathbf{Q}_p \times \mathbf{Q}_p$ be the two dimensional vector space over the filed of p−adic numbers. A $\mathbf{Z}_p$−lattice in V is a $\mathbf{Z}_p$− submodule L of V generated by two linearly independent vectors in V. The standard lattice is $L_0 = \mathbf{Z}_p \times \mathbf{Z}_p$ generated by the basis e_1, e_2. Two lattices L_1, L_2 are called equivalent if there is an element $0 \neq \alpha \in \mathbf{Q}_p$ such that $L_2 = \alpha L_1$. Two equivalence lattices $[L_1]$ and $[L_2]$ are said to be adjacent if there is $L'_i \in [L_i]$ such that $L'_1 \subset L'_2$ and $[L'_2 : L'_1] = p$. Let X denote this graph; then $G = PGL(2, \mathbf{Q}_p)$ acts transitively on the vertices of X. The stabilizer of L_0 is $PGL(2, \mathbf{Z}_p)$ so that the set of vertices is just $PGL(2, \mathbf{Q}_p)/PGL(2, \mathbf{Z}_p)$. G preserves the adjacency so G is a transitive group of automorphisms of X and one has the structure theorem:

Theorem 55 *(Serre) The space X is a $p + 1$ regular tree and*

$$\{A_0(L_0), ... A_p(L_0), A_\infty(L_0)\}$$

are the $p + 1$ representatives of the $p + 1$ vertices adjacent to vertex $[L_0]$. Here
$A_i = \begin{pmatrix} p & i \\ 0 & 1 \end{pmatrix}$ *or $i = 0, ..., p - 1$, and $A_\infty = \begin{pmatrix} 1 & 0 \\ 0 & p \end{pmatrix}$.*

For background on Bruhat-Tits buildings, the reader is directed to Brown (1989) and Garrett (1997).

9.47   Representation Theory

In this section we examine the relationship of Ramanujan graphs and representation theory. Let $G = PGL(2, \mathbf{Q}_p)$ and $K = PGL(2, \mathbf{Z}_p)$ and let Γ be a lattice in $PGL(2, \mathbf{Q}_p)$. Then $\Gamma \backslash G / K$ is a $(p + 1)-$regular graph, in fact a finite graph since every lattice in $PGL(2, \mathbf{Q}_p)$ is cocompact (v., Serre (1980)).

In terms of representation theory, Lubotzky (1994) has shown:

Theorem 56 *(Lubotzky) If Γ is a cocompact lattice in G, then $\Gamma \backslash G / K$ is a Ramanujan graph if and only if no complementary series representations occurs in $L^2(\Gamma \backslash G)$ except for the trivial one corresponding to $\lambda = p + 1$ or $\lambda = -(p + 1)$.*

9.48   Hecke Eigenvalues for Bruhat-Tits Buildings

The Hecke operator

$$T(f) = \sum_{d(y,x)=1} f(y)$$

is related to the combinatorial Laplacian Δ by

$$T = (p + 1)I - \Delta.$$

For the case $X = G/K$ where $G = PGL(2, \mathbf{Q}_p)$ and $K = PGL(2, \mathbf{Z}_p)$, then in terms of spherical functions on X one has:

Theorem 57 *If (H, χ) is an irreducible unitary representation of G of class one with associated spherical function ϕ, then*

$$T\phi = \lambda\phi$$

where $|\lambda| \leq p + 1$. For every such λ, there is a unique spherical function ϕ^λ such that

$$T\phi^\lambda = \lambda\phi^\lambda.$$

Theorem 58 *The spherical functions* ϕ_t *where*

$$T\phi_t = (p^{1-t} + p^t)\phi_t$$

have two forms:

(1) the principal series, $t \in \{\frac{1}{2} + ir | r \in \mathbf{R}\}$ *where* $|\lambda| \leq 2\sqrt{p}$

(2) complementary series, $t \in \{r + \frac{n\pi i}{\log p} | 0 \leq r \leq 1, n \in \mathbf{Z}\}$ *where* $2\sqrt{p} < |\lambda| \leq p + 1$.

9.49 Induced Hecke Operator

For vertex $v \in V(\Gamma \backslash X)$ and edge $e \in E(\Gamma \backslash X)$ let Γ_v and Γ_e denote the stabilizer subgroups. Set $m(v) = |\Gamma_v|^{-1}$ and $m(e) = |\Gamma_e|^{-1}$.

The adjacency operator defined by

$$(Tf)(x) = \sum_{x,y} f(y)$$

where y is adjacent to x induces an operator on $\Gamma \backslash X$ by

$$(T_\Gamma f)(x) = \sum_{e=(x,y) \in E(\Gamma \backslash X)} \frac{m(e)}{m(x)} f(y)$$

where $f : \Gamma \backslash X \to \mathbf{C}$.

9.50 Lattice Models in Function Fields

Let $\mathbf{F}_q$, the finite field with $q = p^n$ elements of characteristic p, and let $A = \mathbf{F}_q[t]$ denote the polynomial ring in one indeterminant t. Its quotient field is $k = \mathbf{F}_q(t) = quot(A)$, i.e., the field of rational functions in t over $\mathbf{F}_q$. k is provided with a degree valuation $v : k \to \mathbf{Z} \cup \{\infty\}$, viz. $v(a/b) = deg(b) - deg(a)$ for $a, b \in A$. The norm at infinity is given by

$$|a/b|_\infty = q^{deg(a) - deg(b)}$$

where a, b are two polynomials in $\mathbf{F}_q[t]$; so $|a|_\infty = q^{-v(a)}$ for $a \in k$. The completion of $\mathbf{F}_q(t)$ with respect to this norm is the field $k_\infty = \mathbf{F}_q((\pi))$ of formal Laurent series in the uniformizer $\pi = t^{-1}$:

$$\sum_{n=-N}^{\infty} a_n t^{-n}$$

where $a_n \in \mathbf{F}_q$. Those series with $N \geq 0$ form the maximal compact subring O of the local integers in k_∞, i.e. $O = O_\infty = \mathbf{F}_q[[\pi]]$, the ring of formal power series in $\pi = 1/t$. The maximal ideal is $m_\infty = \pi O_\infty$ and the residue class is $k(\infty) \simeq \mathbf{F}_q$. Let C denote the completion of the valued field $\bar{k}_\infty$.

Let $G = PGL(2, k_\infty)$ be the group of all 2×2 invertible matrices over k_∞ modulo the scalar matrices and let $K = PGL(2, O)$, a maximal compact subgroup. Consider the homogeneous space

$$\mathcal{T} = G/K$$

on which G acts via isometries.

Let Γ be the subgroup of elements of polynomial entries, $\Gamma = GL(2, \mathbf{F}_q[t])$. It is a discrete subgroup that acts on $\mathcal{T}$. Let

$$F = \Gamma \backslash \mathcal{T}.$$

An $O-$lattice in k_∞^2 is a set

$$L = \{\alpha v_1 + \beta v_2 | \alpha, \beta \in O\}$$

where v_1, v_2 is a basis for k_∞^2. One can associate to L a matrix (v_1, v_2) in $GL(2, k_\infty)$ and different choices of bases v_1, v_2 will give cosets in G/K. Two lattices are called equivalent if $L' = aL$ for some $a \in k_\infty^x$. Thus, there is a natural correspondence between equivalence classes of lattices and points in $\mathcal{T}$.

9.51   Eisenstein Series on Graphs

For each row vector (x, y) in $k_\infty \times k_\infty$, define the height function $ht(x, y) = sup\{|x|_\infty, |y|_\infty\}$. Then K is the group of linear transformations of $k_\infty \times k_\infty$ which preserve ht:

$$ht((x, y)k) = ht(x, y)$$

for every $(x, y) \in k_\infty \times k_\infty$ and $k \in K$.

Define the function

$$\psi_s(g) = |det(g)|_\infty^s ht((0, 1)g)^{-2s}$$

where $s \in \mathbf{C}$ and $g \in G$. Then $\psi_s(g)$ is K right invariant and N left invariant where $N = \{\begin{pmatrix} 1 & x \\ 0 & 1 \end{pmatrix} | x \in G\}$. One checks that ψ_s is an eigenfunction of T:

$$(T\psi_s)(g) = (q^s + q^{1-s})\psi_s(g).$$

Let $\kappa_1, ..., \kappa_\mu$ denote the set of inequivalent cusps and let Γ_{κ_i} denote the stabilizer in Γ of κ_i. There is an element $\tilde{\kappa}_i \in G$ such that $\tilde{\kappa}_i \infty = \kappa_i$.

The Eisenstein series at cusp κ_i is given by

$$E_i(g, s) = \sum_{\gamma \in \Gamma_{\kappa_i} \backslash \Gamma} \psi_s(\tilde{\kappa}_i^{-1} \gamma g).$$

It follows from above that

$$(TE_i)(g, s) = (q^s + q^{1-s})E_i(g, s)$$

for $g \in X = G/K$ where $K = PGL(2, O)$.

Li (1989) has shown for $\Gamma = \Gamma(A)$ the Fourier series of $E_i(g, s)$ at cusp κ_j has the constant term of the form

$$\delta_{ij} q^{ns} + \phi_{ij} q^{n(1-s)}.$$

The matrix $\Phi(s) = (\phi_{ij}(s))$ is called the scattering matrix and $\phi(s) = det\ \Phi(s)$ is the scattering determinant. For the principal congruence group, Li has shown:

Theorem 59 *(Li) The scattering matrix element $\phi_{ij}(s)$ is a rational function in q^{-2s} and for fixed $g \in X$, $E_i(g, s)$ is a rational function in q^{-s}; both functions are holomorphic on $Re(s) \geq 1/2$ except for simple poles at $s = 1 + n\pi i/\log q$ for $n \in \mathbf{Z}$. Furthermore, $\Phi(s)$ is symmetric and satisfies the functional equation*

$$\Phi(s)\Phi(1 - s) = I.$$

9.52 Multiloop Scattering

For related work on scattering theory and Bruhat-Tits spaces, we direct the reader to the work of Chekhov, Mironov and Zabrodin (1989), Chekhov (1995, 1996, 1995, 1999), Freund (1991), Freund and Zabrodin (1993), Novikov (1997), Romanov and Rudin (1995, 1997).

In particular, Freund (1991) has noted the analogy of the scattering matrix for the hyperbolic plane, which is the symmetric space $H_2 = SL(2, \mathbf{R})/SO(2)$, and the p-adic hyperbolic plane, $H_2^{(p)} = SL(2, \mathbf{Q}_p)/SL(2, \mathbf{Z}_p)$. In the first case the S-matrix is given by

$$S^{(\infty)}(\lambda) = \frac{c^{(\infty)}(\lambda)}{c^{(\infty)}(-\lambda)}$$

in terms of Harish-Chandra $c-$function

$$c^{(\infty)}(\lambda) = \frac{1}{\sqrt{\pi}} \frac{\Gamma(\frac{1}{2}i\lambda)}{\Gamma(\frac{1}{2}(i\lambda + 1))}.$$

For the $p-$adic case, Freund (1991) has noted:

Theorem 60 *(Freund) The Harish-Chandra $c-$function is*

$$c^{(p)}(\lambda) = \frac{\zeta_p(i\lambda)}{\zeta_p(i\lambda + 1)}$$

where

$$\zeta_p(s) = (1 - p^{-s})^{-1}.$$

The S-matrix is given by

$$S^{(p)}(\lambda) = \frac{c^{(p)}(\lambda)}{c^{(p)}(-\lambda)}.$$

Chekhov (1995) has extended this result to p-adic multiloop surfaces, graphs X and related reduced graphs X_{red} to show that the S-matrix has the form in terms of Hashimoto-Bass L-functions:

$$C = (\frac{\alpha_+}{\alpha_-})^{|X_{red}|}(\frac{1 - \alpha_-^2}{1 - \alpha_+^2})^{|X_{red}|-|E_{red}|}\frac{L(\alpha_+)}{L(\alpha_-)}.$$

Here $|E|$ and $|X|$ are the total number of edges and vertices. If t is the eigenvalue of the Laplacian on the tree, then $\alpha_\pm = (t/2p) \pm \sqrt{(t^2/2p^2) - (1/p)}$ where $\alpha_+\alpha_- = 1/p$.

For further details, the reader is directed to Chekhov (1995, 1999). In the later reference he presents certain results related to exceptional spectra and Bruhat-Tits spaces.

9.53 Ramanujan Graphs and Function Fields

In the case mentioned above for a lattice

$$\Gamma \subset G = PGL(2, \mathbf{Q}_p)$$

and the $(p + 1)$-regular tree $X_p = G/PGL(2, O_p)$ where O_p are the integers of $\mathbf{Q}_p$, then Γ was a uniform subgroup, $\Gamma\backslash G$ is compact and $\Gamma\backslash X_p$ is a finite $(p + 1)$-regular graph. By the Ramanujan conjecture proven by Deligne, for congruence subgroups Γ of arithmetic groups of G, every eigenvalue λ of $\Gamma\backslash X_p$, which is not $\pm(p+1)$ satisfies $|\lambda| \leq 2\sqrt{p}$; thus these graphs are Ramanujan graphs.

Morgenstern (1993) has extended these results to the function field case. Let $k = \mathbf{F}_q(x)$ denote the quotient field of $\mathbf{F}_q[x]$ with the completion k_x and the integers O_x. Let $G = PGL(2, k_x), K = PGL(2, O_x)$; then $X_x = G/K$ is a $(q + 1)$-regular tree where

$$q + 1 = |\mathbf{P}^1(O_x/m_xO_x)|.$$

Based on the quaternion algebra over k, there is a free group $\Gamma(1)$ of $(q + 1)/2$ generators (v. Morgenstern (1993)). Let $g(x) \in \mathbf{F}_q[x]$ be prime to $x(x - 1)$. Then the finite index sublattice

$$\Gamma(g) = \{t \in \Gamma(1)|t \equiv I \, mod \, g\}$$

of $\Gamma(1)$ is uniform and $X_g = \Gamma(g)\backslash X_x$ is a finite $(q + 1)$-regular graph. Drinfeld's (1988) proof of Ramanujan's conjecture for the function field case implies that X_g is a Ramanujan graph.

Morgenstern has given an explicit construction of these graphs. Let $\mathbf{F}_{q^d} = \mathbf{F}_q[x]/g\mathbf{F}_q[x]$ where g is irreducible of even degree d. Let $\bar{i} \in \mathbf{F}_{q^d}$ be a square root of $\epsilon \in \mathbf{F}_q$. In $PGL(2, \mathbf{F}_{q^d})$ there are exactly $q+1$ matrices, $\xi_1, ..., \xi_{q+1}$, of the form

$$\begin{pmatrix} 1 & \gamma - \delta\bar{i} \\ (\gamma + \delta\bar{i})(x - 1) & 1 \end{pmatrix}$$

where $\gamma, \delta \in \mathbf{F}_q$ and $\delta^2\epsilon - \gamma^2 = 1$.

Theorem 61 *(Morgenstern) Assume x is not a square root in $\mathbf{F}_{q^d}$; then the Cayley graph of $PGL(2, \mathbf{F}_{q^d})$ with respect to the generators $\xi_1, ..., \xi_{q+1}$ is a bipartite $(q+1)$−regular Ramanujan graph.*

9.54 Statistical Properties of Ramanujan Eigenvalues

Terras and her students (v., Terras (1992) and Celniker et al. (1993)) have constructed a class of Ramanujan graphs living on the hyperbolic upper half-plane. Li (1998) has shown that these graphs can be related to Morgenstern's Ramanujan graphs discussed in the last Section. Terras has examined the statistical properties of the eigenvalues of her Ramanujan graphs. If $\{\lambda_i\}$ are the eigenvalues of the adjacency matrix for a Ramanujan graph (v.i.), Terras (1992) has conjectured that $\{\lambda_i/\sqrt{q} | i = 1, ..., q-1\}$ asymptotically has the Sato-Tate distribution, i.e. for $E \subset [-2, 2]$

$$lim_{q\to\infty} \frac{1}{q-1} |\{\lambda_i | \lambda_i/\sqrt{q} \in E\}| = \frac{1}{2\pi} \int_E \sqrt{4 - x^2} dx.$$

For these models, Kuang (1997) has calculated the first and second moments of the asymptotic distribution of the eigenvalues of the adjacency matrices and showed that they asymptotically match those of the Sato-Tate distribution:

$$lim_{q\to\infty} \frac{1}{q-1} \sum_{i=1}^{q-1} \frac{\lambda_i}{\sqrt{q}} = 0$$

and

$$lim_{q\to\infty} \frac{1}{q-1} \sum_{i=1}^{q-1} (\frac{\lambda_i}{\sqrt{q}})^2 = 1.$$

For more details, see Terras (1992).

9.55 Equidistribution for Ramanujan Graphs

Let $X^{p,q}$ denote the Ramanujan graphs constructed by Lubotzky, Phillips and Sarnak (1988). Using the work of Tillich and Zemor (1997) we know that there is an explicit bound on the number of geodesics of length i,

$$O(max(1, p^{i/2}/q^2) q^2 p^{i/2} log(p^i/q^2)).$$

The size of $X^{p,q}$ is $O(q^3)$. Thus, McKay's condition holds in this case and we have the following result of Li (1998):

Theorem 62 *(Li) For the family of k−regular graphs $X^{p,q}$ as prime q tends to ∞, the limit measure is*

$$\mu_{k-1} = \frac{k}{(\sqrt{k-1} + \frac{1}{\sqrt{k-1}})^2 - x^2} \mu_{ST}$$

where $k - 1 = p$.

A similar result holds for Morgenstern's Ramanujan graphs discussed in the last section. As Li (1998) notes, since functions on quaternion graphs are automorphic forms on quaternion groups, which correspond to automorphic forms of GL_2 over **Q**, the above results imply that the eigenvalues of the Hecke operators T_p on cusp forms of weight 2 and level N, where p and N are prime, are equidistributed with respect to μ_p as $N \to \infty$. Note that Serre (1998) showed the same limit holds as $k + N \to \infty$ based on the Selberg trace formula. And as we noted in Chapter 8, Sarnak (1987) showed that the eigenvalues of the Hecke operator T_p on Maass wave forms for the modular group $SL(2, \mathbf{Z})$ are equidistributed with respect to μ_p as the eigenvalue of the Laplacian approaches ∞.

9.56   The Bruhat-Tits Tree for Function Fields

One defines an ordered edge to be a pair (Λ, Λ') of equivalence classes of lattices if there exists $L \in \Lambda, L' \in \Lambda'$ such that $L \subset L'$ and $L'/L \simeq \mathbf{F}_q$. Then, (Λ, Λ') is an edge if and only if its inverse (Λ', Λ) is.

Theorem 63 *(Serre) The graph whose set of vertices is $\mathcal{T}$ and whose edges are the pairs (Λ, Λ') as described form an infinite $(q + 1)$−regular tree. The vertices are given by $V(\mathcal{T}) = G/KZ_\infty$ and the oriented edges are given by $E(\mathcal{T}) = G/\mathcal{I}Z_\infty$ where $\mathcal{I}$ is the Iwahori group $\{\begin{pmatrix} a & b \\ c & d \end{pmatrix} \in K | c \equiv -mod\,\pi\}, \pi = 1/t$ and Z_∞ is the center of $PGL(2, k_\infty)$. The canonical map from $E(\mathcal{T})$ to $V(\mathcal{T})$ associates with each edge e its terminus $t(e)$. G acts on the left on $\mathcal{T}$ as a group of automorphisms. The neighbors of gK are the $q + 1$ cosets $gs_k K, i = 1, ..., q + 1$, given by*

$$\{s_1, ..., s_{q+1}\} = \{\begin{pmatrix} \pi & \beta \\ 0 & 1 \end{pmatrix} | \beta \in \mathbf{F}_q\} \cup \{\begin{pmatrix} 1 & 0 \\ 0 & \pi \end{pmatrix}\}.$$

9.57   Ends and Cusps

An end of a Bruhat-Tits tree is an equivalence class of infinite half lines

$$\bullet - - - \bullet - - - \bullet - - - \quad ...,$$

two of which are identified if they differ in a finite graph. The boundary of the homogeneous tree can be identified with the projective line $\mathbf{P}^1(K)$ or G/B where B is the Borel subgroups with elements $\begin{pmatrix} a & b \\ 0 & 1 \end{pmatrix}$; v. Cartier (1973).

9.58   Modular and Arithmetic Groups

The modular group is defined to be $\Gamma(1) = SL(2, \mathbf{F}_q[t])$, i.e. the matrices of determinant one with polynomial entries. A group Γ is said to be arithmetic if Γ is a subgroup of finite index in $\Gamma(1)$. In this case Γ acts without inversions.

Let $\Gamma(g)$ denote the subgroup of matrices congruent to $\begin{pmatrix} 1 & 0 \\ 0 & 1 \end{pmatrix}\, mod\, g$, where $g \in A$ is nonconstant, i.e.

$$\Gamma(g) = \{A \in \Gamma(1) | A \equiv I\, mod\, g\}.$$

$\Gamma(g)$ is called the principal congruence subgroup of level g. A subgroup Γ is called a congruence subgroup if it contains some $\Gamma(g)$ for $g \in \mathbf{F}_q[t]$.

Serre (1980) showed:

Theorem 64 *(Serre)* $\Gamma(1)$ *is a non-uniform lattice in* G, *i.e.* $\Gamma(1)$ *is discrete and* $\Gamma(1)\backslash G$ *has finite G-invariant measure.*

$\Gamma(g)$ has finite index in $\Gamma(1)$, hence:

Theorem 65 *(Serre)* $\Gamma(g)$ *is a non-uniform lattice in* G *and*

$$X_g = \Gamma(g)\backslash G/K$$

is a $(q+1)$-regular graph.

9.59 Number of Cusps

If $A \in k_\infty$ is written as $\sum_{i=n}^{\infty} a_i t^{-i}$ with $a_n \neq 0$ then $v(A) = n$ and $|A|_\infty = q^{-n}$.
For the principal congruence group $\Gamma(A)$, then number of cusps is given by

$$\sigma(A) = \begin{cases} \frac{1}{q-1}|A|_\infty^2 \prod_{\substack{B,B|A \\ B, monic}} (1 - \frac{1}{|B|_\infty^2}) & \text{if } deg(A) = a \geq 1 \\ 1 & \text{if } A = 1. \end{cases}$$

In comparison for the principal congruence group of level q, in $SL(2,\mathbf{Z})$, the number of inequivalent cusps of $\Gamma(q)$ is given by

$$q^2 \prod_{p|q}(1 - p^{-2}).$$

9.60 Harmonic Analysis on $\Gamma\backslash\mathcal{T}$

Let μ denote the Haar measure on G and normalize it to one on K. This gives rise to an atomic measure that assigns to a vertex s in $\Gamma\backslash\mathcal{T}$ the measure

$$\mu(s) = |\Gamma(s)|^{-1}$$

where $\Gamma(s)$ is the stabilizer of a pre-image of s. Consider the Hilbert space $\mathcal{H} = L^2(\Gamma\backslash\mathcal{T}, \mu)$.

9.61 Hecke Operators for Function Fields

The Hecke algebra is generated by the operator

$$(Tg)(s) = \sum_{d(s,s')=1} g(s')$$

where $g : \mathcal{T} \to \mathbf{C}$ and d is the distance function on the tree. The sum is over s' adjacent to s. T induces an operator on $\mathcal{H}$ which is also denoted by T. Consider the eigenfunctions of T:

$$Tf = \lambda f$$

where $f : \Gamma \backslash \mathcal{T} \to \mathbf{C}$. Specifically T operates as

$$(Tf)(n) = \begin{cases} qf(n-1) + f(n+1) & \text{for } n \geq 1 \\ (q+1)f(1) & \text{for } n = 0. \end{cases}$$

Theorem 66 T *is a self-adjoint operator on F and by positivity of the Laplacian, one has $|\lambda| \leq q + 1$.*

9.62 Hecke Decomposition for Function Fields

Generically, the T−invariant decomposition of $L^2(\Gamma \backslash \mathcal{T}, \mu)$ has the form

$$L^2(\Gamma \backslash \mathcal{T}, \mu) = \mathcal{R} \oplus \mathcal{E} \oplus \mathcal{C}$$

where $\mathcal{R}$ is composed of two one dimensional eigenspaces, viz. the constant functions with eigenvalue $q + 1$ and the alternating functions generated by 1 on even vertices and -1 on odd ones with eigenvalue $-(q+1)$. $\mathcal{E}$ is the continuous spectra defined by the Eisenstein series and are parameterized by the interval

$$[-2\sqrt{q}, 2\sqrt{q}]$$

with multiplity h. $\mathcal{C}$ is the complement of these two spaces and contains the cuspidal functions. More precisely, $f \in L^2(\Gamma \backslash \mathcal{T})$ is said to be cuspidal at a cusp c if

$$\int_{\Gamma_c \backslash H_n} f(s) d\mu(s) = 0$$

where H_n is a horocycle. An eigenfunction is a cusp form if it is cuspidal at all cusps. For a discussion of horocycles, see Cartier (1973).

9.63 Structure Theorem for the Modular Tree

Serre (1980) showed for the modular tree, that $\mathcal{T} = G/K$ is a $q+1$ regular tree and $\Gamma(1) = PGL(2, \mathbf{F}_q[x])$ is a lattice in $G = PGL(2, k)$ where $k = \mathbf{F}_q((1/x))$ which is not co-compact. That is, $\Gamma(1)$ is discrete and finite co-volume and $\Gamma(1) \backslash \mathcal{T}$ and $\Gamma(1) \backslash G$ are infinite.

Theorem 67 *(Serre)* $\mathcal{T}$ *is an infinite* $(q+1)$*-regular tree with typical vertex* $\begin{pmatrix} t^n & x \\ 0 & 1 \end{pmatrix}$ *and the* $q+1$ *adjacent vertices are*

$$\begin{pmatrix} t^{n+1} & x \\ 0 & 1 \end{pmatrix} \quad and \quad \begin{pmatrix} t^{n-1} & \xi t^n + x \\ 0 & 1 \end{pmatrix}$$

where $\xi \in F_q$, $x \in k_\infty$ *and* $n \in \mathbb{Z}$. G *acts on the tree as a group of automorphisms and* $\Gamma \backslash \mathcal{T}$ *is a graph.*

$\Gamma \backslash \mathcal{T}$ *is a graph which is the union of a finite graph* Y *and truncated paths or ends.*

Serre (1980) and Weil (1970) have shown:

Theorem 68 *(Serre and Weil) The quotient graph* $F = \Gamma \backslash \mathcal{T}$ *is given by the cosets of* $\begin{pmatrix} t^n & 0 \\ 0 & 1 \end{pmatrix}$, *where* $n \geq 0$. *In particular* F *is the tree*

$$\bullet \text{-- -- --} \bullet \text{-- -- -- --} \bullet \text{-- -- --} \ldots$$
$$v_0 \qquad\quad v_1 \qquad\quad v_2 \qquad \ldots$$

given by the cosets of $\begin{pmatrix} t^n & x \\ 0 & 1 \end{pmatrix}$ *for* $n \geq 0$. *In other words, the half-line* $h_\infty = (v_0, v_1, v_2, \ldots)$ *of the Bruhat-Tits tree is a fundamental domain for the action of* Γ *on* $\mathcal{T}$. *So each vertex is* Γ*-equivalent to precisely one of the vertices* v_k.

9.64 Harmonic Analysis on the Modular Tree

The spectral theory for the case $\Gamma = \Gamma(1)$ is very simple:

Theorem 69 *(Efrat) For* $\Gamma = \Gamma(1)$ *the Hecke invariant decomposition has the form*

$$L^2(F) = \mathcal{R} \oplus \mathcal{E}$$

Pictorially, this means:

discrete	continuous	discrete
$-(q+1)$	$-2\sqrt{q} \qquad 2\sqrt{q}$	$(q+1)$

The eigenfunctions on F with eigenvalue λ are multiples of the function

$$f_\lambda(n) = \begin{cases} \frac{1}{x_1 - x_2}(\lambda(x_1^n - x_2^n) - q(q+1)(x_1^{n-1} - x_2^{n-1}) & \text{for } n \geq 1 \\ q+1 & \text{for } n = 0 \end{cases}$$

where

$$A = \begin{pmatrix} \lambda & -q \\ 1 & 0 \end{pmatrix}$$

and $x_1, x_2 = \frac{1}{2}(\lambda \pm \sqrt{\lambda^2 - 4q})$ are the characteristic roots of A. The only eigenvalues λ with $|\lambda| > 2\sqrt{q}$ for which $f_\lambda \in L^2(F)$ are $\lambda = \pm(q + 1)$.

Define

$$\tilde{f}_\theta = \frac{x_1 - x_2}{2\sqrt{q}} f_{2\sqrt{q}\cos(\theta)}$$

and extend $\tilde{f}_\theta(n)$ as an odd function of $\theta \in [-\pi, \pi]$. Let

$$F_\psi(n) = \frac{1}{2\pi} \int_{-\pi}^{\pi} \psi(\theta)\tilde{f}_\theta(n)d\theta$$

where $\psi \in L^2([0, \pi])$.

Theorem 70 *(Efrat) Let $\mathcal{E}$ denote the space of functions F_ψ; then $\mathcal{E}$ is a subspace of $L^2(F)$, it is invariant by T and is orthogonal to $\mathcal{R}$.*

Thus, $\Gamma \backslash \mathcal{T}$ in this case has no cusp forms.

9.65 Sarnak's Conjecture in the Function Field Case

Efrat (1989) has extended the result in the last section to show that Sarnak's conjecture is true in the function field case:

Theorem 71 *(Efrat) There exist arithmetic groups of arbitrarily large co-volume which admit no cusp forms.*

9.66 Hecke Congruence Subgroups

For Hecke congruence subgroups, Harder, Li and Weisinger (1980) have shown that the dimension of cusp forms grows asymptotically with co-volume. See also Schleich (1974) and Drinfeld (1982).

9.67 Noncuspidal Forms

As discussed in the early part of this review, for the standard model

$$\Gamma \backslash SL(2, \mathbf{R})/SO(2),$$

where Γ is a cofinite discrete subgroup of $SL(2, \mathbf{R})$, there are only a finite number of noncuspidal independent discrete eigenfunctions. Efrat (1989) has shown that for the function field case, this is no longer true. In fact, the noncuspidal discrete eigenfunctions can be quite abundant:

Theorem 72 *(Efrat) There exist an infinite sequence $q < \mu_1 < \mu_2... \to q + 1$ such that for every $N \geq 1$, the eigenvalues of $\Gamma^{(N)}$ invariant L^2-eigenfunctions are $\pm\mu_1, ... \pm \mu_n$ (or if $q = 2$, $\pm\mu_2, ..., \pm\mu_n$) and the multiplicity of μ_n is given by $(q - 1)q^{N-n}$.*

Here $\Gamma^{(N)}$ is a tower of arithmetic groups, $\Gamma^{(0)} \supset \Gamma^{(1)} \supset ...$ where $\Gamma^{(0)} = \Gamma(t)$, the principal congruence group of level t. And $\Gamma^{(N+1)}$ is given as the kernel of a homomorphism $\chi : \Gamma^{(N)} \to \mathbf{F}_q$. The graph $\Gamma^{(N)} \backslash X$ is the union of a finite graph, a distinguished end and a collection of cusps. For the details, see Efrat (1989).

In terms of representation theory, as noted by Efrat (1991) these results can be restated as:

Theorem 73 *(Efrat) There exist arithmetic groups Γ for which no principal series representations, but arbitrarily many complementary series representations occur discretely in the regular representation of $Aut(\mathcal{T})$ in $L^2(\Gamma \backslash \mathcal{T})$.*

9.68 Dissolving for Bruhat-Tits Buildings

Let $L^2_{dis}(\Gamma \backslash \mathcal{T})$ denote the space of nontrivial square integrable eigenfunctions of invariant operators on $\mathcal{T}$ (i.e., excluding the two dimensional part generated by the constant and alternating functions). $L^2_{dis}(\Gamma \backslash \mathcal{T})$ is a finite dimensional space which contains the subspace $L^2_{cusp}(\Gamma \backslash \mathcal{T})$ of cusp forms. Efrat (1990) has proven an analogue of the Phillips-Sarnak dissolving conjecture:

Theorem 74 *(Efrat) For every $N > 0$, there is a group $\Gamma = \Gamma_N$ such that*
(1) $dim\, L^2_{dis}(\Gamma \backslash \mathcal{T}) = N$;
(2) for every character $\chi \neq 1$ in a neighborhood of the identity character, $dim\, L^2_{dis}(\Gamma \backslash \mathcal{T}, \chi) = 0$.

The character χ is defined as follows. Let $\omega : E(X) \to \mathbf{R}$ be a Γ–automorphic 1-cochain, i.e. $\omega(\gamma y) = \omega(y)$. Define $\sum_{x_1}^{x_2} \omega = \sum_y \omega(y)$, where the sum extends over a path from x_1 to x_2. Set $\psi_\omega(\gamma) = \sum_x^{\gamma x} \omega$ and $\chi_\omega(\gamma) = exp(2\pi i \psi_\omega(\gamma))$. Then $\chi_\omega \in (\hat{\Gamma})_0$, the connected component of identity in the group of unitary characters that are trivial on ends. For more details, see Efrat (1989).

9.69 Cuspidal Eigenfunctions

The function field models also provide the opposite extreme to models described above, viz. examples with arbitrarily many cusp forms:

Theorem 75 *(Efrat) There exists a sequence of groups Γ_N where $vol(\Gamma_N \backslash \mathcal{T})$ is bounded and $dim\, L^2_{dis}(\Gamma_N \backslash \mathcal{T}) \to \infty$, i.e. the models have finite volume and arbitrarily many cusp forms.*

Efrat's approach uses the concept of graphs of groups developed by Bass (1984) and Serre (1980); for a discussion see also Bass and Kulkarni (1990). In particular, Efrat takes $\Gamma_N \simeq (C_q \oplus (C_q)^N \oplus \mathbf{Z}) * V^\infty$ where V^∞ is the group of sequences from the cyclic group C_q of order q which are trivial almost everywhere. The reader is directed to Efrat's paper for the details of the construction.

In contrast to Selberg's full density theorem discussed earlier, Efrat (1990) has shown:

Theorem 76 *(Efrat) For any positive integer N there are arbitrarily large integers M and pairs $\Gamma' \subset \Gamma$ with*

(1) $\dim L^2_{dis}(\Gamma \backslash \mathcal{T}) = N$

(2) $[\Gamma : \Gamma'] = M$

(3) $L^2_{dis}(\Gamma' \backslash \mathcal{T}) = L^2_{dis}(\Gamma \backslash \mathcal{T})$.

9.70   Scheja's Computational Graphs

Scheja (1998) has recently described the zeta functions of the finite part of the quotient $\Gamma(n) \backslash \mathcal{T}$, where $\mathcal{T}$ is the Bruhat-Tits tree over $\mathbf{F}_q((1/x))$ and $\Gamma(n)$ is the principal congruence subgroup of $\Gamma = GL(2, \mathbf{F}_q[x])$ of level $n \in \mathbf{F}_q[x]$.

As we have described above, the Ihara-Hashimoto results show that the inverted zeta function can be computed in terms of the determinant of an operator which is just the generalized Laplacian on the quotient space. Scheja has shown that a similar result holds for the finite part of the Bruhat-Tits tree. Let X be a tree and let G be a group which acts discretely and without inversions on X. Here discrete means that all the stabilizers are finite. Assume that $q(x) + 1$ is the degree of vertex x.

Let $H_\rho(VX) = \{f : VX \rightarrow V_\rho | f(\gamma x) = \rho(\gamma) f(x) \text{ for all } x \in VX, \gamma \in G\}$ and let $H_\rho(EX) = \{f : EX \rightarrow V_\rho | f(\gamma e) = \rho f(e) \text{ for all } e \in EX, \gamma \in G\}$ denote the spaces of G−equivariant functions mapping into the representation space of ρ, V_ρ. The adjacency operators are defined as follows:

$$T(e) = \sum_{o(e')=t(e), e' \neq \bar{e}} e',$$

$$\delta(x) = \sum_{o(e)=x} t(e),$$

and let $Q(x)$ denote the diagonal matrix with values $q(x), x \in VX$. These operators induce the operators T_ρ, δ_ρ and Q_ρ on the spaces of equivariant functions given by

$$(T_\rho f) = \sum_{o(e')=t(e), e' \neq \bar{e}} f(e')$$

$$(\delta_\rho f)(x) = \sum_{o(e)=x} f(t(e))$$

and

$$(Q_\rho f)(x) = q(x) f(x).$$

Define the generalized Laplacian operator

$$\Delta_\rho(u) = I - \delta_\rho u + Q_\rho u^2.$$

Bass (1992) has shown:

Theorem 77 *(Bass)*

$$L(X,G,\rho,u)^{-1} = det(I - uT_\rho) = (1 - u^2)^{\chi_\rho} det(\Delta_\rho(u))$$

where $\chi_\rho = dim_{\mathbf{C}} H_\rho(VX) - \frac{1}{2} dim_{\mathbf{C}} H_\rho(EX)$.

The reader can check that if $\rho = 1$ and $Y = G\backslash X$, then δ_1 is given by

$$(\delta_1 f)(y) = \sum_{o(e)=y} i(e) f(t(e))$$

so δ_1 can be interpreted as the adjacency matrix of the weighted graph Y and $\Delta_1(u)$ is the generalized Laplacian on Y.

Scheja has generalized this construction to produce what are called computational graphs Y_ρ. Viz., let $\{v_i\}_{i \in I}$ be a finite system of representatives of the G−operator on VX. Then

$$H_\rho(VX) = \bigoplus_{i \in I} H_\rho^{(i)}(VX)$$

where

$$H_\rho^{(i)}(VX) = \{f : Gv_i \to V_\rho | f(gx) = \rho(g)f(x) \text{ for all } x \in Gv_i, g \in G\}.$$

Set

$$V_\rho^{G_i} = \{x \in V_\rho | \rho(g)x = x, g' \in G_i\}.$$

Then, $V_\rho^{G_i} \simeq H_\rho^{(i)}(VX)$ where $v \to h_v$ with $h_v(gv_i) = \rho(g)v$ for all $g \in G$ is an isomorphism of vector spaces. If $f_j \in H_\rho^{(i)}(VX) \to H_\rho(VX)$, then the equation

$$Q_\rho f_j = q(v_i) f_j$$

holds. If

$$\delta_\rho f_j = \sum_{k \in J} a_{jk} f_k$$

we put an edge between f_j and $f_k \in VY_\rho$ whenever $a_{jk} \neq 0$. Thus, δ_ρ is the adjacency matrix of Y_ρ and

$$det(\Delta_\rho(u)) = det(I - u\delta_\rho + u^2 Q_\rho)$$

where Δ_ρ is the generalized Laplacian on Y_ρ.

Bass (1992) has shown:

Theorem 78 *(Bass) The L−function of (X,G,ρ) is given by*

$$L(X,G,\rho,u)^{-1} = det(I - uT_\rho) = (1 - u^2)^{\chi_\rho} det(\Delta_\rho(u))$$

where $\chi_\rho = dim_{\mathbf{C}} H_\rho(VX) - \frac{1}{2} dim_{\mathbf{C}} H_\rho(EX)$.

9.71   Functorial Properties of L-Functions

Although we will not cover the detailed properties of L-functions in this review, they have the same functorial properties discussed earlier for zeta functions.

Theorem 79 *(Bass, Hashimoto) If $G' \subset G$ is a subgroup of finite index and ρ' is a representation of G', then*

$$L(X, G', \rho', u) = L(X, G, ind_{G'}^{G} \rho', u).$$

If G' is normal in G, then

$$L(X, G', 1, u) = L(X, G, \rho_{reg}, u)$$

where ρ_{reg} is the regular representation of G/G'.

If the action of G' on X is free, then

$$X \to Y = G' \backslash X$$

is said to be an unramified covering and one sets

$$Z_Y(u) = L(X, G', 1, u).$$

Theorem 80 *(Bass, Hashimoto) If ρ, ρ' are representations of G, then*

$$L(X, G, \rho \oplus \rho', u) = L(X, G, \rho, u)L(X, G, \rho', u).$$

It follows from this theorem that

Theorem 81 *If $G' \subset G$ is a normal subgroup of finite index and G' acts freely on X with quotient $Y = G' \backslash X$, then*

$$Z_Y(u) = L(X, G', 1, u) = \prod_{\rho \in \widehat{G/G'}} L(X, G, \rho, u)^{deg(\rho)}.$$

For further details, see Sunada (1986), Bass (1992) and Hashimoto (1989, 1990).

9.72   Vertices of Type i

Let v_i, resp. e_i, denote the cosets of $\begin{pmatrix} \pi^{-i} & 0 \\ 0 & 1 \end{pmatrix} \, mod \, K$, respectively $\mathcal{I}$. Let Γ_i denote the stabilizer of v_i in Γ; viz.

$$\Gamma_0 = \Gamma_{v_0} = GL(2, \mathbf{F}_q)$$

$$\Gamma_i = \Gamma_{v_i} = \{ \begin{pmatrix} a & b \\ 0 & d \end{pmatrix} \, | a, d \in \mathbf{F}_q^*, deg(b) \leq i \}$$

for $i \geq 1$ and

$$\Gamma_{e_i} = \Gamma \cap \Gamma_{i+1} = \Gamma_i$$

for $i \geq 1$ and

$$\Gamma_{e_0} = \Gamma_0 \cap \Gamma_1.$$

Let Π be a congruence subgroup of Γ. Consider the covering

$$\pi_\Pi : T \to \Pi\backslash T \to \Gamma\backslash T.$$

Elements in the fiber of v_i in each level of this covering are called vertices of type i, i.e. for $e \in E(\Gamma\backslash T)$ and $v \in V(\Pi\backslash T)$, then e is of type i if and only if $\pi_\Pi(e) = e_i$ and v is of type i if and only if $\pi_\Pi(v) = v_i$. Let $V_i = V_i(\Pi\backslash T) = \{v | type(v) = i\}$ and similarly for E_i. Since Γ operates transitively on vertices of type i of , one can show (v. Serre (1980) and Gekeler and Nonnengardt (1995)):

Theorem 82 *(Serre and Gekeler - Nonnengardt) There are bijections*

$$\Pi\backslash\Gamma/\Gamma_i \simeq V_i(\Pi\backslash T) \simeq \Gamma_i\backslash\Gamma/\Pi.$$

Similarly,

$$(\Gamma_i \cap \Gamma_{i+1})\backslash\Gamma/\Pi \simeq E_i(\Pi\backslash T).$$

9.73 Structure Theorem for Arithmetic Groups

Let Γ be an arithmetic subgroup of G. Then Γ acts on the Bruhat-Tits space T with finite stabilizers. Serre's (1980) structure theorem cited above for the modular group extends to any arithmetic subgroups as follows:

Theorem 83 *(Serre) $\Gamma\backslash T$ is the union of a finite graph $X(n)$ and a finite number of half lines, i.e. the ends of $\Gamma\backslash T$. There is a canonical bijection between the sets of ends of $\Gamma\backslash T$ and the orbits $\Gamma\backslash\mathbf{P}^1(K)$ on the projective line $\mathbf{P}^1(K)$ and the cusps of $\bar{M}_\Gamma$, the compactification of the Drinfeld space $\Gamma\backslash\Omega$ where $\Omega = C - K_\infty$.*

Each vertex of type i for $1 \leq i \leq d-1$ has q neighbors of type $i-1$ and 1 neighbor of type $i+1$. While for $i \geq k$, each vertex of type i has exactly one neighbor of type $i-1$ and one of type $i+1$. Thus, the vertices and edges of type greater than or equal to d form the finitely many half lines or cusps.

Let X_d denote the subgraph of T that is induced by all vertices less than d. Then the finite part of $X(n)$ is

$$X(n) = \Gamma(n)\backslash X_d = \Gamma(n)\backslash T - \{cusps\}.$$

Finally, $X(n)$ is combinatorial, i.e. it is a graph, not a multigraph.

For related work on the structure theorem, see Gekeler and Reversat (1996), Lubotzky (1991) and Raghunathan (1989).

The problem with this representation of the graph $X(n)$ is that X_d is not a tree. Scheja has found a remedy to this problem. Let X denote the component of v_0 in X_d. Then

Theorem 84 *(Scheja) Let $\Gamma(n)$ denote the principal congruence subgroup of level n. There are subgroups Θ and $\Theta(n)$ of $\Gamma(1)$ such that there is a sequence of coverings*

$$X \to X(n) = \Theta(n)\backslash X \to \Theta\backslash X$$

where X is a certain subtree of the Bruhat-Tits tree of $GL(2, k_\infty)$. In particular X is the universal unramified covering of $X(n)$ with fundamental group $\Theta(n)$.

Moreover, Scheja has shown that:

Theorem 85 *(Scheja)*

$$\Theta = < GL(2, \mathbf{F}_q), \Gamma_{d-1} >$$

and

$$\Theta(n) = \Theta \cap \Gamma(n).$$

Set

$$G(n) = \Gamma/\Gamma(n)$$

and

$$\Theta_0(n) = \Theta \cap \Gamma_0(n)$$

where $\Gamma_0(n)$ is the nth Hecke congruence group

$$\Gamma_0(n) = \{\begin{pmatrix} a & b \\ c & d \end{pmatrix} | c \equiv 0 \, mod \, n\}.$$

Set $X_0(n) = \Theta_0(n)\backslash X$.

Theorem 86 *(Scheja) $X_0(n)$ is the finite part of the graph $\Gamma_0(n)\backslash X$.*

9.74 Zeta Function of $X(n)$

Since $X \to X(n)$ is an unramified covering, the zeta function is given by

$$Z_{X(n)}(u) = L(X, \Theta(n), 1, u),$$

which in turn can be computed from $L(X, \Theta, \rho, u)$ for all irreducible representations ρ of $\Theta/\Theta(n) \simeq G(n)$; these representations result from the principal series representations or the supercuspidal representations.

Let $A = \mathbf{F}_q[x]$ denote the polynomial ring over $\mathbf{F}_q$. For a field K define

$$B(K) = \{\begin{pmatrix} a & b \\ 0 & d \end{pmatrix} | a, d \in K^z, b \in K\}.$$

Assume that $n \in A$ is prime of degree d do that A/n is a finite field of order d. Representations of $G(n)$ have been studied by Rust (1998).

Theorem 87 *(Scheja) Let $\rho = \rho_\mu$ be induced by the character μ of $B(A/n)$. The nontrivial factors of $L(X, \Theta, ind\,\mu, u)^{-1}$ are computed as the determinants of the generalized Laplacian of Y_ρ where $Y_\rho = X_0(n)$.*

As a factor of $Z_{X(n)}(u)$ the L-function $L(X, \Theta, \rho, u)$ shows up with multiplicity degree ρ for every irreducible representation ρ of $G(n)$. Thus, the behavior of induced representations when restricted to $G(n)$ determines the multiplicity of $L(X, \Theta, \rho, u)$ in $Z_{X(n)}(u)$.

9.75 Scheja's Example

An example developed by Scheja is the case where $n \in A$ is of degree 2. Here the shape of the graph $\Gamma_0(n)\backslash \mathcal{T}$ is

The zeta function for the finite part for this case is

$$Z_{X(n)}(u)^{-1} = -(1 - u^2)^{g(X(n))}(q(q-1)u^2 - 1)(q(q-1)u^4 + qu^2 + 1)^{q^2}$$

$$\times (qu^2 + 1)^{A_q}((q(q-1)u^4 - u^2 + 1)((q-1)u^2 + 1))^{B_q}$$

where $A_q = (q-2)q^2 + 1)/2$ and $B_q = q(q^2 + 1)/2$. The genus is $g(X(n)) = (q^2 + 1)(q^2 - q - 1) + 1$.

9.76 Several Examples of Gekeler

In the case $d = deg(n) = 1$, then $g(\Gamma)$ the genus is zero and the number of cusps is $c(\Gamma) = 2$. In this case $\Gamma(n)\backslash \mathcal{T}$ is isomorphic to the straight line

Consider the case $\Gamma = \Gamma_0(n)$, where $n = T(T^2 + T + 1) \in \mathbf{F}_2[T]$. Then $\Gamma\backslash\mathcal{T}$ has the form

where γ_i are two cycles of length 4. The arrows denote the cusps, and in this example $c(\Gamma) = 4$.

Consider the case $\Gamma = \Gamma_0(n)$ for $n \in A = \mathbf{F}_q[T]$ where $n = T^3$. Then $\Gamma\backslash\mathcal{T}$ has the form

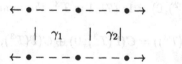

Here $g(\Gamma) = 1$ and $c(\Gamma) = 4$.

Consider the case $n = T^2(T - 1)$ then $\Gamma \backslash \mathcal{T}$ has the form

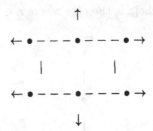

Here $g(\Gamma) = 1$ and $c(\Gamma) = 6$. These are the only two conductors n for which $g = 1$.

For further details see Gekeler (1995).

Li (1978), Weisinger (1977) and Gekeler (1985) have considered several additional examples. Let A be a monic polynomial in $\mathbf{F}_q[T]$ and set

$$\Gamma_1(A) = \{ \begin{pmatrix} P & Q \\ R & S \end{pmatrix} \in \Gamma | P \equiv S \equiv 1 \, mod \, A, R \equiv 0 \, mod \, A \}$$

and

$$\Gamma_0(A) = \{ \begin{pmatrix} P & Q \\ R & S \end{pmatrix} \in \Gamma | R \equiv 0 \, mod \, A \}.$$

Let $C(\Gamma_1(A))$ and $C(\Gamma_0(A))$ denote the cusp forms for these groups. Furthermore, let $C(\Gamma(A), \lambda)$ denote the space of cusp forms with eigenvalue λ. The following cases are known:

Theorem 88 *(Li, Weisinger) (1) If $A = T$, then the number of cusps of $\Gamma_0(T)$ is 2, and $dim \, C(\Gamma_0(T)) = 0$; (2) if $A = T^2$, then the number of cusps of $\Gamma_0(T^2)$ is 3, and $dim \, C(\Gamma_0(T^2)) = 0$; (3) if $A = T^3$, the number of cusps of $\Gamma_0(T^3)$ is 4 and $dim \, C(\Gamma_0(T^3)) = 0$; (4) if $A = T^4$, the number of cusps of $\Gamma_0(T^4)$ is 6 and $dim \, C(\Gamma_0(T^4)) = 0$; (5) if $A = T$, the number of cusps of $\Gamma(T)$ is 3 and $dim \, C(\Gamma(T)) = 0$; (6) if $A = T^2$, the number of cusps of $\Gamma(T^2)$ is 12 and $dim \, C(\Gamma(T^2)) = C(\Gamma(T^2), 0) = 1$; (7) $A = T^3$, the number of cusps of $\Gamma(T^3)$ is 48 and*

$$C(\Gamma(T^3)) = C(\Gamma(T^3), 0) \oplus C(\Gamma(T^3), -\sqrt{2}) \oplus$$

$$C(\Gamma(T^3), \sqrt{2}) \oplus C(\Gamma(T^3), -2) \oplus C(\Gamma(T^3), 2) \oplus C(\Gamma(T^3), -\sqrt{3 + \sqrt{3}})$$

$$\oplus C(\Gamma(T^3), \sqrt{3 + \sqrt{3}}) \oplus C(\Gamma(T^3), -\sqrt{3 - \sqrt{3}}) \oplus C(\Gamma(T^3), \sqrt{3 - \sqrt{3}}),$$

which have dimensions 14,6,6,3,3,4,4,4,4 respectively.

Diagrams for the fundamental domains for these cases are shown in Li (1978) where possible.

Let $K = \mathbf{F}_q(T)$ denote the field of rational functions in T and let n denote a positive divisor of K.

Theorem 89 *(Gekeler, Li, Weisinger) For the case of Hecke congruence group,* $\Gamma_0(n)$, *there are no cusp forms for* $deg(n) \leq 3$. *For the case* $deg(n) = 4$, *the dimension of the space of cusp forms depends on the decomposition of n into prime divisors. E.g., if* $m = p_1 q_1 r_1 s_1$, *then* $dim\, C(n) = 2q$.

There are 11 cases for $deg(n) = 4$; for the remaining cases, see Gekeler (1985).

Consider the case of a prime ideal of the form (π_t) where $\pi_t = T - t$ with $t \in \mathbf{F}_q$. Let H_t denote the Hecke operator. Gekeler has noted that the space of cusp forms has a decomposition

$$C = W_{sp} \oplus W_{\overline{sp}}.$$

Let χ denote the parity character; the map $i : f \to \chi f$ of C is an involution which interchanges W_{sp} and $W_{\overline{sp}}$. Here both spaces have dimension equal to $\frac{1}{2} dim\, C$. There is a natural isomorphism

$$Res_{sp} : W_{sp} \to C_0(\mathbf{P}_1)$$

where the right hand side is the vector space of functions f on $\mathbf{P}_1(\mathbf{F}_q)$ with $\sum f(x) = 0$ (and in certain cases $f(0) = 0$, v. Gekeler (1985)). Here Res_{sp} is the map $f \to f_{\{P_x\}}$. For $x \in \mathbf{F}_q$ there are functions $\phi_x \in W_{sp}$ defined by

$$\phi_x(P_y) = \begin{cases} \delta_{x,y} & y \in \mathbf{F}_q \\ -1 & y = \infty. \end{cases}$$

There functions form a basis for W_{sp} and in terms of this basis the Hecke operator has the form

$$H_t \phi_x = \sum_{y \in \mathbf{F}_q} H_t \phi_x(P_y)\phi_y = \sum H_{x,y} \phi_y.$$

The matrix coefficients satisfy certain properties, e.g.

$$\sum_y H_{x,y} = -\delta_{x,t}.$$

Simple cases have been developed by Gekeler; e.g. for $q = 2$, $T^3 + T + 1$ and $t = 0$, resp. $t = 1$, then

$$H_{x,y} = \begin{pmatrix} -3 & 2 \\ -1 & 1 \end{pmatrix},$$

resp.

$$H_{x,y} = \begin{pmatrix} -2 & -2 \\ 1 & -2 \end{pmatrix}.$$

The operator algebra generated by H_0 and H_1 is $\mathbf{Z}[\sqrt{2}]$.

Gekeler has noted that the cuspforms $f \in W_{sp}$ can also be viewed as vectors $f = (f(0), f(1), ..., f(q))$ where $\sum_{i \leq q} f(i) = 0$. And there is a natural normalization with $f(q) = -1$. E.g., in the case $n = T^3 - 2 \in \mathbf{F}_q$ and $q = 7$, then f has the form $(f(0), ..., f(7)) = (4, 1, 1, -2, 1, -2, -2, -1)$. Other examples of rational eigenvectors of the Hecke algebra for $q \leq 16$ are presented in Gekeler (1985).

9.77   Scheja's Zeta Function

An example of the zeta function for a finite volume graph has been developed by
Scheja (1998b). Scheja's work has shown that the inverted zeta function is given
by

$$(1 - u^2)^{c(\Gamma)+g(\Gamma)-1}(1 - qu^2)^{-c(\Gamma)}det(\Delta(u)).$$

In particular for the case $n = T^3$ discussed above, the zeta function is

$$Z(u) = \frac{(1 - qu^2)^4}{(1 - u^2)^{q+2}(1 - q^2u^2)}.$$

Scheja has conjectured that the zeta function for all T^n have this form, when
adjusted for the number of cusps and the genus. These results have not been
published yet.

9.78   Genus

The genus of $\Gamma(n)\backslash\mathcal{T}$ is the rank of the first homology group $H_1(\Gamma(n)\backslash\mathcal{T}, \mathbf{Z})$, i.e.
the number of independent cycles in the graph $\Gamma(n)\backslash\mathcal{T}$. The genus is unchanged
if we cut off all cusps. The Euler formula states that in terms of the truncated
graph:

$$g(\Gamma(n)\backslash\mathcal{T}) = 1 + |\{ \text{ nonoriented edges}\}| - |\{ \text{ vertices}\}|.$$

For $n \in A$ given by

$$n = \prod_{1\leq i\leq s} f_i^{r_i} = \prod_{1\leq i\leq s} n_i$$

wher f_i are different monic primes, $deg f_i = l_i$ and $deg n = \sum r_i l_i = d$. Set
$q_i = q^{l_i} = |A|/f_i$. Define

$$\phi(n) = \prod_{1\leq i\leq s} q^{r_i-1}(q_i - 1)$$

and

$$\epsilon(n) = \prod_{1\leq i\leq s} q^{r_i-1}(q_i + 1).$$

Gekeler and Nonnengardt (1995) have shown:

Theorem 90 *(Gekeler-Nonnengardt)*

$$g(\Gamma(n)\backslash\mathcal{T}) = 1 + \frac{\phi(n)\epsilon(n)}{q^2 - 1}(q^d - q - 1).$$

And in the case of the Hecke group

$$g(\Gamma_0(n)\backslash\mathcal{T}) = \begin{cases} (q^d - q^2)/(q^2 - 1), & d \text{ even} \\ (q^d - q)/(q^2 - 1), & d \text{ odd}. \end{cases}$$

9.79 Nonuniform Lattices

As discussed above, the Ramanujan graphs had the form $\Gamma\backslash G/K$ where Γ is a cocompact lattice, or so-called uniform lattice in G. In the case $G = PGL(2, F)$ where F is a local non-archimedian field of $char(F) = p > 0$, then G has non-uniform lattices, i.e. discrete subgroups Γ of G such that $\Gamma\backslash G$ has a finite invariant measure but is not compact. In this case Morgenstern (1994a) introduced the concept of a diagram to describe the infinite graphs $\Gamma\backslash G/K$ with finite volume.

Recall Serre's structure theorem states that the diagrams considered there have the form of the disjoint union of a finite graph plus finitely many infinite rays or cusps. Lubotzky (1991) showed that all non-uniform lattices of $PGL(2, \mathbf{F}_q((1/t)))$ have this form. However, Bass and Lubotzky (1998) have shown that this is not true for arbitrary lattices in $G = Aut(X_k)$ where X_k is a $k-$regular tree. In this case the diagrams can have infinitely many cusps.

9.80 Ramanujan Diagrams

Morgenstern (1994) defines a diagram to be an infinite graph with weights for the vertices and edges such that the total volume is finite. More precisely, a diagram is a triple $D = (V, E, w)$ where $Y = (V, E)$ is an undirected countable graph and $w : V \cup E \to \{\frac{1}{n} | n = 1, 2, ...\}$ is a weight function for all $e = (u, v) \in E$ and $1/w(e)$ divides $1/w(u)$ and $1/w(v)$. For $A \subset V$, set $\mu(A) = \sum_{u \in A} w(u)$. Let $\theta(u, v) = w(e)/w(u)$, the so-called entering degree of $e = (u, v)$. D is called k-regular if for all $u \in V$

$$\sum_{(u,v) \in E} \theta(u, v) = k.$$

Let X_k be a $k-$regular tree. Let $G = Aut(X_k)$. If $k = p^r + 1 = q + 1$, then $H = PGL(\mathbf{F}_q(1/t)) \subset G$ and let Γ be a nonuniform lattice in G or H. H is cocompact in G so a lattice in H is automatically a lattice in G.

Let $\Gamma_x = \{\gamma \in \Gamma | \gamma(x) = x\}$ and $\bar{x} = \Gamma x$. Define the structure of a diagram on $\Gamma\backslash X_k$ by $w(\bar{x}) = 1/|\Gamma_x|$ and $w(\bar{e}) = 1/|\Gamma_e|$. As Γ is discrete Γ_e and Γ_x are finite; $\mu(\Gamma\backslash X_k) = \tilde{\mu}(\Gamma\backslash G)$ where $\tilde{\mu}$ is normalized to be one on K, $\tilde{\mu}(K) = 1$.

9.81 Adjacency Operator for Diagrams

The inner product on D is

$$< f, g > = \int f\bar{g}d\mu = \sum_{v \in V} f(v)\bar{g}(v)w(v).$$

Let

$$(Af)(u) = \sum_{(u,v) \in E} \theta(u, v)f(v)$$

Assume that D is k-regular and bipartite, $V = I \cup O$. Let

$$L_0(D) = \{f \in L^2(D)| \sum_{v \in I} f(v)w(v) = \sum_{v \in O} f(v)w(v) = 0\}$$

and let

$$\lambda(D) = \sup Spec\{A|_{L_0^2(D)}\}.$$

Theorem 91 *(Morgenstern) For a k-regular infinite diagram D, one has*

$$\lambda(D) \geq 2\sqrt{k-1}.$$

A k-regular diagram with $\lambda(D) = 2\sqrt{k-1}$ is called a Ramanujan diagram. From the results of Section 81, we see that

Theorem 92 *(Efrat) If $\Gamma = SL(2, \mathbf{F}_q[t])$, then $D = \Gamma \backslash X_{p+1}$ is a Ramanujan diagram.*

Morgenstern has shown:

Theorem 93 *(Morgenstern) Let Γ' be a congruence subgroup of $\Gamma(1)$. Then the diagram*

$$\Gamma' \backslash X_{q+1} = \Gamma' \backslash PGL(2, \mathbf{F}_q((1/t)))/PGL(2, \mathbf{F}_q[[1/t]])$$

is a Ramanujan diagram.

Morgenstern's proof proceeds as follows. By the result of Lubotzky λ is an eigenvalue of Δ if and only if the irreducible unitary class one representation of G, ρ^λ, belongs to $L_2(\Gamma \backslash G)$. By Drinfeld's (1988) theorem, it is a principle series representation, thus $k - 2\sqrt{q} \leq \lambda \leq k + 2\sqrt{q}$.

9.82 Discrete Analogue of Gutzwiller's Trace Formula

Shirai (1997) has presented the following discrete analogue of Gutzwiller's trace formula for graphs. Let X be a countable set, let Δ_X denote the Laplacian on $l^2(X)$ and let V be a real-valued, bounded function; then

$$L = -\Delta_X + V$$

is the discrete Schrödinger operator. It is a linear, bounded self-adjoint operator on $l^2(X)$. Let A be a finite subset of X and consider the eigenvalue problem

$$L\phi(x) = \lambda\phi(x)$$

and the Dirichlet problem

$$L_A\phi(x) = L\phi(x) = \lambda\phi(x)$$

for $x \in X \backslash A$ and

$$L_A\phi(a) = 0$$

for $a \in A$. Let $g_\lambda(x, y)$ and $g_\lambda^A(x, y)$ denote the integral kernels of $(L - \lambda)^{-1}$ and $(L_A - \lambda)^{-1}$. Let G_λ^A denote the $|A| \times |A|$ matrix with elements $g_\lambda(a, b)$ for $a, b \in A$.

Theorem 94 *(Shirai) Let $\lambda_0 = \inf Spec(L)$ and $\lambda_\infty = \sup Spec(L)$. Then, the determinant $\det G_\lambda^A$ is holomorphic for $\lambda \in \mathbf{C}\backslash Spec(L)$ and for $\lambda \in \mathbf{C}\backslash[\lambda_0, \lambda_\infty]$, the determinant $\det G_\lambda^A$ is nonzero.*

As an analogue of the work of Craig (1989) and Gesztesy et al. (1995), Shirai has shown:

Theorem 95 *(Shirai)*

$$\frac{1}{|A|}Tr(L - L_A) = \frac{1}{|A|}\sum_{a \in A} V(a) = \lambda_\infty - 1 - \int_{\lambda_0}^{\lambda_\infty} \theta_A(\lambda)d\lambda$$

where $\theta_A(\lambda)$ is the generalized Krein's spectral shift function.

For a discussion on Krein's spectral shift function, see Gesztesy et al. (1995). In particular, $\theta_A(\lambda)$ is given by

$$\theta_A(\lambda) = \lim_{\epsilon \to 0} \frac{1}{\pi|A|} Im \, log \, det \, G_{\lambda+i\epsilon}^A$$

and one can check that:

Theorem 96 *(Shirai) For almost every $\lambda \in \mathbf{R}, 0 \le \theta_A(\lambda) \le 1, \theta_A(\lambda) \in \{\frac{k}{|A|}, k \in \mathbf{Z}\}$ and*

$$\theta_A(\lambda) = \begin{cases} 0 & \lambda < \lambda_0 \\ 1 & \lambda > \lambda_\infty. \end{cases}$$

Define the distance function on $V(X)$ by

$$d_\lambda(x, y) = -\frac{1}{2}log\frac{g_\lambda(x,y)g_\lambda(y,x)}{g_\lambda(x,x)g_\lambda(y,y)}.$$

One notes that if X is a d regular tree, and V is identically zero, then the spectrum of $-\Delta_X$ is in the interval $[1 - \alpha_d, 1 + \alpha_d]$ where $\alpha_d = 2\sqrt{d-1}/d$ and

$$d_\lambda(x, y) = d(x, y)(-log\,m_d(\lambda)).$$

Here

$$m_d(\lambda) = \frac{d}{2d-2}(1 - \lambda - \sqrt{(1 - \lambda)^2 - \alpha_d^2})$$

and $d(x, y)$ is the geodesic distance function on the tree.

In this case Shirai (1997b) has shown

Theorem 97 *Let X be a d−regular tree and let A be an arbitrary finite subset of X with cardinality N. Then*

$$lim_{|A| \to \infty} \frac{1}{|A|} log \, det \, G_\lambda^A = log \int_{\mathbf{R}} \frac{1}{\lambda - x} n(dx)$$

where $n(dx) = \frac{2}{\pi}\sqrt{1 - \alpha_d^{-2}(x - 1)^2}dx$; that is, the semi-circle law holds for a d−regular tree.

Let σ denote the shift transformation on A and let

$$\Sigma = \{\bar{a} = (a_n)_{n \in \mathbf{N}}, a_n \in A, \sigma a_n \neq a_{n+1}\}.$$

Let $F(n) = \{\bar{a} \in \Sigma | \sigma^n \bar{a} = \bar{a}\}$ and $P(n) = F(n) \backslash \cup_{k|n} F(k)$. Define the equivalence relation $\bar{a} \sim \bar{b}$ if there is a $k \leq n-1$ such that $\sigma^k \bar{a} = \bar{b}$. Set $\Gamma_n = P(n)/ \sim$. An element $\gamma \in \Gamma_n$ is said to be a prime periodic orbit with period n. Let L_γ denote the period of γ and let Γ denote the set of prime periodic orbits. Then Shirai's discrete analogue of the Gutzwiller trace formula is:

Theorem 98 *(Shirai) There exists $\lambda_1 \in \mathbf{R}$ such that for any $\lambda < \lambda_1$*

$$\sum_{x \in X} (g_\lambda(x,x) - g_\lambda^A(x,x)) =$$

$$\sum_{a \in A} \frac{d}{d\lambda} \log g_\lambda(a,a) + \sum_{\gamma \in \Gamma} \frac{dS_\gamma(\lambda)}{d\lambda} \sum_{n \geq 1} exp(-nS_\gamma(\lambda) - n\pi i L_\gamma).$$

Here $S_\gamma(\lambda)$ is the length of a periodic orbit γ with respect to the distance d_λ

The proof arises by decomposing G_λ^A as

$$G_\lambda^A = D_\lambda^A(I + K_\lambda^A)$$

where D_λ^A is the diagonal matrix $(D_\lambda^A)_{a,a} = g_\lambda(a,a)$ and

$$(K_\lambda^A)_{a,b} = \begin{cases} \frac{g_\lambda(a,b)}{g_\lambda(b,b)} & \text{if } a \neq b \\ 0 & \text{if } a = b. \end{cases}$$

And one checks that

$$det(I + K_\lambda^A) = \prod_{\gamma \in \Gamma} (1 - exp(-nS_\gamma(\lambda) - n\pi i L_\gamma)).$$

Shirai (1997) notes the relationship to the Ruelle zeta function. We refer the reader to this paper for the details.

9.83 Selberg Trace Formula

The Selberg trace formula for principal congruence groups has been developed by Nagoshi (1998). Let

$$c(n) \in \mathbf{C}$$

for $n \in \mathbf{Z}$ with $c(n) = c(-n)$ and $\sum_{n \in \mathbf{Z}} q^{|n|/2}|c(n)| < \infty$. Let h designate the Fourier transform $h(r) = \sum c(n)q^{inr}$ where $s = \frac{1}{2} + ir$. Let μ denote the number of inequivalent cusps and let $\{\lambda_1, ... \lambda_M\}$ denote the set of discrete spectrum of T_Γ. Set $\mathcal{P}_\Gamma$ equal to the set of primitive hyperbolic conjugacy classes of Γ.

Theorem 99 *(Nagoshi) Let q be an odd prime power and let* $\Gamma = \Gamma(A)$ *for* $A \in \mathbf{F}_q[t]$ *with* $\deg(A) = a \geq 1$. *Then the Selberg trace formula is*

$$\sum_{n=1}^{M} h(r_n) = C(I) + C(H) + C(P) =$$

$$vol(\Gamma \backslash X)k(0) + \sum_{\{P\} \in \mathcal{P}_\Gamma} \sum_{l=1}^{\infty} \frac{\deg P}{q^{\frac{l \deg P}{2}}} c(l \deg P) +$$

$$(\mu - Tr\Phi(1/2))(\frac{1}{2}c(0) + \sum_{m=1}^{\infty} c(2m)) +$$

$$\frac{1}{4\pi} \int_{-\frac{\pi}{\log q}}^{\frac{\pi}{\log q}} h(r) \frac{\phi'}{\phi} (\frac{1}{2} + ir) dr - \mu(a + \frac{1}{q-1}) c(0).$$

The first term $C(I)$ *is given by*

$$C(I) = \int_{\Gamma(I) \backslash X} k(g, Ig) dg$$

where the kernel is $k(g, g') = k(d(g, g'))$.

9.84 Scattering Determinant

Nagoshi (1998) has developed the following properties of the scattering determinant which follow from Li's theorem discussed above:

Theorem 100 *(Nagoshi)* $\phi(s)$ *is a rational function in* q^{2s} *which can be written as*

$$\phi(s) = c \frac{(q^{2s} - qa_1)...(q^{2s} - qa_m)}{(q^{2s} - qb_1)...(q^{2s} - qb_n)}$$

where c is a constant. The moduli of a_i, b_j *are not equal to one. Furthermore,* $\{q^{2s} - qa_i | a_i \neq 0, i = 1, .., m\}$ *and* $\{q^{2s} - qb_j | b_j \neq 0, j = 1, .., n\}$ *are in one to one correspondence so that* $q^{2s} - qa$ *with* $a \neq 0$ *in the numerator corresponds to* $q^{2s} - qb$, $b = 1/a$ *in the denominator.*

9.85 Selberg Zeta Function

For $\{P\} \in \mathcal{P}_\Gamma$ set $N(P) = sup\{|\lambda_i|^2_\infty\}$ where λ_i is an eigenvalue of the matrix P. For the case $\Gamma \subset PGL(2, \mathbf{F}_q[t])$ then $N(P) = q^{\deg P}$. The Selberg zeta function is defined as before as

$$Z_\Gamma(s) = \prod_{\{P\} \in \mathcal{P}_\Gamma} (1 - N(P)^{-s})^{-1}.$$

Define the determinant for T_Γ by

$$\det(T_\Gamma, s) = \det{}_D(T_\Gamma, s)\det{}_C(T_\Gamma, s)$$

where

$$det_D(T_\Gamma, s) = det_D(1 - T_\Gamma q^{-s} + q^{1-2s}) = \prod_{n=1}^{M}(1 - \lambda_n q^{-s} + q^{1-2s})$$

and

$$det_C(T_\Gamma, s) = \prod_{|b_j|<1}(1 - q^{-2s+1}b_j) \prod_{|b_j|>1}(1 - q^{-2s+1}b_j)^{-1}.$$

Nagoshi has shown the following extension of Ihara's theorem:

Theorem 101 *(Nagoshi) Let q be an odd prime power and set* $\Gamma = \Gamma(A)$ *with* $deg(A) \geq 1$. *Then*

$$Z_\Gamma(s)^{-1} = (1 - q^{-2s})^\chi(1 - q^{-2s+1})^{-\rho} det(T_\Gamma, s)$$

where $\chi = vol(\Gamma\backslash X)$ *and* $\rho = \frac{1}{2}Tr(I_\mu - \Phi(\frac{1}{2}))$.

In particular Nagoshi has shown:

Theorem 102 *(Nagoshi) Let q be an odd prime power and set* $\Gamma = \Gamma(A)$ *with* $deg(A) = a \geq 1$. *Let* $A = A_1^{e_1}A_2^{e_2}...A_l^{e_l}$ *be a decomposition of A into distinct irreducible polynomials where* $deg(A_i) = a_i$ *and* $\sum_{i=1}^{l} = e_i a_i = deg(A)$. *Let* $det'_D(T_\Gamma, s) = \prod_{\lambda_n}(1 - \lambda_n q^{-s} + q^{1-2s})$ *where the product is taken over discrete spectra of* T_Γ *except for the two trivial eigenvalues* $\pm(q+1)$. *Then*

$$Z_\Gamma(s)^{-1} = (1 - q^{-2s})^\chi(1 - q^{-2s+1})^{-\rho} det'_D(T_\Gamma, s) \prod_{|b_j|<1}(1 - q^{-2s+1}b_j).$$

Here

$$\chi = \frac{q^{3a}}{(q+1)(q-1)} \prod_{i=1}^{l}(1 - 1/q^{2a_i})$$

and

$$\rho = \frac{1}{2}\{\frac{q^{2a}}{q-1} \prod_{i=1}^{l}(1 - 1/q^{2a_i}) + q^a \prod_{i=1}^{l}(1 + 1/q^{a_i})\}.$$

Finally,

$$vol(\Gamma\backslash X) = \frac{2q^{3a}}{(q+1)(q-1)^2} \prod_{i=1}^{l}(1 - 1/q^{2a_i}).$$

9.86   Spectral Statistics

As noted above, Jakobson, Miller, Rivin and Rudnick (1998) have studied the level spacing distribution of a generic k-regular graph. They computed the eigenvalues of generic adjacency matrices and unfolded the spectrum using McKay's law. Then, they computed the level spacing distribution and compared it with the GOE distribution. Lafferty and Rockmore (1993) have examined the eigenvalue spacings for adjacency matrices for 4-regular (2-generator) Cayley graphs for three families of groups: $SL(2, \mathbf{F}_p)$, the symmetric groups, S_m for $m = 10$, and large cyclic groups; in addition they examined the eigenvalue spacings for random 4-regular graphs (which are not Cayley graphs). Their numerical work shows strong agreement in the 4-regular Cayley case with the Poisson distribution and in the random case the behavior appears to follow the GOE distribution, which is consistent with the work of Jakobson et al. In a related work, the reader should note Kottos and Smilansky (1998).

Based on his work on the Selberg trace formula for finite volume graphs, Nagoshi (1998) has proposed the following two conjectures regarding the spectral statistics for the case $X(q, A) = \Gamma(q, A)\backslash X_q$ where as above X_q is a $(q+1)-$regular tree. Consider two cases: (1) the degree of A, tends to infinity and (2) q tends to infinity with the degree of A fixed. Normalize the adjacency operator to consider $T' = T/\sqrt{q}$ and define the two measures: $\mu_{ST}(x) = \frac{1}{\pi}\sqrt{1 - \frac{x^2}{4}}dx$ and $\mu_q(x) = \frac{q+1}{(q^{1/2}+q^{-1/2})^2 - x^2}\mu_{ST}(x)$; the reader should recall Sarnak's result in the last chapter, Serre's results in Section 22 and Li's results in Section 53. Then Nagoshi has conjectured that in case (1) in the limit the normalized spectra of $X(q, A)$ is equidistributed with respect to the spectral measure $\mu_q(x)$ on $[-2, 2]$ and in case (2) in the limit the normalized spectra of $X(q, A)$ is equidistributed with respect to the Sato-Tate measure $\mu_{ST}(x)$ on $[-2, 2]$.

Theorem 103 *(Nagoshi) Let q be a fixed odd prime power. For any $\{A_i\}, i = 1, 2, \ldots$ with*

$$A_i \in \mathbf{F}_q[t]$$

such that $\deg(A_i) \to \infty$ as $i \to \infty$, the discrete spectra D'_i of $T' = T/\sqrt{q}$ on $\Gamma(A_i)\backslash X$ are equidistributed with respect to the measure $\mu_q(x)$ on $\Omega = [-2, 2]$; i.e. for $f(x) \in C(\Omega)$

$$\lim_{i\to\infty} \frac{1}{|D'_i|} \sum_{\lambda \in D'} f(\lambda) = \int_\Omega f(x)d\mu_q(x).$$

Theorem 104 *(Nagoshi) For any $\{q_i, A_i\}$ $i = 0, 1, 2, \ldots$ such that $q_i \to \infty$ with $\deg(A_i) \to \infty$ as $i \to \infty$ and $A_i \in \mathbf{F}_{q_i}[t]$, then the discrete spectra D'_i of $T' = T/\sqrt{q_i}$ on $\Gamma(A_i)\backslash X$ are equidistributed with respect to the measure $\mu_\infty(x)$ on $\Omega = [-2, 2]$; i.e. for $f(x) \in C(\Omega)$*

$$\lim_{i\to\infty} \frac{1}{|D'_i|} \sum_{\lambda \in D'} f(\lambda) = \int_\Omega f(x)d\mu_{ST}(x).$$

Nagoshi's proof is very elegant. First, he shows that

Theorem 105 *(Nagoshi) Let $\dim(T)$ denote the number of eigenvalues of T and let $vol(\Gamma\backslash X)$ denote the volume which is normalized so that the Haar measure of K is one; then*

$$\dim(T) = vol(\Gamma\backslash X) + \frac{1}{2}(\mu - Tr\Phi(\frac{1}{2})) + 1 - \sum_{|b_i|<1, b_i \neq 0} 1 - \mu(\alpha + \frac{1}{q-1}).$$

So $dim(T) \sim vol(\Gamma \backslash X)$ as $deg(A) \to \infty$.

The proof here follows from noting that

$$\frac{1}{4\pi} \int_{-\frac{\pi}{\log q}}^{\frac{\pi}{\log q}} \frac{\phi'}{\phi}(\frac{1}{2}+ir)dr = \sum_{|a_i|<1, a_i \neq 0} 1 - \sum_{|b_i|<1, b_i \neq 0} 1 = 1 - \sum_{|b_i|<1, b_i \neq 0} 1$$

and

$$\sum_{|b_i|<1, b_i \neq 0} 1 = o(q^{3a-3}).$$

Let $N_m = \sum_{deg\, P|m, P \in P_\Gamma} deg(P)$. The Selberg zeta function is then

$$Z_\Gamma(u) = exp(\sum_{m=1}^{\infty} \frac{N_m}{m} u^m).$$

The next step is to apply the test functions $c(n) = 1$ for n even and $c(n) = 0$ for n odd to Nagoshi's trace formula. One obtains

$$\sum_r (q^{imr} + q^{-imr}) + 2(q^{\frac{m}{2}} + q^{-\frac{m}{2}}) =$$

$$-vol(\Gamma \backslash X)(q-1)q^{-\frac{m}{2}} + N_m q^{-\frac{m}{2}} + (\sigma - tr\Phi(1/2)) + 2q^{-\frac{m}{2}} - 2\sum_{b_j} b_j^{\frac{m}{2}}.$$

Here the sum on the left hand side if taken over the eigenvalues of T: $\lambda = q^s + q^{1-s}$ for $s = 1/2 + ir$. From the conditions on $\{q_i, A_i\}$, the theorem follows.

9.87   Anderson Localization

Schanz and Smilansky (1999) have presented a quantum graph model which produces Anderson localization properties. Their model is a quantum graph with one dimensional topology. Time evolution is described by the operator

$$U_{j',v';j,v}(k) = e^{ikL_j}(\delta_{+v',v}\delta_{j',j+v}t_{j+\frac{v+1}{2}} + \delta_{-v',v}\delta_{j',j}r_{j+\frac{v+1}{2}})$$

where j runs over the vertices, $v = \pm$ and t,r are the transition and reflection amplitudes. Schanz and Smilansky have shown that k is an eigenvalue of the graph if and only if 1 is in the spectrum of $U(k)$, i.e. $\mathbf{a} = U(k)\mathbf{a}$.

The degree of localization in their work is characterized by the k averaged quantum return probability:

$$\mathcal{P}(n) = < |(U^n)_{0,+;0,+}|^2 >$$

where initially $a_{j,v} = \delta_{j,0}\delta_{v,+}$. The average $< .. >$ is over a large k interval.

Theorem 106 *(Schanz and Smilansky)*

$$P(n) = < |\sum_\lambda A_\lambda exp(ik\mathcal{L}_n)|^2 >$$

where λ runs over all trajectories contributing to the return probability, i.e. all sequences $v_\nu, \nu = 0, 1, ...n-1$ with $\sum_\nu v_\nu = 0$. Here the length of the orbit is given by

$$\mathcal{L}_\lambda = \sum_{\nu=1}^n |x_{j_\nu} - x_{j_{\nu-1}}|,$$

and the product of the transition and reflection amplitudes along the orbits is

$$A_\lambda = \prod_{\nu=1}^n A_{\lambda,v}$$

with

$$A_{\lambda,v} = \begin{cases} t_{j_\nu} & if \ v_{\nu+1} = v_\nu \\ r_{j_\nu} & otherwise. \end{cases}$$

The diagonal approximation yields the standard result for classical diffusion:

Theorem 107 *(Schanz and Smilansky)*

$$P_{cl}(n = 2m) = \frac{1}{2^n} \binom{n-1}{n/2} \approx \frac{1}{\sqrt{2n\pi}}.$$

If the lengths of all the periodic orbits are the same, then

$$P_{per}(n = 2m) = \frac{1}{2^n} \binom{m-1}{\lceil \frac{m-1}{2} \rceil}^2 \approx \frac{1}{n\pi}.$$

If there is a set of disordered bond lengths in the quantum graph,

$$\mathcal{L}_\lambda = 2 \sum_j m_j L_j$$

where $\sum_j m_j = m$ and a family of isometric orbits contains all the orbits with $\mathcal{L} = [m_0, m_{\pm 1}, ...]$. If these are not symmetry related orbits, then one has

$$P(n = 2m) = \sum_{\mathcal{L} \in \mathcal{F}_n} |\sum_{\lambda \in \mathcal{L}} A_\lambda|^2 = \sum_{\mathcal{L} \in \mathcal{F}_n} P(\mathcal{L}).$$

Then, Schanz and Smilansky have shown Anderson localization can be reproduced for this quantum graph model:

Theorem 108 *(Schanz and Smilansky) In the case of disorder,*

$$lim_{m \to \infty} P_{dis}(n = 2m) = \frac{\pi^2}{3} - 3.$$

9.88   Spectral Statistics of Quantum Graphs

Schanz and Smilansky (1999) have studied the spectral statistics of quantum graphs. Consider the 2-star graph. The S matrix in this case has the form

$$S_B(k, \eta) = \begin{pmatrix} e^{2ikL_1} & 0 \\ 0 & e^{2ikL_2} \end{pmatrix} \begin{pmatrix} cos\eta & isin\eta \\ isin\eta & cos\eta \end{pmatrix}$$

where this model has two loops of bond length $2L_1$ and $2L_2$. Here $0 \leq \eta \leq \pi/2$. The eigenvalues of S_B are taken to be $e^{ik(L_1+L_2)}e^{\pm i\lambda/2}$. For a fixed η, the k averaged spacing distribution is given by

$$P(\theta) = \frac{1}{\Delta_k} \int_{k_0-\Delta_k/2}^{k_0+\Delta_k/2} dk \delta(\theta - 2arcos[cos\eta cosk(L_1 - L_2)])$$

$$= \begin{cases} 0 & \text{if } cos(\theta/2) > |cos\eta| \\ \frac{sin(\theta/2)}{\sqrt{cos^2\eta-cos^2(\theta/2)}} & \text{otherwise} \end{cases} .$$

Here θ is the smaller of the intervals between the two eigenphases, so $0 \leq \theta \leq \pi$.

Theorem 109 *(Schanz and Smilansky) The form factor is*

$$K_2(n, \eta) = \frac{1}{2} < |trS_B(\eta)^n|^2 >$$

which for $\eta = \pi/4$ gives

$$K_2(n, \pi/4) = 1 + \frac{(-1)^{m+n}}{2^{2m+1}} \binom{2m}{m} \approx 1 + \frac{(-1)^{m+n}}{2\sqrt{\pi n}}.$$

Schanz and Smilansky also consider the case where η is distributed with measure $d\mu(\eta) = |cos\eta sin\eta|d\eta$. This results in

$$P(\theta) = 2sin^2(\theta/2)$$

and

$$K_2(n) = \begin{cases} \frac{1}{2} & \text{for } n = 1 \\ 1 & \text{for } n \geq 2, \end{cases}$$

which is the CUE result for 2×2 matrices.

Schanz and Smilansky also consider the periodic orbit expansion of the form factor. For the 2-star model this can be expressed in terms of Krawtchouk polynomials. We refer the reader to the paper for details.

9.89   Zeta Functions on Non-uniform Trees

There is other recent work in the general area of Selberg trace formula and zeta functions on trees, including that of Deitmar (1997) and Clair and Mokhtari-Sharghi (1999). Clair and Mokhtari-Sharghi have generalized Bass' work on zeta functions for uniform tree lattices using results from von Neumann algebras, in particular an extension of the determinant of Fuglede and Kadison (1952).

Let Γ be a lattice of a tree X which has uniformly bounded valence. Let $\mathcal{A}$ be a von Neumann *-algebra of bounded operators on a Hilbert space which is closed in the weak or strong topology. $\mathcal{A}$ is equipped with a trace function $Tr_{\mathcal{A}} : \mathcal{A} \to \mathbf{C}$. Let M be a Hilbert $\mathcal{A}$-module and let ρ be a representation of Γ to the unitary operators on M which commute with $\mathcal{A}$. Let $\epsilon \in \mathcal{E}(\Gamma)$ be a hyperbolic end of X. For $g \in \Gamma_\epsilon$, the stabilizer of ϵ, let $\tau_\epsilon(g)$ denote the translation towards ϵ; define $l(\epsilon)$ by $Im(\tau_\epsilon) = l(\epsilon)\mathbf{Z}$. Let Γ_ϵ^0 denote the kernel of τ_ϵ. Set

$$\sigma_\epsilon = \frac{1}{|\Gamma_\epsilon^0|} \sum_{g \in \Gamma_\epsilon, \tau_\epsilon(g) = l(\epsilon)} g.$$

Clair and Mokhtari-Sharghi define the zeta function of Γ by

$$Z_\rho(u) = \prod_{\epsilon \in \Gamma \backslash \mathcal{E}(\Gamma)} Det_{\mathcal{A}}(I - \rho(\sigma_\epsilon)u^{l(\epsilon)}).$$

For a non-uniform lattice one must be careful as there may be infinitely many classes $\epsilon \in \Gamma \backslash \mathcal{E}(\Gamma)$ with the same length.

Theorem 110 *(Clair and Mokhtari-Sharghi) If the total volume of the edges of $\Gamma \backslash X$ is finite and the total volume of the loops of length n in $\Gamma \backslash X$ is finite for each n, then $Z_\rho(u)$ is well defined in $\mathbf{C}[[u]]$ and*

$$Z_\rho(u) = Det_{\mathcal{A}}(I - T_\rho u).$$

If $\sum_{e \in EB} dim_{\mathcal{A}} M^{\Gamma_{\tilde{e}}} < \infty$, then Ihara's formula holds:

$$Det_{\mathcal{A}}(I - T_\rho u) = (1 - u^2)^{-\chi_\rho(X)} Det_{\mathcal{A}}(I - \delta_\rho u + Q_\rho u^2).$$

Let $1 \to \Lambda \to \Gamma \to \pi \to 1$ be an exact sequence. Let ρ be the coset representation of Γ on $M = l^2(\pi)$. Then, $Z_\rho(u)$ can be interpreted as the zeta function of the infinite graph $Y = \Lambda \backslash X$.

Theorem 111 *(Clair and Mokhtari-Sharghi) If Y is a locally finite graph and the discrete subgroup π acts on Y with quotient B having finite volume, then*

$$Z_\pi(Y, u) = \prod_{\gamma \in \pi \backslash P} (1 - u^{l(\gamma)})^{1/|\pi_\gamma|} = (1 - u^2)^{-\chi^{(2)}(Y)} Det_\pi(\Delta_u).$$

Here $\Delta_u = I - \delta u + Qu^2$ acts on $l^2(VY)$, P is the set of free homotopy classes of primitive closed paths in Y, $l(\gamma)$ is the length of the shortest representative of γ. The Euler characteristic $\chi_\rho(X)$ is the L^2 Euler characteristic $\chi^{(2)}(Y)$.

As an example, let $\pi = \mathbf{Z} \times \mathbf{Z} = <a> \times <b>$. Let Y be the Cayley graph of π with generators $(a, 0)$ and $(0, b)$, so Y is an infinite grid. Here π acts freely on all loops and

$$Z_\pi(Y, u) = \prod_{\gamma \in \pi \backslash P} (1 - u^{l(\gamma)}) = \prod_{l=1}^{\infty} (1 - u^l)^{N(l)}$$

so that

$$-log Z_\pi(Y,u) = -\sum_{M=1}^{\infty} \sum_{L|M} N(2L)\frac{L}{M}u^{2M} =$$

$$\sum_{n=1}^{\infty} Tr_\pi(1-\Delta(u))^n + \sum_{M=1}^{\infty}\frac{1}{M}u^{2M}.$$

Here $N(L)$ is the number of translation classes of tail-less, back-less primitive loops in Y of length L. The weighted Laplacian $\Delta(u)$ acts on $l^2(VY) \simeq l^2(\pi)$ and for this case it is a 1×1 matrix:

$$\Delta(u) = (1 - (a + a^{-1} + b + b^{-1})u + 3u^2).$$

Theorem 112 *(Clair and Mokhtari-Sharghi) For this examples one has:*

$$-log Z_\pi(Y,u) = \sum_{M=1}^{\infty} [\frac{1}{M} + \sum_{d=0}^{M}\frac{(-3)^{M-d}}{M+d}\binom{M+d}{M-d}\binom{2d}{d}^{2d}]u^{2M}.$$

In particular the first few values of $N(2L)$ are $0, 2, 4, 26, 152, 1004, ...$

References

R. Abou-Chacra, P. Anderson and D. J. Thouless, A selfconsistent theory of localization, J. Phys. C6 (1973) 1734-1752.

R. Abou-Chacra and D. J. Thouless, A selfconsistent theory of localization: II, J. Phys. C7 (1973) 65-75.

E. Abrahams, Theoretical approaches to the metal-insulator transition in 2D, Ann. Phys. 8 (1999) 1-10.

E. Abrahams, Scaling at the metal-insulator transition in two dimensions (preprint, 1999).

E. Abrahams, P. Anderson, D. Licciardello and R. Ramakrishnan, Phys. Rev. Lett. 42 (1979) 673-676.

V. Acosta and A Klein, Analyticity of the density of states in the Anderson model in the Bethe lattice, J. Stat. Phys. 69 (1992) 277-305.

R. Adams, Sobolev Spaces (Academic Press, New York, 1975).

V. M. Adamyan, Scattering matrices for microschemes, Oper. Theory: Adv. Appl. 59 (1992) 1-10.

J. Aguilar and J.-M. Combes, A class of analytic perturbations for one-body Schrödinger Hamiltonians, Comm. Math. Phys. 22 (1971) 269-279.

G. Ahumada, Fonctions periodiques et formule des traces de Selberg sur les arbres, C. R. Acad. Sci. Paris 305 (1987) 709-712.

M. Aizenman, Localization at weak disorder: some elementary bounds, Rev. Math. Phys. 6 (1994) 1163-1182.

M. Aizenman and G. Graf, Localization bounds for an electron gas, J. Phys. A31 (1998) 6783-6806.

M. Aizenman and S. Molchanov, Localization at large disorder and at extreme energies: an elementary derivation, Comm. Math. Phys. 157 (1993) 245-278.

M. Aizenman, J. Schenker, R. Friedrich and D. Hundertmark, Finite-volume criteria for Anderson localization (preprint, 1999).

R. Akis, D. K. Ferry and J. P. Bird, Magnetotransport fluctuations in regular semiconductor ballistic quantum dots, Phys. Rev. B54 (1996) 17705-17715.

R. Akis, J. P. Bird and D. Ferry, The effects of inelastic scattering in open quantum dots, J. Phys. C8 (1996) L667.

R. Akis, D. Ferry and J. P. Bird, Wave function scarring in open ballistic quantum dots, Jpn. J. Appl. Phys. 36 (1997) 3981-3985.

R. Akis, D. Ferry and J. P. Bird, Wave function scarring effects in open stadium shaped quantum dots, Phys. Rev. Lett. 79 (1997) 123-126.

R. Akis, P. Vasilopoulos and P. Debray, Ballistic transport in electron stub tuners, Phys. Rev. 52 (1995) 2805-2813.

R. Akis, P. Vasilopoulos and P. Debray, Bound states and transmission antiresonances in parabolically confined cross structures: influence of weak magnetic fields, Phys. Rev. B56 (1997) 9594-9602.

E. Akkermans, A. Comtet, J. Desbois, G. Montambaux, and C. Texier, Spectral determinants on quantum graphs (preprint, 1999).

S. Albeverio, F. Gesztesy, R. Høegh-Krohn, and H. Holden, Solvable models in quantum mechanics (Springer-Verlag, New York, 1988).

S. Albeverio and V. Geyler, The band structure of the general periodic Schrödinger operator with point interactions, (preprint, 1999).

S. Albeverio, V. Geyler and O. Kostrov, Quasi-one-dimensional nanosystems in a uniform magnetic field: explicitly solvable model, Rep. Math. Phys. 44 (1999) 1-8.

B. Altshuler and B. Simon, Universalities from Anderson localization to quantum chaos, in Les Houches Session LXI, 1994 (ed., E. Akkermans et al., Elsevier, Amsterdam, 1995).

P. W. Anderson, Absence of diffusion in certain random lattices, Phys. Rev. 109 (1958) 1492-1505.

P. W. Anderson, A question of classical localization. A theory of white paint, Phil. Mag. B53 (1958) 505-509.

T. Ando and H. Tamura, Conductance fluctuations in quantum wires with spin-orbit and boundary-roughness scattering, Phys. Rev. B46 (1992) 2332-2338.

M. Andrews and C. Savage, Bound states of two-dimensional nonuniform waveguides, Phys. Rev. A50 (1995) 4535-4537.

M. Antoine, A. Comtet, and S. Ouvry, Scattering on a hyperbolic torus in a constant magnetic field, J. Phys. A23 (1990) 3699-3710.

M. Ashbaugh and P. Exner, Lower bounds to bound state energies in bent tubes, Phys. Lett. A150 (1990) 183-186.

A. Aslanyan, L. Parnovski and D. Vassiliev, Complex resonances in acoustic waveguides (preprint, 1999).

Y. Avishai and Y. Band, Ballistic conductance of wide orifice, Phys. Rev. B41 (1990) 3253-3255.

Y. Avishai, M, Kaveh and Y. Band, Conductance of Fabry-Pérot and two slit electronic waveguides, Phys. Rev. B42 (1990) 5867-5870.

Y. Avishai, D. Bessis, B. Giraud, and B. Mantica, Quantum bound states in open geometries, Phys. Rev. B44 (1991) 8028-8034.

Y. Avishai and J. Luck, Quantum percolation and ballistic conductance on a lattice of wires, Phys. Rev. B45 (1992) 1074-1095.

J. E. Avron, M. Klein, A. Pnueli, and L. Sadun, Hall conductance and adiabatic charge transport of leaky tori, Phys. Rev. Lett. 69 (1992) 128-131.

J. Avron, A. Raveh and B. Zur, Adiabatic transport in multiply connected systems, Rev. Mod. Phys. 60 (1988) 873-915.

J. Avron and R. Seiler, Quantization of the Hall conductance for general, multi Schrödinger Hamiltonians, Phys. Rev. Lett. 54 (1985) 259-262.

J. Avron, R. Seiler and B. Simon, Homotopy and quantization in condensed matter physics, Phys. Rev. Lett. 51 (1983) 51-53.

J. Avron, R. Seiler and B. Simon, Charge deficiency, charge transport and comparison of dimensions, Comm. Math. Phys. 159 (1994) 399.

J. Avron and L. Sadun, Adiabatic quantum transport in networks with macroscopic component, Ann. Phys. 206 (1991) 440-493.

M. Azbel', Quantum particle in a random potential: exact solution and its application, Phys. Rev. B45 (1992) 4208-4216.

P. Bagwell, Evanescent modes and scattering in quasi-one dimensional wire, Phys. Rev. B41 (1990) 10354-10371.

E. Balslev and A. Venkov, The Weyl law for subgroups of the modular group, Geom. Funct. Anal. 8 (1998) 437-465.

E. Balslev and A. Venkov, Phillips-Sarnak's conjecture for $\Gamma_0(8)$ with primitive character, (preprint, 1999).

E. Balslev and A. Venkov, Stability of character resonances, (preprint, 1999).

E. Balslev and A. Venkov, Phillips-Sarnak's conjecture for Hecke groups with primitivie character, (preprint, 1999).

E. Balslev and A. Venkov, Spectral theory of Laplacians for Hecke groups with primitivie character, (preprint, 1999).

H. Baranger, Multiprobe electron waveguides, filtering and bend resistances, Phys. Rev. B42 (1990) 11479-11495.

H. Baranger, R. Jalabert and A. Stone, Weak localization and integrability in ballistic cavities, Phys. Rev. Lett. 70 (1993) 3876-3879.

H. Baranger, R. Jalabert, and A. Stone, Quantum-chaotic scattering effects in semiconductor microstructures, Chaos 3 (1993) 665-682.

H. U. Baranger and P. A. Mello, Mesoscopic transport through chaotic cavities: a random S-matrix theory approach, Phys. Rev. Lett. 73 (1994) 142-145.

H. Baranger and P. Mello, Effect of phase breaking on quantum transport through a chaotic cavity, Phys. Rev. B51 (1995) 4703-4706.

H. Baranger and A. Stone, Electrical linear-response theory in an arbitrary magnetic field: a new Fermi-surface formation, Phys. Rev. B40 (1989) 8169-8193.

J.-M. Barbaroux, J. M. Combes, P. D. Hislop, Localization near band edges for random Schrödinger operators, Helv. Phys. Acta 70 (1997a) 16-43.

J.-M. Barbaroux, J. M. Combes, P. D. Hislop, Landau Hamiltonians with unbounded random potentials, Lett. Math. Phys. 40 (1997b) 355-369.

J.-M. Barbaroux, J. M. Combes and R. Montcho, Remarks on the relation between quantum dynamics and fractal spectra, J. Math. Anal. Appl. 213 (1997) 698-722.

J.-M. Barbaroux, W. Fischer and P. Müller, Dynamical properties of random Schrödinger operators (preprint, 1999).

H. Bass, Finitely generated subgroups of GL_2, in The Smith Conjecture, (H. Bass and J. Morgan, eds., Academic Press, New York, 1984).

H. Bass, The Ihara-Selberg zeta function of a tree lattice, Inter. J. Math. 3 (1992) 717-797.

H. Bass and R. Kulkarni, Uniform tree lattices, J. Amer. Math. Soc. 3 (1990) 843-902.

H. Bass and A. Lubotzky, Rigidity of group actions on locally finite trees, Proc. London Math. Soc. 69 (1994) 541-575.

C. W. J. Beenakker, Random-matrix theory of quantum transport, Rev. Mod. Phys. 69 (1997) 731-808.

C. W. J. Beenakker, Photon statistics of a random laser, in Diffuse Waves in Complex Media (ed., J. Fouque, Kluwer Academic Press, Dordrecht, 1998) 137-164.

C. W. J. Beenakker, Thermal radiation and amplified spontaneous emission from a random medium, Phys. Rev. Lett. 81 (1998) 1829-1832.

C. W. J. Beenakker and M. Patra, Photon shot noise, Mod. Phys. Lett. B13 (1999) 337-347.

V. Bellani, E. Diez et al., Experimental evidence of delocalized states in random dimer superlattices, Phys. Rev. Lett. 82 (1999) 2159-2162.

J. Bellissard, A. van Elst and H. Schulz-Baldes, The noncommutative geometry of the quantum Hall effect, J. Math. Phys. 35 (1994) 5373-5451.

F. Bentosela, P. Exner and V. Zagrebnov, Mechanism of porous-silica luminescence, Phys. Rev. B57 (1998) 1382-1385.

K.-F. Berggren, C. Besev and Z. Ji, Transition from laminar to vortex flow in a model semiconductor nanostructure, Physica Scr. T42 (1992) 141-148.

K.-F. Berggren, C. Besev and Z. Ji, Spatial distribution of quantum mechanical currents and charge in a model semiconductor nanostructure, in Int. Workshop on Quantum Effect Physics, Luxor, 1992, 25-28.

K.-F. Berggren and Z. Ji, Resonant tunneling via bound states in T-shaped electron waveguide structures, Superlatt. Microst. 8 (1990) 59-61.

K.-F. Berggren and Z. Ji, Resonant tunneling via quantum bound states in a classically unbound system of crossed, narrow channels, Phys. Rev. B43 (1991) 4760-4764.

K.-F. Berggren and Z. Ji, Transition from laminar to vortical current flow in electron waveguides with circular bends, Phys. Rev. B47 (1993) 6390-6394.

K.-F. Berggren and Z. Ji, Quantum chaos in nano-sized billiards in layered two-dimensional semiconductor structures, Chaos 6 (1996) 543-553.

K.-F. Berggren, Z. Ji and T. Lundberg, Origin of conductance fluctuations in large circular quantum dots, Phys. Rev. B54 (1996) 11612-11621.

K.-F. Berggren and C Wang, Different orbitals for different electrons in a system of intersecting quantum wires, Int. J. Q. Chem. 63 (1997) 667-673.

M. Berry, J. Katine, C. Marcus, R. Westervelt and A. Gossard, Weak localization and conductance fluctuations in a chaotic quantum dot, Surf. Sci. 305 (1994) 495-500.

M. V. Berry, Quantum scars of classical closed orbits in phase space, Proc. Roy. Soc. London A423 (1989) 219-231.

C. Besez, Z. Ji, K.-F. Berggren, Quantum ballistic transport in a model semiconductor nanostructure, Cray Channels (1992) 20-25.

J. P. Bird, K. Ishibashi et al., Spectral characteristics of conductance fluctuations in ballistic quantum dots: the influence of finite magnetic field and temperature, Phys. Rev. B 52 (1995) 8295-8304.

J. P. Bird, D. K. Ferry, R. Akis et al., The role of lead openings in regular mesoscopic billiards, Superlatt. Microstr. 20 (1996) 287-295.

J. P. Bird, R. Akis, D. K. Ferry, et al., Periodic conductance fluctuations and lead-induced scarring in open quantum dots, J. Phys: Cond. Mat. 9 (1997) 5935-5950.

J. P. Bird, D. K. Ferry, R. Akis et al., Periodic conductance fluctuations and stable orbits in mesoscopic semiconductor billiards, Europhys. Lett. 35 (1996) 529-534.

J. P. Bird, K. Ishibashi et al, Phase breaking in ballistic quantum dots: a correlation field analysis, Surf. Sci. 361/362 (1996) 730-734.

J. P. Bird, K. Ishibashi et al., Quantum transport in open mesoscopic cavities, Chaos, Solitons & Fractals, 8 (1997) 1299-1324.

J. P. Bird, H. Linke, J. Cooper et al., Phase breaking as a probe of the intrinsic level spectrum of open quantum dots, Phys. Stat. Sol. 204 (1997) 314.

J. P. Bird, A. Micolich, H. Linke et al., Environmental coupling and phase breaking in open quantum dots, J. Phys. C10 (1998) L55.

J. P. Bird, D. Olatona et al., Lead-induced transition to chaos in ballistic mesoscopic billiards, Phys. Rev. B52 (1995) R14336-R14339.

R. Blackenbecler, M. Goldberger and B. Simon, The bound states of weakly coupled long-range one-dimensional quantum Hamiltonians, Ann. Phys. 108 (1977) 69-78.

Ya. Blanter and M. Büttiker, Shot noise in mesoscopic conductors (preprint, 1999).

G. Boebinger, G. Muller, et al., Stable orbit bifurcations in a quantum well in a tilted magnetic field, Surf. Sci. 361-2 (1996) 742-746.

E. Bogomolny, N. Pavloff and C. Schmit, Diffractive corrections in the trace formula for polygonal billiards, (preprint, 1999).

A. Bostrom, IEEE Trans. Microwave Theory Tech. 31 (1983) 752.

E. Brezin, D. Gross, and C. Itzykson, Density of states in the presence of a strong magnetic field and random impurities, Nucl. Phys. B235 (1984) 24-44.

P. Briet, J.-M Combes and P. Duclos, On the location of resonances for Schrödinger operators in the semiclassical limit II, Comm. PDE 12 (1987) 201-222.

K. Broderix, N. Heldt, and H. Leschke, Exact results on Landau-level broadening, J. Phys. A24 (1991) L825-L831.

K. Broderix, D. Hundertmark, W. Kirsch, and H. Leschke, The fate of Lifshits tails in magnetic fields, J. Stat. Phys. 80 (1995) 1-22.

R. Brooks, The spectral geometry of k−regular graphs, J. d'Analyse 57 (1991) 120-151.

R. Brooks, Some relations between spectral geometry and number theory, in Topology 90 (ed., Apanosov et al., Walter de Gruyter, 1992) 97-107.

R. Brooks, Reflections on the first eigenvalue, Texas Tech. Univ. Math. Ser. 19 (1996).

R. Brooks, Isospectral graphs and isospectral surfaces, Sem. Theorie Spec. Geom. 15 (1996) 105-113.

R. Brooks, Some relations between graph theory and Riemann surfaces, Israel Math. Conf. Proc. (1999).

R. Brooks and R. Tse, Isospectral surfaces of small genus, Nagoya Math. J. 107 (1987) 13-24.

P. W. Brouwer, Transmission through a many-channel random waveguide with absorption, Phys. Rev. B57 (1998) 10526-10536.

P. W. Brouwer and C. W. J. Beenakker, Voltage-probe and imaginary potential models for dephasing in a chaotic quantum dot, Phys. Rev. B55 (1997) 4695-4702.

K. Brown, Buildings (Springer-Verlag, New York, 1989).

J. Brüning and V. Geyler, The spectrum of periodic point pertubations and the Krein resolvent formula (preprint, 1998).

J. Brum, Electronic properties of quantum-dot superlattices, Phys. Rev. B43 (1991) 12082-12085.

M. Büttiker, Absence of backscattering in the quantum Hall effect in multi-probe conductors, Phys. Rev. B38 (1988) 9375-9389.

W. Bulla, F. Gesztesy, W. Renger and B. Simon, Weakly coupled bound states in quantum waveguides, Proc. AMS 125 (1997) 1487-1495.

W. Bulla and T. Trenkler, The free Dirac operator on compact and noncompact graphs, J. Math. Phys. 31 (1990) 1157-1163.

P. Buser, Geometry and Spectra of Compact Riemann Surfaces (Birkhauser, Boston, 1992).

M. A. Callan, C. M. Linton and D. V. Evans, Trapped modes in two-dimensional waveguides, J. Fluid Mech. 229 (1991) 51-64.

J. Cambell and W. Jones, Symetrically truncated right-angle corners in parallel-plate and rectangular waveguides, IEEE Trans. MTT 16 (1974) 687-696.

A. Carey, K. Hannabuss and V. Mathai, Quantum Hall effect on the hyperbolic plane in the presence of disorder, Comm. Math. Phys. 190 (1998) 629-673.

J. Carini, J. Londergan, K. Mullen and D. Murdock, Bound states and resonances in quantum wires, Phys. Rev. B46 (1992) 15538-15541.

J. Carini, J. Londergan, K. Mullen and D. Murdock, Multiple bound states in sharply bent waveguides, Phys. Rev. B48 (1993) 4503-4514.

J. Carini, J. Londergan, D. Murdock, D. Trinkle and C. Yung, Bound states in waveguides and bent quantum wires, I. Applications to waveguide systems, Phys Rev. B55 (1997) 9842-9851.

J. Carini, J. Londergan and D. Murdock, Bound states in waveguides and bent quantum wires, II. Electrons in quantum wires, Applications to waveguide systems, Phys Rev. B55 (1997) 9852-9859.

R. Carlson, Hill's equation for a homogeneous tree, Elec. J. Diff. Eqs. 1997 (1997) No. 23, 1-30.

R. Carlson, Adjoint and self-adjoint differential operators on graphs, Elec. J. Diff. Eqs 1998 (1998) No. 6, 1-10.

R. Carmona and J. Lacroix, Spectral theory of random Schrödinger operators (Birkhäuser, Basel-Berlin, 1990).

P. Cartier, Geometrie et analyse sur le arbres, Sem. Bourbaki no. 407 (1971/72) 123-140.

P. Cartier, Harmonic analysis on trees, Proc. Symp. Pure Math. 26 (1973) 419-424.

N. Celniker, S. Poulos, A. Terras, C. Trimble and E. Velasquez, Is there life on finite upper half planes? Cont. Math. 143 (1993) 65-88.

I. Chan, R. Clarke, R. Marcus et al., Ballistic conductance fluctuations in shape space, Phys. Rev. Lett. 74 (1995) 3876-3879.

L. Chekhov, L-functions in scattering on p-adic multiloop surfaces, J. Math. Phys. 36 (1995) 414-425.

L. Chekhov, L-functions in scattering on generalized Cayley trees, J. Tech. Phys. 37 (1996) 301-305.

L. Chekhov, Scattering on univalent graphs from L-function viewpoint, Theor. Math. Phys. 103 (1995) 723-737.

L. Chekhov, Spectral problem on graphs and L-functions, (preprint, 1999).

L. Chekhov, A. Mironov and A. Zabrodin, Multiloop calculations in p-adic string theory and Bruhat-Tits trees, Comm. Math. Phys. 125 (1989) 675-711.

L. Christensson, H. Linke, P. Omling et al., Classical and quantum dynamics of electrons in open, equilateral triangular billiards, Phys. Rev. B57 (1998) 12306-12313.

C. Chu and R. Sorbello, Effects of impurities on the quantized conductance of narrow channels, Phys. Rev. B40 (1989) 5941-5949.

F. Chung, Diameter and eigenvalues, J. AMS 2 (1989) 65-88.

F. Chung, Laplacians of graphs and Cheeger inequalities, Bolyai Soc. Math. Stud. 2 (1993) 157-172.

F. Chung, Eigenvalues of graphs, in Proc. of ICM, Zürich, 1994, (Birkhauser Verlag, Berlin) 1333-1342.

F. Chung and S.-T. Yau, Eigenvalues of graphs and Sobolev inequalities, Comb. Prob. Comp. 4 (1995) 11-25.

B. Clair and S. Mokhtari-Sharghi, Zeta functions of discrete groups acting on trees (preprint, 1999).

R. Clarke, I. Chan, C. Marcus et al., Temperature dependence of phase breaking in ballistic quantum dots, Phys Rev. B52 (1995) 2656-2659.

Y. Colin de Verdiere, Pseudo-Laplacians II, Ann. Inst. Fourier 33 (1983) 87-113

J. Combes and P. Hislop, Localization for some continuous, random Hamiltonians in d-dimensions, J. Func. Anal. 124 (1994) 149-180.

J. Combes and P. Hislop, Landau Hamiltonians with random potentials: localization and the density of states, Comm. Math. Phys. 177 (1996) 603-629.

J. Combes, P. Hislop and E. Mourre, Spectral averaging, perturbation of singular spectra and localization, Trans. Amer. Math. Soc. 348 (1996) 4883-4894.

J. Combes, P. Hislop and E. Mourre, Correlated Wegner inequalities for random Schrödinger operators, Contemp. Math. 217 (1998) 191-203.

J. Combes, P. Hislop and A. Tip, Band edge localization and the density of states for acoustic and electromagnetic waves in random media, Ann. Inst. H. Poincaré, Phys. Theo. 70 (1999) 381-428.

J. M. Combes and L. Thomas, Asymptotic behavior of eigenfunctions for multiparticle Schrödiner operators, Comm. Math. Phys. 34 (1973) 251-270.

A. Connes, Non-commutative differential geometry, Publ. IHES 62 (1986) 257-360.

A. Connes, Noncommutative Geometry (Academic Press, San Diego, 1994).

W. Craig, The trace formula for Schrödinger operators on the line, Comm. Math. Phys. 126 (1989) 379-407.

M. Crommie, C. Lutz and D. Eigler, Imaging standing waves in a two-dimensional electron gas, Nature 363 (1993) 524-527.

M. Crommie, C. Lutz and D. Eigler, Confinement of electrons to quantum corrals on a metal surface, Science 262 (1993) 218-220.

M. Crommie, C. Lutz and D. Eigler, Waves on a metal surface and quantum corrals, Surf. Rev. and Lett. 2 (1995) 127-137.

H. Cycon, R. Froese, W. Kirsch, and B. Simon, Schrödinger Operators with application to quantum mechanics and global geometry (Springer-Verlag, Berlin, 1987).

D. Damanik and P. Stollmann, Multi-scale analysis implies strong dynamical localization, (preprint, 1999).

L. D'Amato, H. Pastawski and J. Weisz, Half-integer and integer quantum-flux periods in magnetoresistance of one-dimension rings, Phys. Rev. B39 (1989) 3554-3562.

S. Datta, Superlatt. Microstr. 6 (1988) 83.

S. Datta and S. Bandyopadhyay, Aharonov-Bohm effect in semiconductor microstructures, Phys. Rev. Lett. 58 (1987) 717-720.

S. Datta, M. Melloch and S. Bandyopadhyay, Novel interference effect between parallel quantum wells, Phys. Rev. Lett. 55 (1985) 2344-2347.

E. B. Davies and L. Parnovski, Trapped modes in acoustic waveguides, Q. J. Mech. Appl. Math. (1997).

S. De Bièvre and J. Pulé, Propagating edge states for a magnetic Hamiltonian, Math. Phys. Elec. J. 5 (1999) Paper 3.

P. Debray, R. Akis et al., Ballistic electronic stub tuner for potential use in analog-to-digital conversion, Appl. Phys. Let. 66 (1995) 3137-3139.

A. Deitmar, Geometric zeta-functions on p-adic groups, (preprint, 1997).

P. Deligne, Cohomologie étale (SGA4$\frac{1}{2}$), in Lecture Notes in Math. 569 (1977).

P. Deligne, La conjecture de Weil, I, Publ. Math. IHES 43 (1974) 273-308.

F. Delyon, Y. Lévy and B. Souillard, Anderson localization for multidimensional systems at large disorder or low energy, Comm. Math. Phys. 100 (1985) 463-470.

F. Delyon, B. Simon and B. Souillard, From power pure point to continuous spectrum in disordered systems, Ann. Inst. H. Poincaré 42 (1985) 283-309.

C. Dembowski, H. D. Gräf, R. Hofferbert, H. Rehfeld, A. Richter and T. Weiland, Anderson localization in a string of microwave cavities (preprint, 1999).

B. Derrida and E. Gardner, Lyapunov exponent of one-dimension Anderson model: weak disorder expansion, J. Phys. 45 (1984) 1283-1295.

J. Deshoulliers and I. Iwaniec, The nonvanishing of Rankin-Selberg zeta functions at special points, Contemp. Math. 53 (1986) 51-95.

E. Diez, A. Sánchez and F. Domínguez-Adame, Intentionally disordered superlattices with high-DC conductance, IEEE J. Q. Elec. 31 (1995) 1919-1926.

V. Dobrosavljevic, E. Abrahams, E. Miranda and S. Chakravarty, Scaling theory of two-dimensional metal-insulator transitions, Phys. Rev. Let. 79 (1997) 455-458.

F. Domínguez-Adame, A. Sánchez and E. Diez, Extended states and dynamical localization in semiconductor superlattices, J. Appl. Phys. 81 (1997) 777-780.

T. Dorlas, N. Macris and J. Pulé, Localization in a single-band approximation to random Schrödinger operators in a magnetic field, Helv. Phys. Acta 68 (1995) 329-364.

T. Dorlas, N. Macris, J. Pulé, Localization in single Landau bands, J. Math. Phys. 37 (1996) 1574-1595.

T. Dorlas, N. Macris and J. Pulé, The nature of the spectrum for a Landau Hamiltonian with delta impurities, J. Stat. Phys. 87 (1997) 847-875.

T. Dorlas, N. Macris, J. Pulé, Characterization of the spectrum of the Landau Hamiltonian with delta impurities, Comm. Math. Phys. 204 (1999) 367-396.

H. von Dreifus and A. Klein, A new proof of localization in the Anderson tight binding model, Comm. Math. Phys. 124 (1989) 285-299.

H. von Dreifus and A. Klein, Localization for random Schrödinger operators with correlated potentials, Comm. Math. Phys. 140 (1991) 133-147.

V. G. Drinfeld, Number of two-dimensional irreducible representations of the fundamental group of a curve over a finite field, Func. Anal. Appl. 14 (1982) 294-295.

V. G. Drinfeld, The proof of Peterson's conjecture for $GL(2)$ over a global field of characteristic p, Func. Anal. and its Appl. 22 (1988) 28-43.

P. Duclos, On a global approach to the location of quantum resonances, in Proc. of Sym. PDE and Math. Phys. (Birkhauser, Basel, 1992) 39-49.

P. Duclos and P. Exner, Curvature vs. thickness in quantum waveguides, Czech. J. Phys. 41 (1991) 1009-1018; erratum 42 (1992) 344.

P. Duclos and P. Exner, Curvature-induced bound states in quantum waveguides in two and three dimensions, Rev. Math. Phys. 7 (1995) 73-102.

P. Duclos, P. Exner and D. Krejcirik, Locally curved quantum layers (preprint, 1999).

P. Duclos, P. Exner and B. Meller, Exponential bounds on curvature-induced resonances in a two-dimensional Dirichlet tube, Hel. Phys. Acta 71 (1998) 133-162.

P. Duclos, Pl. Exner and P. Stovicek, Curvature-induced resonances in a two-dimensional Dirichlet tube, Ann. Inst. H. Poincare Phys. Theo. 62 (1995) 81-101.

D. Dunlap, H.-L. Wu and P. Phillips, Phys. Rev. Lett. 65 (1990) 88-91.

G. Dunne and R. Jaffe, Bound states in twisted Aharonov-Bohm tubes, Ann. Phys. 223 (1993) 180-196.

L. Eaves, T. Fromhold et al., Quantum chaology in semiconductor heterostructures, Phys. Scrip. T68 (1996) 51-55.

I. Efrat, Cusp forms and higher rank, (thesis, New York University, 1983).

I. Efrat, The Selberg trace formula for $PSL_2(R)^n$, Mem. Amer. Math. Soc. 65 (1987) 359.

I. Efrat, On the existence of cusp forms over functions fields, J. Reine Angew. Math. 399 (1989) 173-187.

I. Efrat, Spectral deformations of automorphic functions over graphs of groups, Invent. Math. 102 (1990) 447-462.

I. Efrat, On the discrete spectrum of certain discrete groups, Bull. AMS 24 (1991) 125-129.

I. Efrat, Automorphic spectra on the tree of PGL_2, L'Enseign. Math. 37 (1991) 31-43.

D. V. Evans, Trapped acoustic modes, IMA J. Appl. Maths. 49 (1992) 45-60.

D. V. Evans and C. M. Linton, Trapped modes in open channels, J. Fluid Mech. 225 (1991) 153-175.

D. V. Evans and C. M. Linton, Acoustic resonance in ducts, J. Sound Vib. 173 (1994) 85-94.

D. V. Evans, C. M. Linton and F. Ursell, Trapped mode frequencies embedded in continuous spectrum, Q. J. Mech. Appl. Math. 46 (1993) 253-274.

D. V. Evans, M. Levitin and D. Vassiliev, Existence theorems for trapped modes, J. Fluid Mech. 261 (1994) 21-31.

D. V. Evans and R. Porter, Trapped modes about multiple cylinders in a channel, J. Fluid Mech. 339 (1997) 331-356.

D. V. Evans and R. Porter, Trapped modes embedded in the continuous spectrum, Q. J. Mech. Appl. Math. 51 (1998) 263-274.

D. V. Evans and R. Porter, Trapping and near-trapping by arrays of cylinders in waves, J. Eng. Math. 35 (1999) 149-179.

D. V. Evans and R. Porter, Near-trapping of waves by circular arrays of vertical cylinders, Appl. Ocean Res. 19 (1997) 83-99.

D. V. Evans and R. Porter, Trapped modes about multiple cylinders in a channel, J. Fluid. Mech. 339 (1997) 331-356.

P. Exner, A model of resonance scattering on curved quantum wires, Ann. Physik 47 (1990) 123-138.

P. Exner, Bound states and resonances in quantum wires, in "Order, Disorder and Chaos in Quantum Systems", (Birkhauser, Basel, 1990) 65-85.

P. Exner, Bound states in curved quantum waveguides of a slowly decaying curvature, J. Math. Phys. 34 (1989) 23-28.

P. Exner, Bound states in quantum waveguides of a slowly decaying curvature, J. Math. Phys. 34 (1993) 23-28.

P. Exner, Lattice Kronig-Penny models, Phys. Rev. Lett. 74 (1995) 3503-3506.

P. Exner, Contact interactions on graph superlattices, J. Phys. A29 (1996) 87-102.

P. Exner, Weakly coupled states on branching graphs, Lett. Math. Phys. 38 (1996) 313-320.

P. Exner, A duality between Schrödinger operators on graphs and certain Jacobi matrices, Ann. Inst. H. Poincaré 66 (1997) 359-371.

P. Exner, Magnetoresonances on a lasso graph, Found. Phys. 27 (1997) 171-190.

P. Exner, Point interactions in a tube (preprint, 1999).

P. Exner and R. Gawlista, Band spectra of rectangular graph superlattices, Phys. Rev. B53 (1996) 7275-7286.

P. Exner, R. Gawlista, P. Seba and M. Tater, Point interactions in a strip, Ann. Phys. 252 (1996) 133-179.

P. Exner, A. Joye and H. Kovarik, Edge currents without edges, (preprint, 1999).

P. Exner and E. Krejcirik, Quantum waveguides with a lateral semitransparent barrier: spectral and scattering properties, J. Phys. A32 (1999) 4475-4494.

P. Exner and P. Seba, A new type of quantum interference transistors, Phys. Lett. A129 (1988) 477-480.

P. Exner and P. Seba, Bound states in curved quantum waveguides, J. Math. Phys. 30 (1989) 2574-2580.

P. Exner and P. Seba, Free quantum motion on a branching graph, Rep. Math. Phys. 28 (1989) 2574-2580.

P. Exner and P. Seba, Electrons in semiconductor microstructures: a challenge to operator theorists, in "Schrödinger Operators, Standard and Non-Standard", (World Scientific, Singapore, 1989) 85-106.

P. Exner and P. Seba, Trapping modes in a curved electromagnetic waveguide with perfectly conducting walls, Phys. Lett. A144 (1990) 347-350.

P. Exner and P. Seba, Resonance statistics in a microwave cavity with a thin antenna, Phys. Lett. A228 (1997) 146-150.

P. Exner, P. Seba and P. Stovicek, On existence of a bound state in an L-shaped waveguide, Czech. J. Phys. B39 (1989) 1181-1191.

P. Exner, P. Seba, M. Tater and D. Vanek, Bound states and scattering in quantum waveguides coupled laterally through a boundary window, J. Math. Phys. 37 (1996) 4867-4887.

P. Exner and E. Šerešová, Appendix resonances on a simple graph, J. Phys. A27 (1994) 8296-8278.

P. Exner and M. Tater, A one-band model for a weakly coupled quantum-wire resonator, Phys. Rev. B50 (1994) 18350-18354.

P. Exner and M. Tater, Evanescent modes in a multiple scattering factorization, Czech. J. Phys. 48 (1998) 617-624.

P. Exner and S. Vugalter, Bound states in a locally deformed waveguide: the critical case, Lett. Math. Phys. 39 (1997) 57-69.

P. Exner and S. Vugalter, Asymptotic estimates for bound states in quantum waveguide coupled laterally through a narrow window, Ann. Inst. H. Poincare 65 (1996) 109-123.

P. Exner and S. Vugalter, Bound-state asymptotic estimates for window-coupled Dirichlet strips and layers, J. Phys. A30 (1997) 7863-7878.

W. Faris, A localization principle for multiplicative perturbations, J. Func. Anal. 67 (1986) 105-114.

W. Faris, Localization for a random discrete wave equation, in Random Media (ed. G. Papanicolaou, Springer-Verlag, New York, 1987) 121-128.

F. Faure, Topological properties of quantum periodic Hamiltonians (preprint, 1999).

B. Felderhof, Transmission and reflection of waves in one-dimension disordered array, J. Stat. Phys. 43 (1986) 267-279.

K. Feng and W.-C. W. Li, Spectra of hypergraphs and applications, J. Number Theory 60 (1996) 1-22.

D. K. Ferry, J. P. Bird, R. Akis, et al., Quantum transport in single and multiple quantum dots, Jpn. J. Appl. Phys. 36 (1997) 3944-3950.

A. Figà-Talamanca and C. Nebbia, Harmonic Analysis and Representation Theory for Groups Acting on Homogeneous Trees (London Math. Soc. LNS 162, Cambridge Univ. Press, 1991).

A. Figotin, High-contrast photonic crystals, in Diffuse Waves in Complex Media (ed., J. Fouque, Kluwer Academic Press, Dordrecht, 1998) 109-136.

A. Figotin and Y. Godin, The computation of spectra of some 2D photonic crystals, J. Comp. Phys. 136 (1997) 585-598.

A. Figotin and I. Khalfin, Bound states of a one-band model for 3D periodic medium, J. Comp. Phys. 138 (1997) 153-170.

A. Figotin and A. Klein, Localization phenomenon in gaps of the spectrum of random lattice operators, J. Stat. Phys. 75 (1994a) 997-1021.

A. Figotin and A. Klein, Localization of electromagnetic an acoustic waves in random media: lattice models, J. Stat. Phys. 76 (1994b) 985-1003.

A. Figotin and A. Klein, Localization of classical waves I: acoustic waves, Comm. Math. Phys. 180 (1996) 439-482.

A. Figotin and A. Klein, Localization of classical waves II: electromagnetic waves, Comm. Math. Phys. 184 (1997) 411-441.

A. Figotin and A. Klein, Localized classical waves created by defects, J. Stat. Phys. 86 (1997) 165-177.

A. Figotin and A. Klein, Localization of light in lossless inhomogeneous dielectrics, J. Opt. Soc. Am. A15 (1998) 1423-1435.

A. Figotin and A. Klein, Midgap defect modes in dielectric and acoustic media, SIAM J. Appl. Math. 58 (1998) 1748-1773.

A. Figotin and P. Kuchment, Band-gap structure of spectra of periodic dielectric and acoustic media I. scalar model, SIAM J. Appl. Math. 56 (1996) 68-88.

A. Figotin and P. Kuchment, Band-gap structure of spectra of periodic dielectric and acoustic media II. photonic crystals, SIAM J. Appl. Math. 56 (1996) 1561-1620.

W. Fischer, T. Hupfer, H. Leschke and P. Müller, Existnce of the density of states for multi-dimensional continuum Schrödinger operators with Gaussian random potentials, Comm. Math. Phys. 190 (1997) 133-141.

W. Fischer, H. Leschke and P. Müller, Towards localisation by Gaussian random potentials in multidimension continuous space, Lett. Math. Phys. 38 (1996) 343-348.

W. Fischer, T. Hupfer, H. Leschke and P. Müller, Rigorous results on Schrödinger operators with certain Gaussian random potentials in multi-dimensional continuous space, in Differential equations, asymptotic analysis, and mathematical physics (ed. M. Demuth and B. Schulze, Akademie Verlag, Berlin, 1997) 105-112.

J. Flores, Transport in models with correlated diagonal and off-diagonal disorder, J. Phys. Cond. Matt. 1 (1989) 8471-8479.

D. Foata and D. Zeilberger, A combinatorial proof of Bass's evaluations of the Ihara-Selberg zeta function for graphs, Trans. AMS 351 (1999) 2257-2274.

P. Freund, Scattering on p-adic and on adelic symmetric spaces, Phys. Lett. 257B (1991) 119-125.

J. Fröhlich, G. Graf and J. Walcher, On the extended nature of edge states of quantum Hall Hamiltonians (preprint, 1999).

J. Fröhlich, G. Graf and J. Walcher, Extended quantum Hall edge states: general domains (preprint, 1999).

J. Fröhlich and T. Spencer, Absence of diffusion in the Anderson tight binding model for large disorder or low energy, Comm. Math. Phys. 88 (1983) 151-184.

T. Fromhold, L. Eaves et al., Magnetotunneling spectra of a quantum well in the regime of classical chaos, Phys. Rev. Lett. 72 (1994) 2608-2611.

B. Fuglede and R. V. Kadison, Determinant theory in finite factors, Ann. Math. 55 (1952) 520-530.

Y. Fyodorov and H.-J. Sommers, Statistics of resonance poles, phase shifts and time delays in quantum chaotic scattering: random matrix approach for systems with broken time-reversal invariance, J. Math. Phys. 38 (1997) 1918-1981.

P. Garrett, Buildings and Classical Groups (Chapman and Hall, 1997).

P. Gaspard and S. A. Rice, Exact quantization of the scattering from a classically chaotic repellor, J. Chem. Phys. 90 (1989) 2255-2262.

V. Geiler, The two-dimensional Schrödinger operator with a uniform magnetic field, and its perturbation by periodic zero-range potentials, St. Petersburg Math. J. 3 (1992) 489-532.

V. Geiler and V. Margulis, Point perturbation-invariant solutions of the Schrödinger equation with a magnetic field, Math. Notes 60 (1996) 575-580.

E.-U. Gekeler, Automorphe Formen über $F_q(T)$ mit kleinem Führer, Abh. Math. Sem. Univ. Hamburg 55 (1985) 111-146.

E.-U. Gekeler, Analytical construction of Weil curves over function fields, J. Theorie Nombres de Bordeaux 7 (1995) 27-49.

E.-U. Gekeler, On the cuspidal divisor class group of a Drinfeld modular curve, Docu. Math. 2 (1997) 351-374.

E.-U. Gekeler and U. Nonnengardt, Fundamental domains of some arithmetically groups over function fields, Int. J. Math. 6 (1995) 689-708.

E.-U. Gekeler and M. Reversat, Jacobians of Drinfeld modular curves, J. Reine Angew. Math. 476 (1996) 27-93.

S. Gelbart and H. Jacquet, A relation between automorphic representations of $GL(2)$ and $GL(3)$, Ann. Sci. Ecole Norm. Sup 11 (1978) 471-542.

N. Gerasimenko, Inverse scattering problem on a noncompact graph, Theo. and Math. Phys. 75 (1988) 460-470.

N. Gerasimenko and B. S. Pavlov, Scattering problem on noncompact graphs, Teor. Mat. Fiz. 74 (1988) 345-359.

F. Germinet and S. De Bièvre, Dynamical localization for discrete and continuous random Schrödinger operators, Comm. Math. Phys. 194 (1998) 323-341.

F. Gesztesy, H. Holden and B. Simon, Absolute summability of the trace relation for certain Schrödinger operators, Comm. Math. Phys. 168 (1995) 137-161.

V. Geyler and I. Popov, The spectrum of a magneto-Bloch electron in a periodic array of quantum dots: explicitly solvable model, Z. Phys. B93 (1994) 437-439.

V. Geyler and I. Popov, Periodic array of quantum dots in a magnetic field: irrational flux, Z. Phys. 98 (1995) 473-477.

V. Geyler, I. Popov and S. Popova, Transmission coefficient for ballistic transport through quantum resonator, Rep. Math. Phys. 40 (1997) 531-538.

V. Geyler and I. Popov, Eigenvalues imbedded in the band spectrum for the periodic array of quantum dots, Rep. Math. Phys. 39 (1997) 275-281.

S. Girvin, The quantum Hall effect: novel excitations and broken symmetries, in Les Houches, 1998 (Springer Verlag, New York, 1999).

S. Girvin and R. Prange, editors, The Quantum Hall Effect (Springer Verlag, New York, 1987).

C. Godsil and B. Mohar, Walk-generating functions and spectral measures of infinite graphs, Lin. Alg. Appl. 107 (1988) 191-206.

G. Goldoni, F. Rossi and E. Molinari, Quantum interference in nanometric devices: ballistic transport across arrays of T-shaped quantum wires, App. Phys. Lett. 71 (1997) 1519-1521.

G. Goldoni, F. Rossi, E. Molinari and A. Fasolino, Band structure and optical anisotropy in V-shaped and T-shaped semiconductor quantum wires, Phys. Rev. B55 (1997) 7110-7123.

Ya. Gol'dsheid, S. Molchanov and L. Pastur, Pure point spectrum of stochastic one dimensional Schrödinger operators, Func. Anal. Appl. 11 (1977) 1-10.

C. I. Goldstein, Eigenfunction expansions associated with the Laplacian for certain domains with infinite boundaries, Trans. Amer. Math. Soc. 135 (1969) 1 -50.

C. Goldstein, Scattering theory in waveguides, in "Scattering Theory in Mathematical Physics", (Reidel, Dordrecht, 1974) 35-51.

J. Goldstone and R. Jaffe, Bound states twisting tubes, Phys. Rev. B45 (1992) 14100-14107.

C. Gordon and E. Wilson, Isospectral deformations of compact solvmanifolds, J. Diff. Geom. 19 (1984) 241-256.

G. Graf, Anderson localization and the space-time characteristic of continuum states, J. Stat. Phys. 75 (1994) 337-346.

J. Gratus, C. Lambert et al., Quantum mechanics on graphs, J. Phys. A27 (1994) 6881-6892.

M. Groves, Examples of embedded eigenvalues for problems in acoustic waveguides, Math. Methods. Appl. Sci 21 (1998) 479-488.

M. Grundmann and D. Bimberg, Formation of quantum dots in two-fold cleaved edge overgrowth, Phys. Rev. B55 (1997) 4054-4056.

I. Guarneri, On an estimate concerning quantum diffusion in the presence of fractal spectrum, Europhys. Lett. 21 (1993) 729-733.

M. C. Gutzwiller, Stochastic behavior in quantum scattering, Physica 7D (1983) 341-355.

M. C. Gutzwiller, Chaos in Classical and Quantum Mechanics (Springer-Verlag, New York, 1990).

B. I. Halperin, Adv. Chem. Phys. 13 (1967) 123.

B. I. Halperin, Quantized Hall conductance, current-carrying edges states and the existence of extended states in a two-dimensional disordered potential, Phys. Rev. B25 (1982) 2185-2188.

G. Harder, W. Li and R. Weisinger, Dimensions of spaces of cusp forms over function fields, J. f. reine and ange. Math. 319 (1980) 73-103.

P. De la Harpe, A. Robertson and A. Valette, On the spectrum of the sum of generators for a finitely generated group, Israel J. Math. 81 (1993) 65-96.

K. Hashimoto, Zeta functions of finite graphs and representations of p-adic groups, in Adv. Studies in Pure Math. 15 (1989) 211-280.

K. Hashimoto, On zeta and L-functions of finite graphs, Int. J. Math. 1 (1990) 381-396.

K. Hasimoto, Artin type L-functions and the density theorem for prime cycles on finite graphs, Inter. J. Math. 3 (1992) 809-826.

K. Hashimoto and A. Hori, Selberg-Ihara's zeta function for p-adic groups, in Adv. Studies in Pure Math. 15 (1989) 171-210.

Y. Hatsugai, Phys. Rev. Lett. 71 (1993) 3697-3700.

Y. Hatsugai, K. Ishibashi and Y. Morita, Sum rule of the Hall conductance in random quantum phase transition, Phys. Rev. Lett. 83 (1999) 2246-2249.

R. Haug, K. Lee and J. Hong, in Nanostructures: Fabrication and Physics (ed., S. Berger et al, Materials Research Society, Pittsburgh, 1990).

S. He and J. Maynard, Detailed measurements of inelastic scattering in Anderson localization, Phys. Rev. Lett. 57 (1986) 3171-3174.

S. He and X. Xie, A new liquid phase and metal-insulator transition in Si MOSFETs (preprint, 1997).

H. Hegger, B. Huckestein, K. Hecker et al., Fractal conductance fluctuations in gold-nanowires, Phys. Rev. Lett. 77 (1996) 3885-3888.

D. Hejhal, The Selberg trace formula for $PSL(2, \mathbf{R})$, Vol. 2, Lecture Notes in Math. 1001 (1983).

D. Hejhal, Eigenvalues of the Laplacian for Hecke triangle groups, Mem. AMS 469 (1992).

J. Hersch, Sur la frequence fondamentale d'une membrane vibrante: evaluations par defaut et principle de maximum, ZAMP 11 (1960) 387-413.

M. Hilke, Localization properties of the periodic random Anderson model, J. Phys. A30 (1997) L367-L371.

M. Hilke and J. Flores, Delocalization in continuous disordered systems, Phys. REv. B55 (1997) 10625.

M. Hilke, D. Shahar, S. Song, D. Tsui, and Y. Xie, Experimental phase diagram of integer quantized Hall effect (preprint, 1999).

M. Hilke, D. Shahar, S. Song, D. Tsui, Y. Xie and D. Monroe, The quantized Hall insulator: a new insulator in two-dimensions, Nature 395 (1998) 675-677.

Y. Hirayama, Y. Tokura, A. Wieck et al., Transport characteristics of a window-coupled in-plane-gated wire system, Phys. Rev. B48 (1993) 7991-7998.

P. D. Hislop, Some recent results on random operators, (preprint, 1998).

H. Holden and F. Martinelli, On absence of diffusion near the bottom of the spectrum for a random Schrödinger operator on $L^2(\mathbf{R}^\nu)$, Comm. Math. Phys. 93 (1984) 197-217.

J. Howland, Resonances near an embedded eigenvalue, Pac. J. Math. 55 (1974) 1.

B. Huckestein, Scaling theory of the integer quantum Hall effect, Rev. Mod. Phys. 67 (1995) 357-396.

B. Huckestein, R. Ketzmerick and C. Lewenkopf, Quantum transport through ballistic cavities: soft vs. hard quantum chaos (preprint, 1999).

K. Hugill et al., Quantum-well-wire growth by molecular-beam epitaxy: a computer simulation study, J. Appl. Phys. 66 (1989) 3415-3417.

M. Huxley, Introduction to Kloostermania, in Banach Center Publ. 17 (ed., H. Iwaniec,, Warsaw, 1983).

M. Huxley, Scattering matrices for congruence subgroups, in Modular Forms (ed. R. Rankin, Ellis Horwood Ltd, 1984).

Y. Ihara, On discrete subgroups of the two by two projective linear group over p–adic fields, J. Math. Soc. Japan 18 (1966) 219-235.

Y. Ihara, Discrete subgroups of $PL(2, k_p)$, in Proc. Symp. Pure Math. IX (1968) 272-278.

K. Ishii, Prog. Theo. Phys. Suppl. 53 (1973) 77.

H. Ishio, Quantum transport and classical dynamics in open billiards, J. Stat. Phys. 83 (1996) 203-214.

H. Ishio and J. Burgdorfer, Quantum conductance fluctuations and classical short-path dynamics, Phys. Rev. B51 (1995) 2013-2017.

H. Ishio and K. Nakamura, Quantum transport in open billiards: dependence on the degree of opening, J. Phys. Soc. Japan 61 (1992) 2649-2654.

T. Itoh, N. Sano and A. Yoshii, Effects of width increase in the ballistic quantum wire, Phys. Rev. B45 (1992) 14131-14135.

H. Iwaniec, Introduction to the Spectral Theory of Automorphic Forms (Revista Matematic Iberoamericana, 1995).

F. Izrailev and A. Krokhin, Localization and mobility edge in one-dimensional potentials with correlated disorder, (preprint, 1999).

F. Izrailev, S. Ruffo and L. Tessieri, Classical representation of the 1D Anderson model, J. Phys. A31 (1998) 5263.

C. Jacoboni, P. Casarini and A. Ruini, Quantum Transport in Ultrasmall Devices (ed. D. Ferry et al., Plenum Press, New York, 1995) 181-190.

D. Jakobson, S. Miller, I. Rivin and Z. Rudnick, Eigenvalue spacings for regular graphs, in Emerging Applications of Number Theory, IMA Vol 109 (Springer, New York, 1999) 317-327.

D. Jakobson and S. Zelditch, Classical limits of eigenfunction for some completely integrable systems, in Emerging Applications of Number Theory, IMA Vol 109 (Springer, New York, 1999) 329-354.

V. Jaksic and Y. Last, Corrugated surfaces and A.C. spectrum, (preprint, 1999).

V. Jaksic and S. Molchanov, Localization of surface spectra, (preprint, 1999).

V. Jaksic and S. Molchanov, On the surface spectrum in dimension two, (preprint, 1999).

R. Jalabert, H. Baranger and A. Stone, Conductance fluctuations in the ballistic regime: a probe of quantum chaos? Phys. Rev. Lett. 65 (1990) 2442-2445.

R.A. Jalabert, J-L. Pichard and W. W. J. Beenakker, Universal quantum signatures of chaos in ballistic transport, Europhys. Lett. 27 (1994) 255-260.

R. A. Jalabert, A. D. Stone and Y. Alhassid, Statistical theory of coulomb blockade oscillations: quantum chaos in quantum dots, Phys. Rev. Lett. 68 (1992) 3468-3471.

A. Jensen and T. Kato, Spectral properties of Schödinger operators and time decay of the wave functions, Duke Math. J. 46 (1979) 583-611.

Z. Ji, Semicond. Sci. Tech. 7 (1992) 198.

Z. Ji, Ballistic transport through a double bend in an electron waveguide, J. Appl. Phys. 73 (1993) 4468-4472.

Z. Ji and K.-F. Berggren, Quantum bound states in narrow ballistic channels with intersections, Phys. Rev. B45 (1992) 6652-6658.

Z. Ji and K.-F. Berggren, Transition from chaotic to regular behavior of electrons in stadium-shaped quantum dot in a perpendicular magnetic field, Phys. Rev. B52 (1995) 1745-1750.

Z. Ji and K.-F. Berggren, Influence of potential fluctuation on quantum transport through chaotic cavities, Phys. Rev. B52 (1995) R11607-R11610.

S. John, The localization of light and other classical waves in disordered media, Comm. Cond. Mat. Phys. 14 (1988) 193-230.

S. John, Localization of light, Physics Today (May) (1991) 32-40.

S. John, The localization of light, in Photonic Band Gaps and Localization, (ed. C. Soukoulis, Plenum, New York, 1993) p. 1.

R. Johnston and H. Kunz, The conductance of a disordered wire, J. Phys. C16 (1983) 3895-3912.

D. S. Jones, The eigenvalues of $\nabla^2 u + \lambda u = 0$ when the boundary conditions are given on semi-infinite domains, Proc. Camb. Phil. Soc. 49 (1953) 668-684.

J. Journe, A. Soffer, and C. Sogge, $L^p \to L^{p'}$ estimates for time dependent Schrödinger equations, Bull AMS 23 (1990) 519-524.

R. Joynt and R. Prange, Conditions for the quantum Hall effect, Phys. Rev. B29 (1984) 3303-3317.

C. Judge, The Roelcke-Selberg conjecture for surfaces with cone points (preprint, 1991).

C. Judge, On the existence of Maass cusp forms on hyperbolic surfaces with cone points, J. AMS 8 (1995) 715-759.

C. Judge and R. Phillips, Dissolving a cusp form in the presence of multiplicities, Duke Math. J. 88 (1997) 267-280.

W. Kang, H. Stormer, K. Baldwin, L. Pfeiffer and K. West, Tunneling between the edges of two lateral quantum Hall systems (preprint, 1999).

N. Katz and P. Sarnak, The spacing distributions between zeros of zeta functions (preprint, 1996).

N. Katz and P. Sarnak, Zeros of zeta functions and symmetry, Bull. AMS 36 (1999) 1-26.

N. Katz and P. Sarnak, Random Matrices, Frobenius Eigenvalues and Monodromy, (AMS, Providence, 1999).

F. Kassubek, C. Stafford and H. Grabert, Force, charge and conductance of an ideal metallic nanowire (preprint, 1998).

H. Kesten, Symmetric random walks on groups, Trans. AMS 92 (1959) 336-354.

R. Ketzmerick, Fractal conductance fluctuations in generic chaotic cavities, Phys. Rev. B54 (1996) 10841-10844.

D. Khmelnitskii, Quantum Hall effect and additional oscillations of conductivity in weak magnetic fields, Phys. Lett 106A (1984) 182-183.

D. Khmelnitskii, Hel. Phys. Acta 65 (1992) 164.

G. Kirczenow, Bend resistance and junction resonances in narrow quantum conductors, Solid State Comm. 71 (1989) 469-472.

G. Kirczenow, Resonant conduction in ballistic quantum channels, Phys. Rev. B39 (1989) 10452-10455.

G. Kirczenow, Mechanism of the quenching of the Hall effect, Phys. Rev. Lett. 62 (1989) 2993-2996.

A. Kirilenko, V. Litvinov and L. Rudj, Truncated bend of a rectangular waveguide in the H-plane, Sov. J. Radiotech. and Elec. 24 (1979) 1032-1052.

A. Kirilenko, L. Rudj and V. Shestopalov, Wave scattering on a waveguide bend, ibid. 19 (1974) 687-696.

W. Kirsch: Random Schrödinger operators: a course, in Lecture Notes in Physics 345 (1989).

W. Kirsch, S. Kotani and B. Simon, Absence of absolutely continuous spectrum for some one-dimensional random but deterministic Schrödinger operators, Ann. Inst. H. Poincaré Phys. Theor. 42 (1985) 383-406.

W. Kirsch, M. Krishna and J. Obermeit, Anderson model with decaying randomness: mobility edge (1999)

W. Kirsch and S. Molchanov, Random Schrödinger operators with potentials having stationary increments (preprint, 1998)

W. Kirsch, P. Stollmann, G. Stolz, Localization for random perturbations of periodic Schrödinger operators, Random Oper. Stochastic Equations 6 (1998) 241-268.

W. Kirsch, P. Stollmann, G. Stolz, Localization for random Schrödinger operators with long range interactions, (preprint 1998).

W. Kirsch, P. Stollmann, G. Stolz, Anderson localization for random Schrödinger operators with long range interactions, Comm. Math. Phys. 195 (1998) 495-507.

S. Kivelson, D. Lee and S. Zhang, Global phase diagram in the quantum Hall effect, Phys. Rev. B46 (1992) 2223-2238.

M. Klaus, On the bound state of Schrödinger operators in one dimension, Ann. Phys. 108 (1977) 288-300.

M. Klaus, Some applications of the Birman-Schwinger principle, Helv. Phys. Acta 55 (1982) 49-68.

F. Kleespies and P. Stollmann, Lifshitz asymptotics and localization for random quantum waveguides, Rev. Math. Phys. (to apprear).

A. Klein, Absolutely continuous spectrum in random Schrödinger operators, Proc. Symp. Pure Math. 59 (1996) 139-147.

A. Klein, Extended states in the Anderson model on the Bethe lattice, Adv. in Math. 133 (1998) 163-184.

A. Klein, Absolutely continuous spectrum in the Anderson model on the Bethe lattice, Math. Res. Lett. 1 (1994) 399-407.

A. Klein, Spreading of wave packets in the Anderson model on the Bethe lattice, Comm. Math. Phys. 177 (1996) 755-773.

A. Klein, Localization of light in randomized periodic media, in Diffuse Waves in Complex Media (ed., J. Fouque, Kluwer Academic Press, Dordrecht, 1998) 73-92.

A. Klein and J. Perez, On the density of states for random potentials in the presence of a uniform magnetic field, Nucl. Phys. B251 (1985) 199.

K. von Klitzing, G. Dorda and M. Pepper, New method for high accuracy determination of the fine structure constant based on quantized Hall resistance, Phys. Rev. Lett. 45 (1980) 494-497.

F. Klopp, Localization for semi-classical continuous random Schrödinger operators II: the random displacement model, Helv. Phys. Act. 66 (1993) 810-841.

F. Klopp, Localization for some continuous random Schrodinger operators, Comm. Math. Phys. 167 (1995) 553-569.

F. Klopp, Localization pour des operateurs de Schródinger aleatoires dans $L^2(\mathbf{R}^d)$: un model semi-classique, Ann. Inst. Fourier 45 (1995) 265-316.

F. Klopp, A low concentration asymptotic expansion for the density of states of a random Schrödinger operator with Poisson disorder, J. Func. Anal. 145 (1997) 267-295.

F. Klopp, Internal Lifshits tails for random perturbations of periodic Schrödinger operators, Duke Math. J. 98 (1999) 335-396.

F. Klopp and L. Pastur, Lifshitz tails for random Schrödinger operators with negative singular Poisson potential (1999).

V. Kostrykin and R. Schrader, Kirchhoff's rule for quantum wires, J. Phys. A32 (1999) 595-630.

V. Kostrykin and R. Schrader, Scattering theory approach to random Schrödinger operators in one dimension, Rev. Math. Phys. 11 (1999) 187-242.

S. Kotani, Lyapunov exponent and spectra for one-dimensional random Schrödinger operators, in Contemp. Math. 50 (AMS, Providence, 1986).

S. Kotani and B. Simon, Localization in general one-dimensional systems II., Comm. Math. Phys. 112 (1987) 103-119.

T. Kottos, F. Izrailev and A. Politi, Finite-length Lyapunov exponents and conductance for quasi-1D disordered solids (preprint, 1998).

T. Kottos and U. Smilansky, Quantum chaos on graphs, Phys. Rev. Lett. 79 (1997) 4794-4797.

T. Kottos and U. Smilansky, Periodic orbit theory and spectral statistics for quantum graphs, Ann. Phys. 273 (1999) 1.

T. Kottos and U. Smilansky, Chaotic scattering on graphs (preprint, 1999).

L. P. Kouwenhoven, C. M. Marcus, P. L. McEuen et al., Electron Transport in Quantum Dots, in NATO ASI Conference Proceedings, (ed. L. P. Kouwenhoven et al. Kluwer, Dordrecht, 1997).

S. Kravchenko, G. Kravchenko, J. Furneaux, V. Pudalov and M D'Iorio, Possible metal/insulator transition at $B = 0$ in two dimensions (preprint, 1994).

S. Kravchenko, W. Mason, J. Furneaux, J. Caulfield, J. Singleton and V. Pudalov, Temperature induced transitions between insulator, metal and quantum Hall states (preprint, 1994).

M. Krishna, Anderson model with decaying randomness - extended states, Proc. Indian Acad. Sci. 100 (1990) 285-294.

M. Krishna and K. Sinha, Spectra of Anderson type models with decaying randomness, (preprint, 1999).

R. Kronig and W. Penney, Quantum mechanics of electrons in crystal lattices, Proc. Roy. Soc. London 130A (1931) 499-513.

J. Kuang, Eigenfunctions of the finite Poincaré plane (preprint, 1997); and On eigenvalues related to finite Poincaré planes, Finite Fields App. 3 (1997) 151-158.

U. Kuhl, F. Izrailev, A. Krokhin and H.-J. Stöckmann, Experimental observation of the mobility edge in a waveguide with correlated disorder, (preprint, 1999).

U. Kuhl and H.-J. Stöckmann, Microwave transmission spectra in regular and irregular one-dimensional scattering arrangements (preprint, 1999).

A. Kumar and P. Bagwell, Resonant tunneling in a quasi-one-dimensional wires: influence of evanescent modes, Phys. Rev. B43 (1991) 9012-9020.

A. Kumar and P. Bagwell, Evaluation of the quantized ballistic conductance with increasing disorder in narrow-wire arrays, Phys. Rev. B44 (1992) 1747-1753.

H. Kunz, The quantum Hall effect for electrons in a random potential, Comm. Math. Phys. 112 (1987) 121-145.

H. Kunz and B. Souillard, Sur le spectre des operateurs aux differences finies aleatoires, Comm. Math. Phys. 78 (1980) 201-246.

H. Kunz and B. Souillard, The localization transition on the Bethe lattice, J. Phys. (Paris) Lett. 44 (1983) 411-414.

C. Kunze, Leaky and mutually coupled quantum wires, Phys. Rev. B48 (1993) 14338-14346.

J. Lafferty and D. Rockmore, Fast Fourier analysis for SL_2 over a finite field and related numerical experiments, Exp. Math. 1 (1992) 115-139.

J. Lafferty and D. Rockmore, Numerical investigation of the spectrum for certain families of Cayley graphs, in DIMACS Ser. Disc. Math. Theo. Comp. Sci. 10 (1993) 63-73.

J. Lafferty and D. Rockmore, Spectral techniques for expander codes, (preprint, 1997).

J. Lafferty and D. Rockmore, Level spacings for Cayley graphs, in Emerging Applications of Number Theory, IMA Vol 109 (Springer, New York, 1999) 373-386.

L. Landau, Diamagnetismus der metalle, Z. Phys. 64 (1930) 629-637.

L. D. Landau and E. M. Lifshitz, Quantum Mechanics (1974).

R. Landauer, Electrical resistance of disordered one-dimensional lattices, Phil. Mag. 21 (1970) 863-867.

R. Laughlin, Quantized Hall conductivity in two dimensions, Phys. Rev. B23 (1981) 5632-5633.

R. Laughlin, Levitation of extended-state bands in a strong magnetic field, Phys. Rev. Lett. 52 (1984) 2304-2307.

P. Lee and A. Stone, Universal conductance fluctuations in metals, Phys. Rev. Lett 55 (1985) 1622-1625.

F. Leighton, Finite common coverings of graphs, J. Comb. Th. B33 (1982) 231-238.

M. Leng and C. Lent, Magnetic edge states in a 1D channel with a periodic array of antidots, Superlatt. Microst. 11 (1992) 351-355.

M. Leng and C. Lent, Recovery of quantum ballistic conductance in periodically modulated channel, Phys. Rev. Lett. 71 (1993) 137-140.

C. S. Lent, Transmission through a bend in an electron waveguide, Appl. Phys. Lett. 56 (1990) 2554-2556.

C. C. Lent and D. J. Kirkner, Then quantum transmitting boundary method, J. Appl. Phys. 67 (1990) 6353-6359.

Y. Levinson, M. Lubin and E. Sukhorukov, Short-range impurity in a saddle-point potential: conductance of a microjunction, Phys. Rev. B45 (1992) 11936-11943.

W. Li, On modular functions in characteristic p, Trans. Amer. Math. Soc. 246 (1978) 231-259.

W. Li, Eisenstein series and decomposition theory over function fields, Math. Ann. 240 (1979) 115-139.

W.-C. W. Li, Character sums and abelian Ramanujan graphs, J. Numb. Theory, 41 (1992) 199-217.

W-C. W. Li, A survey of Ramanujan graphs, in Arithmetic, Geometry and Coding Theory (de Gruyter, Berlin, 1996) 127-143.

W.-C. W. Li, Number Theory with Applications (World Scientific, Singapore, 1996).

W. Li, Elliptic curves, Kloosterman sums and Ramanujan graphs, Study. Adv. Math. 7 (1998) 179-190.

W. Li, Eigenvalues of Ramanujan graphs, in Emerging Applications of Number Theory, IMA Vol 109 (Springer, New York, 1999) 387-403.

C.-T. Liang, J. Frost, M. Pepper, D. Ritchie and G. Jones, Experimental studies of T-shaped quantum dot transistors: phase-coherent electron transport, Solid State Comm. 105 (1998) 109-111.

I. M. Lifshitz, Structure of the energy spectrum of impurity bands in disordered solid solutions, Sov. Phys. JETP 17 (1963) 1159-1170.

I. M. Lifshitz, Energy spectrum structure and quantum states of disordered condensed systems, Sov. Phys. Uspekhi 7 (1965) 549-573.

I. M Lifshitz, Theory of fluctuation levels in disordered systems, Sov. Phys. JETP 26 (1968) 462.

I. M. Lifshitz, S. Gredeskul and L. Pastur, Introduction to the Theory of Disordered Systems (Wiley, New York, 1988).

K. Lin and R. Jaffe, Bound states and threshold resonances in quantum wires with circular bends, (preprint, 1996).

H. Linke, L. Christensson, P. Omling and P. Lindelof, Stability of classical electron orbits in triangular electron billiards, Phys. Rev. B56 (1997) 1440-1446.

H. Linke, J. P. Bird, J. Cooper et al., Nonequilibrium electrons in a ballistic quantum dot, Phys. Stat. Sol. 204 (1997) 318.

C. M. Linton and D. V. Evans, The interaction of waves with arrays of vertical circular cylinders, J. Fluid Mech. 215 (1990) 549-569.

C. M Linton and D. V. Evans, The interaction of waves with a row of circular cylinders, J. Fluid Mech. 251 (1993) 687-708.

W. Lippert, The measurement of sound reflection and transmission at right-angled bends in rectangular tubes, Acustica 4 (1954) 313-319.

W. Lippert, Wave transmission around bends of different angles in rectangular ducts, Acustica 5 (1955) 274-278.

D. Liu, X. Xie and Q. Niu, Weak field phase diagram for an integer quantum Hall liquid (preprint).

D. Liu, X. Xie, S. Das Sarma and S. Zhang, Electron localization in a 2D system with random magnetic flux (preprint).

W. Lu, L. Viola, K. Pance, M. Rose and S. Sridhar, Microwave study of quantum n-disk scattering (preprint, 1999).

A. Lubotzky, Lattices in rank one Lie groups over local fields, Geom. Func. Anal. (GAFA) 1 (1991) 405-431.

A. Lubotzky, Discrete Groups, Expanding Graphs and Invariant Measures (Birkhäuser Verlag, Basel, 1994).

A. Lubotzky, R. Phillips and P. Sarnak, Ramanujan graphs, Combinatorica 8 (1988) 261-277.

W. Luo, Z. Rudnick, and P. Sarnak, On Selberg's eigenvalue conjecture, Geom. Func. Anal. 5 (1995) 387-401.

W. Luo and P. Sarnak, Number variance for arithmetic hyperbolic surfaces, Comm. Math. Phys. 161 (1994) 419-432.

A. MacDonald and P. Streda, Quantized Hall effect and edge currents, Phys. Rev. B29 (1984) 1616-1619.

N. Macris, P. Martin and J. Pulé, On edge states in semi-infinite quantum Hall systems, J. Phys. A32 (1999) 1985-1996.

N. Macris and J Pulé, Density of states of random Schrödinger operators with a uniform magnetic field, Lett. Math. Phys 24 (1992) 307.

N. Makarov and I. Yurkevich, Localization of 2D electrons by coherent scattering, Sov. Phys. JETP 69 (1989) 628-629.

F. Mancoff, R. Clarke, C. Marcus et al., Magnetotransport of a 2DES in a spatially random magnetic filed, Phys. Rev. B51 (1995) 13269.

F. Mancoff, L. Zielinski, C. Marcus et al., Shubnikov-de Haas oscillations in a two-dimensional electron gas in a spatially random magnetic field, Phys. Rev. B53 (1995) R7599-R7602.

H. Maniar and J. N. Newman, Wave diffraction by long arrays of cylinders, J. Fluid Mech. 339 (1997) 309-330.

M. Marcolli and V. Mathai, Twisted higher index theory on good orbifolds, I: Noncommutative Bloch theory (preprint, 1999).

M. Marcolli and V. Mathai, Twisted higher index theory on good orbifolds, II: Fractional quantum numbers (preprint, 1999).

C. Marcus, A. Rimberg et al., Conductance fluctuations and chaotic scattering in ballistic microstructures, Phys. Rev. Lett. 69 (1992) 506-509.

C. Marcus, R. Westervelt et al., Conductance fluctuations and quantum chaotic scattering in semiconductor microstructures, Chaos 3 (1993) 643.

C. Marcus, R. Westervelt et al., Phase breaking in ballistic quantum dots: experiment and analysis based on chaotic scattering, Phys. Rev. B48 (1993) 2460-2464.

C. Marcus, S. Patel et al., Quantum chaos in open vs. closed quantum dots: signatures of interacting particles, Chaos, Solitons & Fractals, 8 (1997) 1261.

G. Margulis, Explicit group theoretic constructions of combinatorial schemes and their application to the design of expanders and concentrators, Prob. Infor. Trans. 23 (1988) 39-46.

A. Martinez, Complex interaction in phase space, Math. Nach. 167 (1994) 203-254.

J. Martorell, S. Klarsfeld, D. Sprung and H. Wu, Analytical treatment of electron wave propagation in two-dimensional structures, Solid State Comm. 78 (1991) 13-18.

J. Masek and B. Kramer, Coherent-potential approach for the zero-temperature DC conductance of weakly disordered narrow systems, J. Phys. C1 (1989) 6395.

H. Matsumoto, On the integrated density of states for the Schrödinger operators with certain random electromagnetic potentials, J. Math. Soc. Japan 45 (1993) 197-214.

J. Maynard, Acoustic Anderson localization, in Random Media and Composites (e. V. Kohn and G. Milton, SIAM, Philadelphia, 1988) 206-207.

P. McCann, Geometry and the integer quantum Hall effect, in Geometric Analysis and Lie Theory in Mathematics and Physics (e. A. Carey and M. Murray, Cambridge University Press, Cambridge, 1998) 132-208.

P. McEuen, B. Alpehnaar and R. Wheeler, Surf. Sci. 229 (1990) 312.

P. McEuen et al., Disorder, pseudospins and backscattering in carbon nanotubes (preprint, 1999).

M. Mciver and C. Linton, On the non-existence of trapped modes in acoustic waveguides, Q. J. Mech. Appl. Math. 48 (1995) 543-555.

P. McIver, C. M. Linton and M. McIver, Construction of trapped modes for wave guides and diffraction gratings, Proc. Roy. Soc. A454 (1998) 2593-2616.

B. McKay, The expected eigenvalue distribution of a large regular graph, Lin. Alg. Appl. 40 (1981) 203-216.

H. McKean, Selberg's trace formula as applied to a compact Riemann surface, Comm. Pure App. Math. 25 (1972) 207-227.

K. McCormick, M. Woodside, et al., Scanned potential microscopy of edge and bulk currents in the quantum Hall regime, Phys. Rev. B59 (1999) 4654.

R. Mehran, Calculation of microstrip bends and Y-junctions with arbitrary angle, IEEE Trans. MTT 26 (1978) 400-405.

P. A. Mello and A. D. Stone, Maximum entropy model for quantum mechanical interference effects in metallic conductors, Phys. Rev. B44 (1991) 3559-3576.

J.-F. Mestre, La méthode des graphes, in Proc. Int. Conf. on Class Numbers and Fundamental Units of Algebraic Number Fields (Katata, 1986) 217-242.

A. Micolich, R. Taylor, R. Newbury, J. Bird et al., Geometry-induced fractal behavior in a semiconductor billiard, J. Phys. Cond. Matt. 10 (1998) 1339-1347.

A. Micolich, R. Taylor, R. Newbury, J. Bird et al., Geometry-induced fractal behavior: fractional Brownian motion in a ballistic mesoscopic billiard, Physica B 249-251 (1998) 343-347.

J. Miller and B. Derrida, Weak disorder expansion for the Anderson model on a tree, J. Stat. Phys. 75 (1994) 357-389.

N. Minami, Local fluctuation of the spectrum of a multidimensional Anderson tight binding model, Comm. Math. Phys. 177 (1996) 709-725.

A. Mirlin and Y. Fyodorov, Localization transition in the Anderson model on the Bethe lattice: spontaneous symmetry breaking and correlation functions, Nucl. Phys. B366 (1991) 507-532.

B. Mohar, The spectrum of an infinite graph, Lin. Alg. Appl. 48 (1982) 245-256.

B. Mohar, Isoperimetric inequalities, growth and the spectrum of graphs, Lin. Alg. Appl. 103 (1988) 119-131.

B. Mohar, Laplace eigenvalues of graphs – a survey, Disc. Math. 109 (1992) 171-183.

B. Mohar and W. Woess, A survey on spectra of infinite graphs, Bull. London Math. Soc. 21 (1989) 209-234.

S. Molchanov, The local structure of the spectrum of the one-dimensional Schrödinger operator, Comm. Math. Phys. 78 (1981) 429-446.

E. Montroll, Quantum theory on a network, J. Math. Phys. 11 (1970) 635-648.

M. Morgenstern, Ramanujan graphs and diagrams function field approach, DIMACS Ser. Disc. Math. Comp. Sci. 10 (1993) 111-116.

M. Morgenstern, Ramanujan diagrams, SIAM J. Disc. Math. 7 (1994a) 560-570.

M. Morgenstern, Existence and explicit constructions of $q+1$ regular Ramanujan graphs for every prime power q, J. Comb. Theory B62 (1994b) 44-62.

M. Morgenstern, Natural bounded concentrators, Combinatorica 15 (1995) 111-122.

Y. Morita, K. Ishibashi and Y. Hatsugai, Transitions from the quantum Hall state to the Anderson insulator: fate of delocalized states (preprint, 1999).

A. Mosk, T. Nieuwenhuizen, and C. Barnes, Theory of semi-ballistic wave propagation, Phys. Rev. B53 (1996) 15914-15931.

N. Mott and W. Twose, The theory of impurity conduction, Adv. Phys. 10 (1961) 107-163.

E. Mourre, Absence of singular continuous spectrum for certain self-adjoint operators, Comm. Math. Phys. 78 (1981) 391-408.

G. Müller, G. Boebinger, et al, Precursors and transitions to chaos in a quantum well in a tilted magnetic field, Phys. Rev. Lett. 75 (1995) 2875-2878.

W. Müller, Spectral theory for Riemannian manifolds with cusps and a related trace formula, Math. Nachr. 111 (1983) 197-288.

W. Müller, The point spectrum and spectral geometry for Riemannian manifolds with cusps, Math. Nachr. 125 (1986) 243-257.

W. Müller, Spectral geometry and scattering theory for certain complete surfaces of finite volume, Inv. Math. 109 (1992) 265-305.

M Murty and V. Murty, Non-vanishing of L—functions and Applications (Birkhäuser, Boston, 1997).

H. Nagoshi, The Selberg zeta functions over function fields, (preprint, 1999).

H. Nagoshi, The distribution of eigenvalues of arithmetic infinite graphs (preprint, 1999).

H. Nagoshi, e-mail, 1998.

K. Nakamura and H. Ishio, Quantum transport in open billiards: comparison between circle and stadium, J. Phys. Soc. Japan 61 (1992) 3939-3944.

S. Nakamura, Shape resonances for distortion analytic Schrödinger operators, Comm. PDE 14 (1989) 1383-1419.

S. Nakamura, On Martinez' method in phase space tunneling, Rev. Math. Phys. 7 (1995) 431-441.

S. Nakamura and J. Bellissard, Low energy bands do not contribute to the quantum Hall effect, Comm. Math. Phys. 131 (1990) 283-305.

S. Nakao, On the spectral distribution of the Schrödinger operator with random potential, Japan. J. Math. 3 (1977) 111-139.

E. Narimanov and A. Stone, Magnetunneling through a quantum well in a tilted field, (1997).

E. Narimanov and A. Stone, Origin of strong scarring in quantum wells in a tilted magnetic field, (1997).

E. Narimanov, A. Stone and B. Boebinger, Some classical theory of magneto transport through a chaotic quantum well (1997).

L. Nedelec, Sur les resonances de l'operateur de Dirichlet dans un tube, Comm. PDE 22 (1997) 143-163.

K. Nikolić and A. MacKinnon, Conductance and conductance fluctuations of narrow disordered quantum wires, Phys. Rev. B50 (1994) 11008-11017.

Q. Niu, D. J. Thouless and Y. S. Wu, Quantized Hall conductance as a topological invariant, Phys. Rev. B31 (1985) 3372-3377.

J. Nixon, J. Davies and H Baranger, Conductance of quantum point calculated using realistic potentials, Superlatt. Microst. 9 (1991) 187-190.

S. Northshield, Two proofs of Ihara's theorem, in Emerging Applications of Number Theory, IMA Vol 109 (Springer, New York, 1999) 469-478.

S. Novikov, The Schrödinger operators on graphs and topology, Russ. Math. Sur. 52 (1997) 178-179.

S. Novikov, Discrete Schrodinger operators and topology, (preprint, 1999).

Y. Ochiai, Y. Okubo, A. Widjaja et al., Backscattering of ballistic electrons in a corrugated gate quantum wire, Phys. Rev. B56 (1997) 1073.

H. Okada, T. Hashizume, and H. Hasegawa, Transport characterization of Schottky in-plane gate $Al_{0.3}Ga_{0.7}As/GaAs$ quantum wire transistors realized iby in-situ electrochemical process, Jpn. J. Appl. Phys. 34 (1995) 6971-6976.

A. Okiji, H. Kasai and A. Nakamura, Ballistic transport in mesoscopic systems, Prog. Theo. Phys. Suppl. 106 (1991) 209-224.

A. Okiji, H. Kasai and K. Mitsutake, Conductance oscillations in a quantum wire with a ring geometry, J. Phys. Soc. Japan 61 (1992) 1717-1723.

Y. Okubo, J. Bird et al., Magnetically induced suppression of phase breaking in ballistic mesoscopic billiards, Phys Rev. B55 (1997) 1368-1371.

Y. Okubo, Y. Ochiai, D. Vasileska et al., Stability of regular orbits in ballistic quantum dots, Phys. Lett. A236 (1997) 120.

G. I. Ol'shanskii, Classification of irreducible representations of groups of automorphisms of Bruhat-Tits trees, Func. Anal. and its Appl. 11 (1977) 26-34.

R. Parker, Resonance effects in water shedding from parallel plates: some experimental observations, J. Sound Vib. 4 (1966) 62-72.

R. Parker, Resonance effects in water shedding from parallel plates: calculation of resonant frequencies, J. Sound Vib. 5 (1966) 233-242.

R. Parker and S. Stoneman, The excitation and consequences of acoustic resonances in enclosed fluid flow around solid bodies, Proc. Inst. Mech. Engrs. 203 (1989) 9-19.

M. Pascaud and G. Montambaux, Thermodynamics and transport in mesoscopic disordered networks, (preprint, 1998).

M. Pascaud and G. Montambaux, Persistent currents on graphs, Phys. Rev. Lett. 82 (1999) 4512-4515.

L. Pastur, Spectra of random selfadjoint operators, Russ. Math. Surv. 28 (1973) 1-67.

L. Pastur, Behavior of some Wiener integrals as $t \to \infty$ and the density of states of Schrödinger equations with random potential, Theor. Math. Phys. 32 (1977) 615.

L. Pastur and A. Figotin, Spectra of Random and Almost-Periodic Operators (Springer-Verlag, New York, 1992).

M. Patra and C. W. J. Beenakker, Excess noise for coherent radiation propagating through amplifying random media (preprint, 1999).

L. Pauling, The diamagnetic anisotropy of aromatic molecules, J. Chem. Phys. 4 (1936) 673-677.

F. Peeters, Bound and resonant states in quantum wire structures, in "Science and Engineering of One-and Zero-Dimensional Semiconductors", (ed. S. Beaumont and C. Torres, Plenum, New York, 1990) 107.

F. Peeters, Superlatt. Microstr. 6 (1989) 217.

M. Persson, J. Pettersson, et al., Conductance oscillations related to the eigenenergy spectrum of a quantum dot in weak magnetic fields, Phys. Rev. B52 (1995) 8921-8933.

L. Pfeiffer et al., Quantum wire exciton lasers (preprint, 1998).

P. Phillips and H.-L. Wu, Localization and its absence: a new metallic state for conducting polymers, Science 252 (1991) 1805-1812.

R. Phillips, Scattering theory for the Hilbert modular group, IMRN Inter. Math. Res. Not. 4 (1996) 161-200.

R. Phillips and P. Sarnak, On cusp forms for cofinite subgroups of $PSL(2, \mathbf{R})$, Inv. Math. 80 (1985a) 339-364.

R. Phillips and P. Sarnak, The Weyl theorem and the deformation of discrete groups, Comm. Pure App. Math. 38 (1985b) 853-866.

R. Phillips and P. Sarnak, Spectrum of Fermat curves, Geom. Func. Anal. 1 (1991) 79-146.

R. Phillips and P. Sarnak, Automorphic spectrum and Fermi's Golden Rule, J. D'Anal. Math. 59 (1992) 179-187.

R. Phillips and P. Sarnak, Perturbation theory for the Laplacian on automorphic functions, J. AMS 5 (1992) 1-32

R. Phillips and P. Sarnak, Cusp forms for character varieties, Geom. Func. Anal. 4 (1994) 93-118.

A. Pizer, Ramanujan graphs and Hecke operators, Bull. AMS 23 (1990) 127-137.

A. Pnueli, Scattering matrices and conductances of leaky tori, Ann. Phys. 231 (1994) 56-83.

G. Polya and S. Szego, Isoperimetric Inequalities in Mathematical Physics (Princeton University Press, Princeton, 1951).

A. N. Popov, On the existence of eigenoscillations of a resonator open to a waveguide, Sov. J. Tech. Phys. (Zh. Tekh. Fiz.) 56 (1986) 1916-1922.

I. Yu. Popov, Extension theory and localization of resonances for domains of trap type, Math. USSR Sbornik 71 (1992) 209-234.

I. Yu. Popov, The resonator with narrow slit and the model based on the operator extensions theory, J. Math. Phys. 33 (1992) 3794-3801.

I. Yu. Popov, On the point and continuous spectra for coupled quantum waveguides and resonators,

I. Popov and S. Popova, Zero-width slit model and resonances in mesoscopic systems, Europhys. Lett. 24 (1993) 373-377.

I. Popov and S. Popova, Model of zero-width gaps and resonances effects in a quantum waveguide, Tech. Phys. 39 (1994) 11-15.

I. Popov and S. Popova, On the mesoscopic gate, Acta Phys. Polon. 88 (1995) 1113-1117.

I. Yu. Popov and S. L. Popova, Eigenvalues and bands imbedded in the continuous spectrum for the system of resonators and waveguide: solvable model, Phys. Lett. A222 (1996) 286-290.

S. Popova, The possibility of making nanoelectronic devices using mesoscopic Fresnel zone plates, Tech. Phys. Lett. 19 (1993) 508-509.

W. Porod, Z. Shao and C. Lent, Transmission resonances and zeros in quantum waveguides with resonantly coupled cavities Appl. Phys. Lett. 61 (1992) 1350-1352.

W. Porod, Z. Shao and C. Lent, Resonance-antiresonance line shape for transmission in quantum waveguides with resonantly coupled cavities, Phys. Rev. B48 (1993) 8495-8498.

R. Prange, Quantized Hall resistance and the measurement of the fine-structure constant, Phys. Rev. B23 (1981) 4802-4805.

P. Price, Transmission and reflection peaks in ballistic transport, Appl. Phys. Lett. 62 (1993) 289-290.

R. Quenell, Spectral diameter estimates for k-regular graphs, Adv. Math. 106 (1994) 122-148.

M. S. Raghunathan, Discrete subgroups of algebraic groups over local fields of positive characteristics, Proc. Indian Acad. Sci. 99 (1989) 127-146.

B. Randol, Small eigenvalues of the Laplace operator on compact Riemann surfaces, Bull. AMS 80 (1974) 996-1000.

D. Ravenhall, H. Wyld, and R. Schult, Quantum Hall effect at a four-terminal junction, Phys. Rev. Lett. 62 (1989) 1780-1783.

M. Reed and B. Simon, Methods of Modern Mathematical Physics (Academic, New York, 1978).

F. Rellich, Über das asymptotische Verhalten der Lösunger von $\Delta u + \lambda u = 0$ in unendlichen Gebieten, Jahresber. Deutsch. Math.-Verein. 53 (1943) 157-165.

W. Renger and W. Bulla, Existence of bound states in quantum waveguides under weak conditions, Lett. Math. Phys. 35 (1995) 1-12.

P. Richens and M. V. Berry, Pseudointegrable systems in classical and quantum mechanics, Physica D2 (1981) 495-512.

A. Richter, Playing billiards with microwaves - quantum manifestations of classical chaos, in Emerging Applications of Number Theory (ed. D. Hejhal, et al., Springer, New York, 1998) 479-523.

T. Richter and R. Seiler, Geometric properties of transport in quantum Hall systems (preprint).

R. Romanov and G. Rudin, Scattering of the Bruhat-Tit's tree, I. Phys. Lett. A198 (1995) 113-118.

R. Romanov and G. Rudin, Scattering on p-adic graphs, Comp. Math. Appl. 34 (1997) 587-597.

J.-P. Roth, Le spectre du Laplacien sur un graphe, Lecture Notes in Mathematics 1096 (1983) 521-539.

Z. Rudnick and P. Sarnak, The behavior of eigenstates of arithmetic hyperbolic manifolds, Comm. Math. Phys. 161 (1994) 195-213.

K. Ruedenberg and C. Scherr, Free-electron network model for conjugated systems, I. Theory, J. Chem. Phys. 21 (1953) 1565-1581.

I. Rust, Arithmetically defined representations of groups of type $SL(2, \mathbf{F}_q)$, (preprint, 1998).

A. Sachrajda, R. Ketzmerick et al., Fractal conductance fluctuations in a soft wall stadium and a Sinai billiard, Phys. Rev. Lett. 80 (1998) 1948-1951.

L. Sadun and J. Avron, Adiabatic curvature and the S-matrix, Comm. Math. Phys. 181 (1996) 685-702.

D. Saraga and T. Monteiro, Semiclassical Gaussian matrix elements for chaotic quantum wells (preprint, 1999).

P. Sarnak, Additive number theory and Maass forms, LNM 1052 (1982) 286-309.

P. Sarnak, On cusp forms, in The Selberg trace formula and related topics, Cont. Math. 53 (1986) 393-397.

P. Sarnak, Statistical properties of eigenvalues of the Hecke operators, in Analytic Number Theory and Diophantine Problems (e., A. Adolphson, J. Conrey, A. Ghosh, and R. Roger, Prog. in Math. 70, Birkhäuser, Boston, 1987) 321-331

P. Sarnak, Arithmetic Quantum Chaos, First Annual R. A. Blyth Lectures (1993); Isr. Math. Conf. Proc. 8 (1995) 183-256.

I. Satake, Theory of spherical functions on reductive algebraic groups over $p-$adic fields, IHES Publ. Math. 18 (1963).

I. Satake, Spherical functions and Ramanujan conjecture, in Proc. Symp. Pure Math. IX (1967) 258-264.

H. Schanz and U. Smilansky, Periodic-orbit theory of Anderson localization on graphs (preprint, 1999).

O. Scheja, On zeta functions of arithmetically defined graphs (preprint, 1998a).

O. Scheja, e-mail to author (1998b)

T. Schleich, Einige bemerkungen zur spectralzerlegung der Hecke-algebra für die $PGL(2)$ über funktionenkörpern, Bonner Math. Schr. 71 (1974).

C. Schmit, Quantum and classical properties of some billiards on the hyperbolic plane, in Chaos and Quantum Physics (ed., M.-J. Giannoni, A. Voros and J. Zinn-Justin, Elsevier, New York, 1991) 333-369.

H. Schomerus, K. Frahm, M. Patra and C. W. J. Beenakker, Quantum limit of the laser linewidth in chaotic cavities and statistics of residues of scattering matrix poles (preprint, 1999).

R. L. Schult, D. G. Ravenhall and H. W. Wyld, Quantum bound states in a classically unbounded system of crossed wires, Phys. Rev. B39 (1989) 5476-5479.

R. Schult, H. Wyld and D. Ravenhall, Quantum Hall effect and general narrow-wire circuits, Phys. Rev. B41 (1990) 12760-12780.

H. Schulz-Baldes and J. Bellissard, Anomalous transport: a mathematical framework, Rev. Math. Phys. 10 (1998) 1-46.

H. Schulz-Baldes, J. Kellendonk and T. Richter, Simultaneous quantization of edge and bulk Hall conductivity (preprint, 1999).

M. Scrowston, The spectrum of a magnetic Schrödinger operator with randomly located delta impurities (preprint).

P. Seba, Random matrix theory and mesoscopic fluctuations, Phys. Rev. B53 (1996) 13024-13028.

A. Selberg, Harmonic analysis and discontinuous groups in weakly symmetric Riemannian spaces with applications to Dirichlet series, J. Indian Math. Soc. 20 (1956) 47-87.

A. Selberg, On the estimation of Fourier coefficients of modular forms, AMS Proc. Symp. in Pure Math. VIII (1965) 1-15.

A. Selberg, Remarks on the distribution of poles of Eisenstein series, in Collected Papers (Springer-Verlag, New York, 1989) Vol. 2 15-46.

A. Selberg and S. Chowla, On Epstein's zeta-function, J. reine und angew. Math. 227 (1967) 86-110.

J.-P. Serre, Arbres, Amalgames, SL_2, Astérisque, no 46 (1977).

J.-P. Serre, Trees (Springer-Verlag, New York, 1980).

J.-P. Serre, Répartition asymptotique des valeurs propres de l'opérateur de Hecke T_p, J. Amer. Math. Soc. 10 (1997) 75-102.

D. Shahar, D. Tsui, M. Shayegan, R. Bhatt and J. Cunningham, Universal conductivity at the quantum Hall liquid to insulator transition, Phys. Rev. Lett. 74 (1995) 4511-4514.

Z. Shao, W. Porod and C. Lent, Transmission resonances and zeros in quantum waveguide systems with attached resonators, Phys. Rev. B49 (1994) 7453-7465.

B. Shapiro, Quantum conduction on a Cayley tree, Phys. Rev. Lett. 50 (1983) 747-750.

D. Sheng and Z. Weng, Delocalization of electrons in a random magnetic field, Phys. Rev. Lett. 75 (1995) 2388-2391.

D. Sheng and Z. Weng, Phase diagram of integer quantum Hall effect (preprint, 1999).

T. Shirai, A trace formula for discrete Schrödinger operator, Pub. Res. I. Math. Sci. 34 (1998) 27-41.

T. Shirai, A factorization of determinant related to some random matrices, J. Stat. Phys. 90 (1998) 1449-1459.

A. Shudo, Y. Shimizu, P. Seba, et al., Statistical properties of spectra of pseudointegrable systems, Phys. Rev. E49 (1994) 3748-3756.

B. Simon, The bound state of weakly coupled Schrödinger operators in one and two dimensions, Ann. Phys. 97 (1976) 279-288.

B. Simon, Schrödinger semigroups, Bull. AMS 7 (1982) 447-526.

B. Simon, Lifshits tails for the Anderson model, J. Stat. Phys. 38 (1985) 65-76.

B. Simon, L^p norms of the Borel transform and the decomposition of measures, Proc. AMS 123 (1995) 3749-3755.

B. Simon, Kotani theory for one-dimensional stochastic Jacobi matrices, Comm. Math. Phys. 89 (1983a) 227-234.

B. Simon, Equality of the density of states in a wide class of tight-binding Lorentzian random models, Phys. Rev. B27 (1983b) 3859-3860.

B. Simon, Absence of ballistic motion, Comm. Math. Phys. 134 (1990) 209-212.

B. Simon, Operators with singular continuous spectra, VI: Graph Laplacians and Laplace-Beltrami operators, Proc. AMS 124 (1996) 1177-1182.

B. Simon and T. Wolff, Singular continuous spectrum under rank one perturbations and localization for random Hamiltonians, Comm. Pure Appl. Math. 39 (1986) 75-90.

M. Skriganov, The spectrum band structure of the three-dimensional Schrödinger periodic potential, Invent. Math. 80 (1985) 107-121.

F. Sols, Scattering, dissipation and transport in mesoscopic systems, Ann. Phys. 214 (1992) 386-438.

F. Sols and M. Macucci, Circular bends in electron waveguides, Phys. Rev. B41 (1990) 11887-11891.

F. Sols, M. Macucci, U. Ravaioli and K. Hess, On the possibility of transistor action based on quantum interference phenomena, Appl. Phys. Lett. 54 (1989) 350-352.

F. Sols, M Macucci, U. Ravaioli and K. Hess, Theory for a quantum modulated transistor action, J. Appl. Phys. 66 (1989) 3892-3906.

F. Sols, M. Macucci, U. Ravaioli and K. Hess, in "Nanostructure Physics and Fabrication", ed. M. A. Reed and W. P. Kirk, Academic Press, 1989) p. 157.

A. Soffer and M. Weinstein, Time dependent resonance theory, Geom. Func. Anal. 8 (1998) 1086-1128.

T. Spencer, Localization for random and quasiperiodic potentials, J. Stat. Phys. 51 (1988) 1009-1019.

D. Sprung, H. Wu and J. Martorell, Understanding quantum wires with circular bends, J. Appl. Phys. 71 (1992) 515-517.

D. Sprung, H. Wu and J. Martorell, Scattering by a finite periodic potential, Am. J. Phys. 61 (1993) 310-316.

S. Sridhar, Experimental observations of scarred eigenfunctions of chaotic microwave cavities, Phys. Rev. Lett. 67 (1991) 785-788.

H. Stark and A. Terras, Zeta functions of finite graphs and coverings, Adv. in Math. 121 (1996) 124-165.

H.-J. Stöckmann, M. Barth, U. Dörr, U. Kuhl nd H. Schanze, Microwave studies of chaotic billiards and disordered systems, (preprint, 1999).

P. Stollmann, Wegner estimates and localization for continuum Anderson models with some singular distributions (1999a).

P. Stollmann, Localization for random perturbations of anisotropic periodic media, Israel J. Math. 107 (1998) 125-139.

P. Stollmann, Caught by Disorder: Lectures on Bound States in Random Media (preprint, 1999c).

G. Stolz, Localization for random Schrödinger operators with Poisson potential, Ann. Inst. H. Poincare 63 (1997) 297-314.

M. Stoytchev and A. Genack, Measurment of the probability distribution of total transmission in random waveguides, Phys. Rev. Lett. 79 (1997) 309-312.

T. Sunada, Riemannian coverings and isospectral manifolds, Ann. Math. 121 (1985) 169-186.

T. Sunada, L-functions in geometry and some applications, in Lecture Notes in Math. 1201 (1986) 266-284.

T. Sunada, Fundamental groups and Laplacians, in Lecture Notes in Math. 1339 (1987) 248-277.

A. Szafer and A. D. Stone, Theory of quantum conduction through a constriction, Phys. Rev. Lett. 62 (1989) 300-303.

Y. Takagaki and D. Ferry, Conductance of quantum wave-guides with a rough boundary, J. Phys.: Condens. Matter 4 (1992a) 10421-10432.

Y. Takagaki and D. Ferry, Double quantum point contacts in series, Phys. Rev. B45 (1992b) 13494-13498.

Y. Takagaki and D. Ferry, Conductance of quantum point contacts in the presence of disorder, Phys. Rev. B46 (1992c) 15218-15224.

H. Tamura and T. Ando, Conductance fluctuations in quantum wires, Phys. Rev. B44 (1991) 1792-1800.

R. Taylor, A. Micolich, R. Newbury, T. Fromhold, Correlation analysis of self-similarity in semiconductor billiards, Phys. Rev. B56 (1997) R12733- R12736.

R. Taylor, A. Micolich, R. Newbury, J. Bird, T. Fromhold et al., Exact and statistical self-similarity in magnetoconductance fluctuations: a unified picture, Phys. Rev. B58 (1998) 11107-11110.

R. Taylor, R. Newbury, A. Sachrajda et al., Self-similar magnetoresistance of a semiconductor Sinai billiard, Phys. Rev. Lett. 78 (1997) 1952-1955.

E. Tekman and P. Bagwell, Fano resonances in quasi-one-dimensional electron waveguides, Phys. Rev. B48 (1993) 2553-2559.

E. Tekman and S. Ciraci, Novel features of quantum conduction in a constriction, Phys. Rev. B39 (1989) 8772-8775.

A. Terras, Fourier Analysis on Finite Groups and Applications (Cambridge University Press, Cambridge, 1999).

A. Terras, Survey of spectra of Laplacians on finite symmetric spaces, Exp. Math. 5 (1996) 15-32.

A. Terras, A survey of discrete trace formulas, in Emerging Applications of Number Theory, IMA Vol 109 (Springer, New York, 1999) 643-681.

L. Tessieri and F. Izrailev, One-dimensional tight-binding models with correlated diagonal and off-diagonal disorder, (preprint, 1999).

S. Tessmer, P. Glicofridis et al., Surface charge accumulation imaging of a quantum Hall liquids, Nature (1998) 51-54.

D. J. Thouless, Electrons in disordered systems and the theory of localization, Phys. Rep. 13C (1974) 93-142.

D. J. Thouless, Localization and the two-dimensional Hall effect, J. Phys. C14 (1981) 3475-3480.

D. J. Thouless, Edge voltages and distributed currents in the quantum Hall effect, Phys. Rev. Lett. 71 (1993) 1879-1882.

D. J. Thouless, Topological interpretation of quantum Hall conductance, J. Math. Phys. 35 (1994) 5362-5372.

D. J. Thouless, Topological Quantum Numbers in Nonrelativistic Physics (World Scientific, Singapore, 1998).

D. J. Thouless, M. Kohmoto, M. Nightingale and M. den Nijs, Quantized Hall conductance in a two-dimensional periodic potential, Phys. Rev. Lett. 49 (1982) 405-408.

J.-P. Tillich and G. Zemor, Optimal cycle codes constructed from Ramanujan graphs, SIAM J. Dis. Math. 10 (1997) 447-459.

G. Timp, H. Baranger, et al., Propagation around a bend in a multichannel electron waveguide, Phys. Rev. Lett. 60 (1988) 2081-2084.

G. Timp and R. Howard, Quantum mechanical aspects of transport in nano-electronics, Proc. IEEE 79 (1991) 1188-1207.

A. Tip, Absolute continuity of the integrate density of states of the quantum Lorentz gas for a class of repulsive potentials, J. Phys. A27 (1994) 1057-1069.

F. Toyama and Y. Nogami, Effect of a trapped electron on the transmission characteristic of a quantum wire, Phys. Lett. A196 (1994) 237-241.

C. Tsao and W. Gambling, Curvilinear optical fibre waveguide: characterization of bound modes and radiative field, Proc. Roy. Soc. London A425 (1989) 1-16.

V. Twersky, Multiple scattering or radiation by an arbitrary configuration of parallel cylinders, J. Acous. Soc. Am. 24 (1952) 42-46.

N. Ueki, On spectra of random Schrödinger operators with magnetic fields, Osaka J. Math. 31 (1994) 177-187.

S. E. Ulloa, E. Castano and B. Kirczenow, Ballistic transport in a novel one-dimensional superlattice, Phys. Rev. B41 (1990) 12350-12353.

F. Ursell, Trapping modes in the theory of surface waves, Proc. Camb. Phil. Soc. 47 (1951) 347-358.

F. Ursell, Mathematical aspects of trapping modes in the theory of surface waves, J. Fluid Mech. 183 (1987) 421-437.

F. Ursell, Trapped modes in a circular cylindrical acoustic waveguide, Proc. Roy. Soc. Lond. A435 (1991) 575-589.

T. Usuki, M. Saito et al., Numerical analysis of ballistic-electron transport in magnetic fields by using a quantum point contact and a quantum wire, Phys. Rev. B52 (1995) 8244-8255.

K. Vacek, H. Kasai and A. Okiji, Ballistic transport in the bent quantum wire, J. Phys. Soc. Japan 61 (1992) 27-31.

K. Vacek, A. Okiji and H. Kasai, Magnetotransport in bent quantum wires and scattering in bent potential wells, Tech. Rep. Osaka Univ. 42 (1992) 225-240.

K. Vacek, A. Okiji and H. Kasai, Ballistic transport in quantum wires with periodic bend structure, Solid State Comm. 85 (1993) 507-511.

K. Vacek, A. Okiji and H. Kasai, Multichannel ballistic magnetotransport through quantum wires with double circular bends, Phys. Rev. B47 (1993) 3695-3705.

A. Valette, Can one hear the shape of a group? Est. Rend. Sem. Mat. Fis. Milano LXIV (1994) 31-44.

B. A. van Tiggelen, Localization of waves, in Diffuse Waves in Complex Media (ed., J. Fouque, Kluwer Academic Press, Dordrecht, 1998) 1-60.

G. Vattay, J. Cserti, G. Palla and G. Szalka, Diffraction in the semiclassical description of mesoscopic devices, Chaos, Solitons & Fractals, (to appear).

A. Venkov, Spectral theory of automorphic functions and its applications, (Kluwer, Dordrecht, 1990).

A. Venkov, On essentially cuspidal noncongruence subgroups of $PSL(2, \mathbf{Z})$, J. Funct. Anal. 92 (1990) 1-7.

A. B. Venkov and A. M. Nikitin, The Selberg trace formula, Ramanujan graphs, and some problems of mathematical physics, St. Petersburg Math. J. 5 (1994) 419-484.

A. Venkov and P. Zograf, On analogues of the Artin factorization formulas in the spectral theory of automorphic functions, Math. USSR-Izv. 21 (1983) 435-444.

J. Vidal, R. Mosseri and J. Bellissard, Spectrum and diffusion for a class of tight-binding models on hypercubes, J. Phys. A32 (1999) 2361-2367.

P. de Vries, D. van Coevorden and A. Lagendijk, Point scatterers for classical waves, Rev. Mod. Phys. 70 (1998) 447.

K. Wakabayashi and M. Sigrist, Zero-conductance resonances due to flux states in nanographite ribbon junctions (preprint, 1999).

C.-K. Wang, Quantum bound-states in a ballistic quantum channel with a multiple double-bend discontinuity, Semicond. Sci. Tech. 10 (1995) 1131-1138.

C.-K. Wang, K.-F. Berggren and Z. Ji, Quantum bound states in a double-bend quantum channel, J. Appl. Phys. 77 (1995) 2564-2571.

J. Wang and H. Guo, Time-dependent transport in two-dimensional quantum-wire structures, Phys. Rev. B48 (1993) 12072-12075.

J. Wang, Y.-J. Wang and H. Guo, Many-electron effects on transport through 2-dimension quantum structures, J. Appl. Phys. 75 (1994) 2721-2723.

Y. Wang, N. Zhu, J. Wang and H. Guo, Resonance states of open quantum dots, Phys. Rev. B53 (1996) 16408-16413.

W. Wang, Developpement asymptotique de la densite d'etats pour l'operateur de Schrödinger aleatoire avec champ magnetique, Sem. Equations Der. Part. 1992-1993, Ecole Polytech., Expose XVIII.

W. Wang, Microlocalization, percolation and Anderson localization for the magnetic Schrödinger operator with a random potential, J. Func. Anal. 146 (1997) 1-26.

Z. Wang, M Fisher, S. Girvin and J. Chalker, Short-range interactions and scaling near integer quantum Hall transitions (preprint).

C. Webster, Harmonic analysis and representation theory of the automorphism groups of homogeneous trees (thesis, 1996).

R. Weder, Spectral and scattering theory in deformed optical waveguides, J. Reine Ang. Math. 390 (1988) 130-169.

R. Weder, Spectral and Scattering Theory for Wave Propagation in Perturbed Stratified Media (Springer-Verlag, New York, 1991).

B. J. van Wees, H. van Houten, C. W. J. Beenakker, J. B. Williamson, L. P. Kouwenhoven, et al., Quantized conductance in a two-dimensional electron gas, Phys. Rev. Lett. 60 (1988) 848-850.

F. Wegner, Bounds on the density of states for disordered systems, Z. Phys. 44 (1981) 9-15.

W. Wegscheider, L. Pfeiffer, et al., Lasing from excitons in quantum wires, Phys. Rev. Lett. 71 (1993) 4071-4074.

W. Wegscheider, G. Schedelbeck et al., Atomically precise GaAs/AlGaAs quantum dots fabricated by twofold cleaved edge overgrowth, Phys. Rev. Lett. 79 (1997) 1917-1920.

A. Weil, On the analogue of the modular group in characteristic p, in Functional Analysis, Proc. in honor of M. Stone, (Springer, New York, 1970) 211-223.

J. Weisinger, Some results on classical Eisenstein series and modular forms over functions fields (thesis, Harvard, 1977).

A. Weisshaar, J. Lary, S. Goodnick and V. Tripathi, Analysis of discontinuities in quantum waveguide structures, Appl. Phys. Lett. 55 (1989) 2114-2116.

A. Weisshaar, J. Lary, S. Goodnick and V. Tripathi, SPIE Proc. 1284 (1990) 45.

A. Weisshaar, J. Lary, S. Goodnick and V. Tripathi, Analysis and modeling of quantum waveguide structures and devices, J. Appl. Phys. 70 (1991) 355-366.

D. A. Wharam, T. J. Thornton, R. Newbury, M. Pepper et al., One-dimensional transport and quantization of the ballistic resistance, J. Phys. C21 (1988) L209-L214.

D. Wiersma, B. Bartolini, A. Lagendijk, and R. Righini, Localization of light in a disordered medium, Nature 390 (1997) 671.

C. Wilcox, Scattering theory for the d'Alembert equation in exterior domains (Springer-Verlag, Berlin, 1975).

P. Wilkinson, T. Fromhold, L. Eaves et al., Observation of 'scarred' wavefunctions in a quantum well with chaotic electron dynamics, Nature 380(1996) 608-610.

P. Wilkinson, T. Fromhold, L. Eaves et al., Evidence for periodic "scar" patterns in the wavefunctions of a chaotic quantum well, Surf. Sci. 361-2 (1996) 696-699.

S. Wolpert, The spectrum of a Riemann surface with a cusp, Taniguchi Symposium Lecture (1989).

S. Wolpert, Disappearance of cusp forms in special families, Ann. Math. 139 (1994) 239-291.

H. Wu and D. Sprung, Quantum probability flow patterns, Phys. Lett. A183 (1993) 413-417.

H. Wu and D. Sprung, Theoretical study of multi-bend quantum wires, Phys. Rev. B47 (1993) 1500-1505.

H. Wu and D. Sprung, Ballistic transport: a view from the quantum theory of motion, Phys. Lett. A196 (1994) 229-234.

H. Wu and D. Sprung, Validity of the transfer-matrix method for a two-dimensional electron waveguide, Appl. Phys. A58 (1994) 581-587.

H. Wu, D. Sprung and J. Martorell, Electronic properties of a quantum wire with arbitrary bending angle, J. Appl. Phys. 72 (1992) 151-154.

H. Wu, D. Sprung, and J. Martorell, Effective one-dimensional square well for two- dimensional quantum wires, Phys. Rev. B45 (1992) 11960-11967.

H. Wu, D. Sprung, and J. Martorell, A comparison of confining potentials in the quantum wire problem, Appl. Phys A56 (1993) 127-131.

H. Wu, D. Sprung and J. Martorell, Periodic quantum wires and their quasi-one-dimensional nature, J. Phys. D 26 (1993) 798-803.

H. Wu, D. Sprung, J. Martorell and S. Klarsfeld, Quantum wire with periodic serial structure, Phys. Rev. B44 (1991) 6351-6360.

J. C. Wu, M. N Wybourne, W. Yindeepol, A. Weisshaar, and S. M. Goodnick, Interference phenomena due to a double bend in a quantum wire, Appl. Phys. Lett. 59 (1991) 102-104.

J. Xia, Geometric invariants of the quantum Hall effect, Comm. Math. Phys 119 (1988) 29-50.

X. Xie and D. Liu, Transition from integer quantum Hall state to insulator

H. Xu, Z. Ji and K.-F. Berggren, Electron transport in finite one-dimensional quantum-dot arrays, Superlatt. Microstr. 12 (1992) 237-242.

K. Yang and R. Bhatt, Current carrying states in a random magnetic field (preprint).

K. Yang and R. Bhatt, Floating of extended states and localization transition in a weak magnetic field (preprint).

K. Yang and R. Bhatt, Quantum Hall - insulator transitions in lattice models with strong disorder (preprint).

X. Yang, H Ishio and J. Burgdorfer, Statistics of magnetoconductance in ballistic cavities, Phys. Rev. B52 (1995) 8219-8225.

W. Yindeepol, A. Chin, A. Weisshaar, et al., in Nanostructures and Mesoscopic Systems (ed. W. Kirk and M. Reed, Academic Press, Boston, 1992) 139-149.

J. Zak, Magnetic translation group, Phys. Rev. A134 (1964) 1602-1607; magnetic translation group II: irreducible representations, ibid. 1607-1611.

S. Zelditch, Quantum dynamics from the semi-classical viewpoint (preprint, 1999).

Z. Zhang, C. Wong et al., Observation of localized electromagnetic waves in three-dimensional networks of waveguides, Phys. Rev. Lett. 81 (1998) 5540-5543.

I. Zozoulenko and K.-F. Berggren, Ballistic weak localization in regular and chaotic quantum-electron billiards, Phys. Rev. B54 (1996) 5823-5828.

I. Zozoulenko and K.-F. Berggren, Weak localization and chaos in ballistic quantum dots, Physica Scr. T69 (1997) 345-347.

I. Zozoulenko and K.-F. Berggren, Quantum scattering, resonant states and conductance fluctuations in an open square electron billiard, Phys. Rev. B56 (1997) 1-11.

I. Zozoulenko, R. Schuster and K.-F. Berggren and K. Ensslin, Ballistic electrons in an open square geometry: selective probing of resonant-energy states, Phys. Rev. B55 (1997) R10209-R10212.

Index